FRACTIONAL AND SUBFRACTIONAL HORSEPOWER ELECTRIC MOTORS

LEESON
G-Series
GEARMASTER
ROBBINS & MYERS
Electra Motors
DRESSER
Anaheim, CA
Power Transmission Division
Dresser Industries Inc.
Westinghouse
AC Motor
MOTOR MUST BE GROUNDED
Westinghouse Electric Corp.
EG&G ROTRON
prestolite
PRESTOLITE, TOLEDO, OHIO
UNIVERSAL ELECTRIC CO. OWOSSO, MICH.
IMPEDANCE PROTECTED
JARD
EASTERN AIR DEVICES
Northland
a Scott Fetzer company
Leland
Electrosystems Inc
ASHLAND ELEC PROD INC
RELIANCE
DUTY MASTER A.C. MOTOR
SINGER
CARTER
REGENT MOTOR
WELCO INDUSTRIES INC.
CINCINNATI, OHIO
Marathon Electric
DC MOTOR
Smith
Franklin Electric
1081 POOL MOTOR
THERMALLY PROTECTED
WARNING
GENERAL ELECTRIC
A-C MOTOR
THERMALLY PROTECTED
AMERICAN BOSCH
AIRPAX
NORTH AMERICAN PHILIPS CONTROLS CORP.
Cheshire Division
FASCO INDUSTRIES INC.
OZARK MO. DIV.
EMERSON
OIL BURNER
DUTY
WARNING
DC SERVO MOTOR
MADE IN U.S.A.
IMC PART NO.
SERIAL NO.
CUST. PART NO.
INERTIAL MOTORS CORP.
280 N. BROAD ST.
DOYLESTOWN, PA. 18901
PHONE NO. (215) 345-1010
TELEX NO. 83-1633
EASTERN AIR DEVICES
Dayton
DAYTON ELECTRIC MFG. CO. CHICAGO 60648 U.S.A.
HOWELL ELECTRIC MOTORS
IMC MAGNETICS CORP.
VOLTAGE 115
HORSEPOWER RATING - 1/6
FREQUENCY 50-60 Hz
FULL LOAD SPEED-3200 R.P.M.
FULL LOAD POWER-200 WATTS
(MUST BE USED WITH 2.5 UF
RUNNING CAPACITOR)
DUMORE CORPORATION
GENERAL ELECTRIC
A-C MOTOR
THERMALLY PROTECTED
Delco
PORTER
PEERLESS
PERMANENT MAGNET D.C. SERVOMOTOR
G.K. HELLER CORP.
GKH
FLORAL PARK NEW YORK
FOR USE WITH G.K.H. CONTROLLERS
BISON

FRACTIONAL AND SUBFRACTIONAL HORSEPOWER ELECTRIC MOTORS

AVAILABLE TYPES, BASIC OPERATING PRINCIPLES, SELECTION, AND MAINTENANCE

CYRIL G. VEINOTT, D.Eng.
Independent Consultant
Fellow IEEE, Tesla Medalist

JOSEPH E. MARTIN, B.S.E.E.
Engineering Manager Industry Standards
Emerson Motor Division

FOURTH EDITION

McGRAW-HILL BOOK COMPANY

New York St. Louis San Francisco Auckland Bogotá Hamburg
Johannesburg London Madrid Mexico Montreal New Delhi
Panama Paris São Paulo Singapore Sydney Tokyo Toronto

Library of Congress Cataloging-in-Publication Data

Veinott, Cyril G. (Cyril George)
Fractional and subfractional horsepower electric motors.

Includes index.
1.Electric motors, Fractional horsepower.
I.Martin, Joseph E. II.Title.
TK2537.V38 1986 621.46′2 85-12487
ISBN 0-07-067393-4

1234567890 DOC/DOC 898765

ISBN 0-07-067393-4

The editors for this book were Betty Sun and Susan B. West, the designer was Naomi Auerbach, and the production supervisor was Sally Fliess. It was set in Caledonia by University Graphics, Inc.

Printed and bound by R. R. Donnelley & Sons Company.

WARNING

Many of the connections and constructions described and illustrated in this book are covered by letters patent or an application for letters patent. No attempt is made in this book to indicate whether any construction or connection is patented or is common property. Servicepersons, of course, can repair a patented device or apparatus that was purchased from a duly authorized seller. Manufacturers, however, are warned to check the patent situation before incorporating into their own designs any of the types of construction or connections given in this book. Publication in this book does not imply that a device or system described herein is public property available to all who care to use it. Neither the publishers nor the authors will assume any responsibility for damages arising out of patent litigation or suits involving any apparatus described in this book.

To
DOROTHY VEINOTT
and
MARILYN MARTIN,

without whose patient tolerance and wholehearted support this volume could never have been prepared, the authors lovingly dedicate this book

CONTENTS

LIST OF TABLES

PREFACE

It was with a deep sense of responsibility that the authors undertook the revision of this volume, to bring it as up to date as possible. To make room for new material, some old material, now of less importance, has been omitted.

The authors believe that the need for the information presented in this volume is now of greater value and importance than ever before, not only because of the continued growth of this industry, but even more so because exploding electrical and electronic technology has crowded out the study of electric machines in so many of our universities and colleges.

In this edition, as in the first three editions, the authors' objectives are to tell as simply and as fully as possible the story of this giant industry, to explain the basic underlying principles of why small electric machines work and why they do what they do, to provide information that will help the would-be user to select and apply the right machine to his or her task as well as to connect and service and repair it, and to write an easy-to-understand text that is completely free of mathematics but is supplemented with numerous references at the end of each chapter that readers can use to pursue further any needs not covered.

Students and instructors in technical institutes and trade schools should find this volume most helpful in enhancing their practical knowledge. University and college faculty as well as graduate and undergraduate students should also find much reference material in this book to aid them with various projects which involve the use of small motors. This field is now so broad that a reference such as this can be of great help.

The authors have kept in mind the needs of the many persons employed in the production of these machines—design engineers and quality control and laboratory as well as shop personnel—and have tried to present material that will be helpful to all. We believe that this volume contains the broadest and most comprehensive up-to-date data base on small electric machines to be found anywhere.

The authors wish to thank the many persons and companies who have assisted us in providing the latest information on so many subjects. It is their cooperation that has made it possible to include so much information in this volume. Specifically, of the companies who participated, 76 are listed in Table 1-1 and an additional 12 in Table 15-1. A great deal of material was taken from NEMA and from IEEE, as well as from Underwriters Laboratories. Just a few of the individuals who helped were J. Lloyd, D. Felt, K. Theis, K. Hoemann, N. Hargis, and K. Giffard.

Finally, both authors want to express their deepest appreciation to the Emerson Motor Division for the encouragement and support it has given to coauthor Martin, who has gathered most of the updating information; his participation has meant much to coauthor Veinott.

Cyril G. Veinott
Joseph E. Martin

FRACTIONAL AND SUBFRACTIONAL HORSEPOWER ELECTRIC MOTORS

CHAPTER

1

THE STORY BEHIND THE NAMEPLATE

Manufactured goods contain a label or marking to identify the manufacturer of the product and the size or rating and to provide information concerning use or maintenance. This type of identification is particularly important for the electric motor.

The motor manufacturer marks a product by means of ink stamping, indenting, embossing, or affixing a label or other form of nameplate thereon. The nameplate contains the principal information needed to put the motor into service; it describes the motor briefly; it contains a remarkable wealth of information written in a language of its own. To some, little of the information contained thereon is intelligible; to others, most of the information is meaningful; but usually only to the motor manufacturer is all the information significant. The message is coded, in a sense; part of the code is common to the industry, and part is the creation of the particular manufacturer to aid in identifying the motor completely, when it becomes necessary to do so.

If it is necessary to write to the manufacturer for information or to order parts, the first and unalterable rule should always be, *Copy the entire nameplate reading.* It is essential that the manufacturer have the complete reading of the nameplate in order to positively identify the motor. Even when it is known that a single number on the plate provides total identification, it is best to send the entire nameplate reading, for it often happens that an error is made in recording or transmitting the important number.

Later, we shall examine a typical nameplate, but first let us define the terms fractional horsepower and subfractional horsepower motors.

1-1 Fractional and Subfractional Horsepower Motors. These terms are both defined in the Glossary. Curiously enough, the definition of fractional horsepower is not simple.

(1) Fractional horsepower motor. A fractional horsepower motor is any motor built in one of the frame sizes shown in Fig. 19-1 (see Chap. 19), even though the actual rating may be in excess of 1 hp. When these frame sizes were originally developed, all motors built in them having four poles or more were rated at less than 1 hp, but advancing technology has now given us ratings in excess of 1 hp in these frame sizes. The term fractional horsepower relates more to the physical size of the machine than to the actual power output rating. Common ratings are given in Chap. 19 in Table 19-1.

(2) Subfractional horsepower motor. This term has not been formally defined. By common consent it has come to mean any motor smaller than a fractional horsepower motor, or any motor rated at less than 1/20 hp. Common ratings, given in millihorsepower, are shown in Chap. 19 in Table 19-2. Many of the subfractional horsepower motors are rated in watts.

1-2 Manufacturer. Now, let us examine the typical nameplate, shown in Fig. 1-1. The manufacturer's name is listed in a prominent location, together with the address, which in this case is near the top center of the plate. Names and addresses of leading manufacturers are given in Table 1-1. This table and the notes that follow it give extensive information about the products made by each manufacturer and the way they are identified.

1-3 CSA Monogram. The CSA monogram, appearing in the upper left-hand corner of Fig. 1-1, indicates that the design and construction of this motor have been certified by the Canadian Standards Association.[1,*] The monogram is often followed by a file number under which the motor was originally investigated and is periodically reexamined.

1-4 Thermally Protected. A large percentage of fractional horsepower motors are now provided with built-in thermal protection, a subject more fully discussed in Arts. 17-16 to 17-21. The use of such pro-

*For numbered references, see References at end of this chapter.

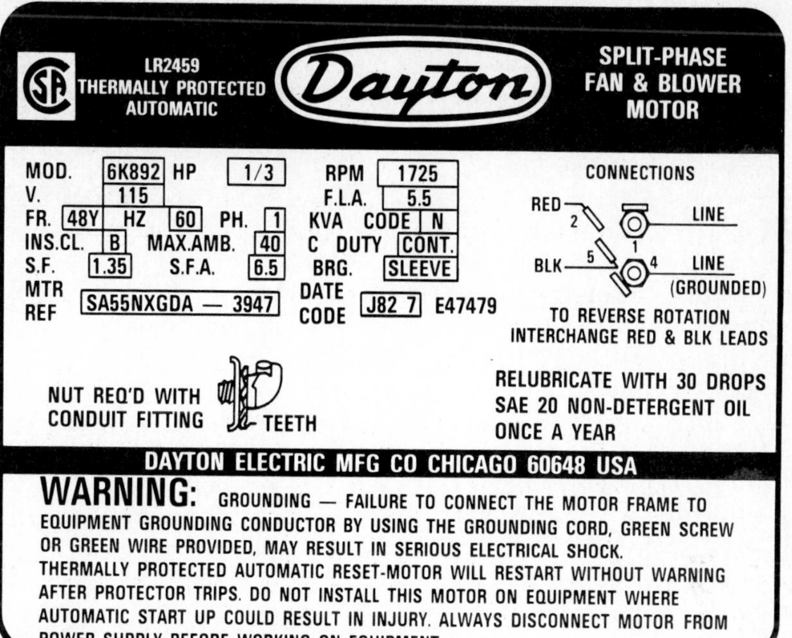

Fig. 1-1 A motor nameplate.

tection, in whole or in part, is indicated on the motor nameplate by the words "thermally protected," as required by the National Electrical Code,[2] National Electrical Manufacturers Association (NEMA),[3] Underwriters Laboratories (UL),[4] and Canadian Standards Association (CSA).[5] Thermally protected may be abbreviated TP on subfractional horsepower motors. Built-in thermal protection was incorporated into a line of motors in the early thirties by Westinghouse. In 1939 the Underwriters Laboratories issued their requirements for such equipment and began the examination and listing of these devices applied to specific motors. Current requirements are given in Art. 17-17 and more fully in UL 547[4] and CSA C22.2 no. 77[5]. Individual manufacturers may use their own systems of marking thermally protected motors to indicate the kinds of protection and the type of thermal protector.

1-5 Type and Application. Motors fall into a number of different types of classifications, such as split-phase, capacitor-start, and universal. Each type of motor has its own particular set of characteristics. The nameplate of Fig. 1-1 indicates a split-phase motor. Manufacturers usually assign certain letters, or combinations of letters, to denote the motor type. How to read and interpret the type-letter designations of many manufacturers of motors is shown in Table 1-1.

(continued on page 17)

TABLE 1-1 Type-Letter Designations of Fractional and Subfractional Horsepower Motors

		AC motors					
		Single-phase					
			Capacitor types				
Manufacturer's name and address	Scope*	Split-phase	Capacitor-start	Permanent-split	Two-value	Split-phase-capacitor	Shaded-pole
Aerovox Inc., Motor Division 740 Belleville Avenue New Bedford, MA 02741	S						S1, S2
Airpax North American Philips Controls Corp. Cheshire Industrial Park Cheshire, CT 06410	S, G						
American Bosch Electrical Products P.O. Box 2228 McCrary Road Columbux, MS 39701	F, G						
Ametek Inc., Lamb Electric Div. Kent, OH 44240	F, G	I, IW	CAS	PSC, CA, FCA			
Ashland Electric Products Div. 32-02 Queens Blvd. Long Island City, NY 11101	S, F, G	†	†	†	†	†	
Baldor Electric Company Fort Smith, AR 72902	F	S (H)	L (CI)	C	LC		
Barber-Colman Company, Colman AC Motors, Inc. P.O. Box 350 Reedsburg, WI 53959	S, G			DE, KE (AE, OE)			AA, DA, KA, BB, KB, OB, AX, BX, DX, EX, FX, KX (BA, AF DE, KF)
Bison Gear and Engineering Corp. 2424 Wisconsin Avenue Downers Grove, IL 60615	G	†	†	†	†		†
Bodine Electric Company 2500 West Bradley Place Chicago, IL 60618	S, F, G	SI	CS	CI, LI	CCI, DI		SP
Brailsford and Company 670 Milton Road Rye, NY 10580	S, G						
Brevel Motors, Inc. Broad and 16th Streets Carlstadt, NJ 07072	S, F, G		†	†	†		†
Carter Motor Company 2711 West George Street Chicago, IL 60618	S, F, G						

Note: Letters in parentheses represent obsolete types.
*S, subfractional horsepower; F, fractional horsepower; G, gearmotors.
†Type is built, but there are no specific type letters assigned.

1-, 2-, 3-phase synchronous		Polyphase	AC or dc universal	DC motors				
Reluctance	Hysteresis	Squirrel-cage	Uncom-pensated	Brushless	Shunt	Compound	Series	Permanent-magnet
†				†				†
								†
		P	CS		S	C		†
†	†	†			†	†	†	
(MS)		M (MM, GM)	U		D	D	D	P
		†						†
SY, YC, YP, DY, CCY	HY (CH, PH)	PP	SE		SH	CO	SE	PM
				†				†
†								†
			CUA, CUB, MUB, MUC, MUD		CSA, CSB, MSB, MSC, MSD	CSA, CSB, MSB, MSC, MSD	CUA, CUB, MUB, MUC, MUD	CTA, CPA, CPB

TABLE 1-1 Type-Letter Designations of Fractional and Subfractional Horsepower Motors (*Continued*)

		AC motors					
		Single-phase					
			Capacitor types				
Manufacturer's name and address	Scope*	Split-phase	Capacitor-start	Permanent-split	Two-value	Split-phase-capacitor	Shaded-pole
Century Electric, Inc. 1831 Chestnut Street St. Louis, MO 63166	F	SP, SPM, SPS	CS, CSX, CSS	C, CX	CP, CPX	CM, CSM, SPM, CPM, CXPM	
Dayton Electric Manufacturing Co. 5959 West Howard Street Chicago, IL 60648	S, F, G	4K, 5K, 6K, 2Z, 4Z	4K, 5K, 6K, 2Z, 4Z	3M, 5K, 6K, 2Z, 4Z	4K, 5K, 6K	4K, 5K, 6K	3M, 5K, 6K, 2Z, 4Z
Delco Products, Division of General Motors P.O. Box 1042 Dayton, OH 45401	F						
Doerr Electric Corporation P.O. Box 67 Cedarburg, WI 53012	F	F, N, CO, C, X, FU, K	F, N, CO, C, X, FU, K		F, N, FU, T	F, N, CO, C, X, FU	
Dumore Corporation 1300 17th Street Racine, WI 53403	S, F, G						
Eastern Air Devices 1 Progress Drive Dover, NH 03820	S, F, G			†			
ECM Motor Company 1301 East Tower Road Schaumberg, IL 60196	S, G			C			M, T
EG & G Rotron, Custom Division Hasbrook Lane Woodstock, NY 12498	S, F			†			
Electra Motors/Dresser 1110 North Lemon Street Anaheim, CA 92801	F, G		†		†		
Electro-Craft Corporation 1600 Second Street South Hopkins, MN 55343	S, F						
Emerson Motor Division P.O. Box 3946 St. Louis, MO 63136	S, F	S	C (KS)	K	T (KK)	R (RK)	F
Fasco Industries, Inc. Ozark, MO 65721	S, F	25, 40, 44, 46, 60, 65	†	10, 18, 24, 26, 28, 62, 64, 74, 76, 82, 89			8, 21, 51, 63
Franklin Electric Company, Inc. 400 East Spring Street Bluffton, IN 46714	S, F, G	24, 140–143, 440–443, 640–643	110–113, 410–413, 21, 610–613	150–153, 25, 450–458, 650–653, 827, 828, 873, 874	22, 120–123, 420–423, 620–623	145–148, 445–448	825, 826, 841, 842, 871, 872 (821, 822, 861, 862) (AK, AL, AP, AW, AY)

Note: Letters in parentheses represent obsolete types.
*S, subfractional horsepower; F, fractional horsepower; G, gearmotors.
†Type is built, but there are no specific type letters assigned.

1-, 2-, 3-phase synchronous		Polyphase	AC or dc universal	DC motors				
Reluctance	Hysteresis	Squirrel-cage	Uncompensated	Brushless	Shunt	Compound	Series	Permanent-magnet
		SC, SCT, SCM			DN	DN	DN	
		2N, 3N, 3Z	2M, 5K, 3Z, 4Z					2M
		†						
FS, NS, FUS, COS, CS, XS		F, FU, CO, C, X, FU						
			†		†		†	†
†	†	†		†	†	†		
								D
	†	†		†				†
		†						
								†
		P						
64		12, 14, 16, 34, 36, 94, 96						
		23 130–133, 430–433, 630–633,						

TABLE 1-1 Type-Letter Designations of Fractional and Subfractional Horsepower Motors (*Continued*)

Manufacturer's name and address	Scope°	AC motors					
		Single-phase					
			Capacitor types				
		Split-phase	Capacitor-start	Permanent-split	Two-value	Split-phase-capacitor	Shaded-pole
Gearmaster 1809 South Route 31 McHenry, IL 60050	F, G	S	C	K			
General Electric Company 2000 Taylor Street Fort Wayne, IN 46804	S, F	KH, KHM, XB	KC, KCM, XC	KCP, KPM, XE	KC, XC	KHC	KSP, KSB, KSM, KSR, KSS
General Industries Company P.O. Box 4002 Elyria, OH 44036	S, G						AA, LM, FF
Gleason-Avery Division, John G. Rubino, Inc. 45 Aurelius Avenue Auburn, NY 13201	S, G						2506–2532, 500, 700, 800,
Globe Motor Division, TRW Inc. 2275 Stanley Avenue Dayton, OH 45404	S, F, G			†			
GSE/Voorlas P.O. Box 400 Carlisle, PA 17013	S, F						†
G. K. Heller Corporation 36 Mayflower Place Floral Park, NY 11001	S, F, G						
Hoover Electric Company 2100 South Stoner Avenue Los Angeles, CA 90025	S, F, G			†			
Howell Electric Motors, Division SFM Corporation 900 North Avenue Plainfield, NJ 07061	F, G	†	†	†	†	†	
IMC Magnetics Corporation 570 Main Street Westbury, NY 11590	S, F, G			†	†		
Indiana General Motor Products 1168 Barranca Drive El Paso, TX 79935	S, F						
Inertial Motors Corporation 280 North Broad Street Doylestown, PA 18901	S, F						
International Components Corp. 1740 Marshfield Chicago, IL 60062	S, G						

Note: Letters in parentheses represent obsolete types.

°S, subfractional horsepower; F, fractional horsepower; G, gearmotors.

†Type is built, but there are no specific type letters assigned.

1-, 2-, 3-phase synchronous		Polyphase	AC or dc universal	DC motors				
Reluctance	Hysteresis	Squirrel-cage	Uncompensated	Brushless	Shunt	Compound	Series	Permanent-magnet
		P			M		M	
SC, SK, SH, SCP, SCR		K, KM, KS, XA	BA, P		BC	BC	BC	BP
								†
B-30-CS2, B-30-CS3, 500								
†	†	†	†	†	†	†	†	†
			†					†
			GT, K		2T, 6T, MS, NH, 56			T2, 2P, 8P, 12P, P45
		†			†	†	†	†
†	†	†						
†	†	†			†	†	†	†
								†
								†
								†

TABLE 1-1 Type-Letter Designations of Fractional and Subfractional Horsepower Motors (*Continued*)

		AC motors					
		Single-phase					
			Capacitor types				
Manufacturer's name and address	Scope*	Split-phase	Capacitor-start	Permanent-split	Two-value	Split-phase-capacitor	Shaded-pole
Janette Div., Eastern Air Devices 1 Progress Drive Dover, NH 03820	F, G	SS, SU, SA	KU, KA, KS, KC	KC, KS, KA, KU	KC, KS KA, KU		
Japanese Products Corporation 7 Westchester Plaza Elmsford, NY 10523	S, F, G			†			
Jard Company, Inc. Bowen Road Bennington, VT 05201	S						M, MT, MF, MTF
The Jensen Company 110 North Missouri Avenue Sedalia, MO 65301	S, F						
Jones Instrument Corporation 432 Fairfield Avenue Stamford, CT 06904	S, F, G						
Leeson Electric Corporation 2100 Washington Street Grafton, WI 53024	F	S	C	P	K		
Leland Electrosystems, Inc. 740 East National Road Vandalia, OH 45377	F, G		†	†			
Marathon Electric Manufacturing Corp. P.O. Box 1407 Wausau, WI 54401	F	SS, SSF, SD, ST	SCS, SCSE, CD, CT	SC, SCE	B, SC	SSP	
Minnesota Electric Technology, Inc. 352 South Main Winnebago, MN 56098	F, G						
Motor Appliance Corporation 555 Spirit of St. Louis Blvd. P.O. Box 22 Chesterfield, MO 63017	F	S	L	C			
Motor Products Owosso Corporation 201 South Delaney Road Owosso, MI 48867	S, F, G						
Motor Specialty Inc. 2801 Lathrop Avenue Racine, WI 53405	S, F, G						

Note: Letters in parentheses represent obsolete types.
*S, subfractional horsepower; F, fractional horsepower; G, gearmotors.
†Type is built, but there are no specific type letters assigned.

1-, 2-, 3-phase synchronous		Polyphase	AC or dc universal	DC motors				
Reluctance	Hysteresis	Squirrel-cage	Uncom-pensated	Brushless	Shunt	Compound	Series	Permanent-magnet
		PU, PA, PS, PC, PD, PE			DU, DA, DS, DC, DD, DE	DU, DA, DS, DC, DD, DE	DU, DA, DS, DC, DD, DE	
†	†			†				†
			†		†		†	†
			†		†			†
		T						
		†			†	†	†	†
		TS, QS, TSE, QSE, TD, TT, TF, QD, QT			DS	DP	DX	
								32A, 32B, 32C, 560DP, 56TE, 32GMS, 32GMR
								PE, PF, PU, PR
			†		†	—	†	†

TABLE 1-1 Type-Letter Designations of Fractional and Subfractional Horsepower Motors (*Continued*)

		AC motors					
		Single-phase					
			Capacitor types				
Manufacturer's name and address	Scope*	Split-phase	Capacitor-start	Permanent-split	Two-value	Split-phase-capacitor	Shaded-pole
Motor Technology, Inc. 2796 Culver Avenue Dayton, OH 45429	S, F, G						
Northfield Electric Motor Co., Inc. 5317 Ravenswood Avenue Chicago, IL 60640	S, F						
Northland, Div. Scott & Fetzer 968 Bradley Street Watertown, NY 13601	S, F						
Ohio Electric Motors Paint Fork Road Barnardsville, NC 28709	F						
Olsen Machine and Tool Company, Inc. 225 Wales Street North Abington, MA 02172	S						†
Oriental Motor Company, Ltd. 2701 Toledo Street Torrance, CA 90503	S, F, G			K, J			
Oster 5055 North Lydell Avenue Milwaukee, WI 53217	S, F						†
Pacific Scientific, Motor and Control Division P.O. Box 106 Rockford, IL 61105	F						
The Pittman Corporation 343 Godshall Drive Harleysville, PA 19438	S, F, G						
H. K. Porter Company, Inc. 1401 West Market Street Warren, OH 44485	F	S	CT, CY (CES)	C	CTC, CYC (CEC)		
Prestolite Motor Division 511 Hamilton Street Toledo, OH 43964	S, F						
Reliance Electric Rt. 1 Madison, IN 47250	F, G	SP	CS	CL	CH	†	

Note: Letters in parentheses represent obsolete types.
*S, subfractional horsepower; F, fractional horsepower; G, gearmotors.
†Type is built, but there are no specific type letters assigned.

1-, 2-, 3-phase synchronous		Polyphase	AC or dc universal	DC motors				
Reluctance	Hysteresis	Squirrel-cage	Uncompensated	Brushless	Shunt	Compound	Series	Permanent-magnet
			†			†	†	CIR, CIL, CMR, CML, DMR, DML, FIR, FIL, HIR, HIL, GMR, JMR
			†		†	†	†	†
			†				†	†
							†	†
J, K, L	K	K		†				†
			†					
				†				†
				†				†
		P	UD	UD	D	D	D	DPM
						†	†	†
CSR	PMR	P, PM			T, TR	T, TR	T, TR	T, TR

TABLE 1-1 Type-Letter Designations of Fractional and Subfractional Horsepower Motors (*Continued*)

Manufacturer's name and address	Scope*	AC motors					
		Single-phase					
			Capacitor types				
		Split-phase	Capacitor-start	Permanent-split	Two-value	Split-phase-capacitor	Shaded-pole
RMR Corporation P.O. Box 469 Elkton, MD 21921	S, F			C, G, N, S			T, R, Y, H, L
Robbins and Myers 1345 Lagonda Avenue Springfield, OH 45501	S, F, G	KS, SP	KL, CS	KP, KPT, PSC	KK		K
Rotating Components Division, Aeroflex Laboratories, Inc. Service Road Plainview, NY 11805	S, G			†			
Sheller Globe Corporation 989 Athens Street Gainesville, GA 20501	S, F			4J, 6E			2B, 2N, 4G, 4T
The Singer Company, Kearfott Div. 1225 McBride Avenue Little Falls, NY 07057	S, F, G			†			
Singer, Controls Division 9655 West Soreng Avenue Schiller Park, IL 60176	S, G						
The Singer Company, Motor Products Division P.O. Box 35 Pickens, SC 29671	F, G						
Smith, A.O., Corp. 531 North Fourth Street Tipp City, OH 45371	F	S, SP	C, CS	F, PSC	K		
UMC Electronics, Inc., Beau Motors Division 460 Sackett Point Road North Haven, CT 06473	F		†				
Universal Electric Company 300 East Main Street Owosso, MI 48867	F	AG, AH, JG, JH, CG, CH, DG, DH, HG, HH	AP, AN, JP, JN, CP, CN, DP, DN, HP, HN	AE, AF, JE, JF, CE, CF, DE, DF, HE, HF	AE, AF, JE, JF, CE, CF, DE, DF, HE, HF		AA, AB, JA, JB, CA, CB, DA, DB, HA, HB (EA, BA, BB)
Von Weise Gear Company St. Clair, MO 63077	S, G						†
Voorlass Manufacturing Company 1711 South Street Racine, WI 53404	S, F						†
Welco Industries, Inc. 9027 Shell Road Cincinnati, OH 45236	S, F	†	†	†	†	†	
Westinghouse Electric Corporation, Small Motor Division P.O. Box 566 Lima, OH 45802	F	FH, FHT	FJ, FZ	FL, FLL	FT	FHT	FE

Note: Letters in parentheses represent obsolete types.
*S, subfractional horsepower; F, fractional horsepower; G, gearmotors.
†Type is built, but there are no specific type letters assigned.

1-, 2-, 3-phase synchronous		Polyphase	AC or dc universal	DC motors				
Reluctance	Hysteresis	Squirrel-cage	Uncompensated	Brushless	Shunt	Compound	Series	Permanent-magnet
SKP, SKS, SKL, SKK, SL		L, 3P						†
†	†	†						
	†	†		†	†			†
	†							
			†				†	†
		P						
	†			†				†
		AL, AM, JB, JM, CL, CM, DL, DM, HL, HM	VZ, WZ, YY, YZ		VZ, WZ, YY, YZ (WY, ZZ)	VZ, WZ, YY, YZ (WY, ZZ)	VZ, WZ, YY, YZ (WY, ZZ)	YW
								†
			†					†
†		†	†		†	†	†	†
(FBH, FBJ, FBL, FBT, FBS)		FB	(ADS)		(FK)	(FK)	(FK)	

NOTES FOR TABLE 1-1

Aerovox Inc., Motor Division. Makes C-frame (skeleton-type) shaded-pole motors only.

Ametek Inc., Lamb Electric Div. Motors are now identified by a model number. Formerly they were identified by an instruction sheet (IS) number. Either number is stamped on the nameplate if there is one; if not, it is stamped on the motor housing. Ametek Inc., Lamb Electric, also make a brushless dc motor and indirect drive, and direct-drive blowers are also manufactured under the trade name Windjammer.

Ashland Electric Products Div. Makes dual- and triple-speed hysteresis synchronous motors for 60-Hz, 400-Hz, and other frequencies as required. Special motors designed for lasers and optical mirrors.

Bison Gear and Engineering Corp. In addition to a line of gearmotors, also offers a line of gear reducers.

Bodine Electric Company. Has a complete line of metric motors and gearmotors, designed specifically for operation on 50 Hz. Also makes a line of speed-control systems.

Brailsford and Company. Specializes in brushless dc motors of very small size rated from 1 to 10 W, for instruments and timing applications.

Brevel Motors, Inc. Motors are more broadly classified by series rather than by type.

Series 712 are synchronous and asynchronous skeleton-type two-pole shaded-pole motors.

Series 724, 726, and 729 are permanent-magnet designs.

Series 700 are gearmotors.

Carter Motor Company. Makes dc to ac converters that are rated for both 50- and 60-Hz output

Century Electric, Inc. Has made motors under the name Gould Century.

Types CS and CP are high-torque motors.

Types CSX and CPX are medium-torque motors.

Types CX and AM are fan-duty motors.

Types CSS and SPS are for special service.

Types CSM, CPM, CXPM, SPM, and SCM are multipole, multispeed motors.

Dayton Electric Manufacturing Co. Makes a line of speed-control systems.

Doerr Electric Corporation. Type letters in the table refer to construction.

F = totally enclosed, fan-cooled

N = totally enclosed, nonventilated

CO = open, fan-duty, air-over

C = enclosed, fan-duty, air-over

X = open, fan-cooled

FU = TEFC explosion proof

Synchronous motors are designated by adding the suffix letter S to any type letter(s).

Eastern Air Devices. Makes a line of gear reducers, including right-angle reducer with two output shafts.

Electra Motors/Dresser. Makes a line of right-angle and parallel-shaft gear reducers for rigid-base and NEMA C-frame mounting.

Emerson Motor Division. The appearance of a second letter in the model number, before the first number, designates a line of motors for a particular application or a family of motors with a common mechanical feature.

Fasco Industries, Inc. Has separate engineering and manufacturing facilities at Ozark, Eldon, and Cassville, Missouri.

Franklin Electric Company, Inc. Makes a line of submersible pumps and motors. Air-moving motors are engineered and manufactured at Jacksonville, Arkansas.

General Electric Company. Third letter in dc types designates the type of winding, e.g., BCE is a compensated shunt-wound motor. Makes a line of speed-control systems.

Globe Motor Division, TRW Inc. A diversified line of miniature motors, mostly built to military specifications, with many civilian counterparts. Type letters usually denote physical size and mechanical construction as well as electrical type, giving rise to too many combinations to list in the table. Line of motors includes 400- as well as 60-Hz motors, inverter-driven ("brushless dc") ac motors, and brushless dc permanent-magnet motors. Line also includes pumps, fans, linear and rotary actuators, using miniature motors. Makes compensated universal motors.

GSE/Voorlas. Makes a line of blowers with motors up to 3.2 hp for specialized applications requiring high pressures.

Hoover Electric Company. Specializes in custom-built dc and 400-Hz motors for aircraft service and oil-filled motors for deep-submergence applications.

Japanese Products Corporation. Makes a speed-control system which utilizes a feedback-regulated closed loop in which an ac induction motor, tachometer-generator, and solid-state electronic package are integrated to control both speed and torque.

Leland Electrosystems, Inc. Makes two-phase motors, model Q. Second letter in model number indicates

S = standard performance

H = special performance

Third letter, *D*, when present, designates a brake. Mechanical description designators include

O = open

T = totally enclosed, nonventilated

F = totally enclosed, fan-cooled

P = partial motor

G = geared head

Marathon Electric Manufacturing Corp. Makes a line of speed-control systems.

Motor Products Owosso Corporation. Assumed the dc product line of Controls Co. of America. Has since dropped the brush-type dc motors, but carries the Redmond trademark.

Oriental Motor Company, Ltd. Makes a line of speed-control systems featuring a feedback speed-control motor for high efficiency.

Oster. Manufactures a line of speed-control systems for use with its motors in small appliances.

H. K. Porter Company, Inc. (formerly Peerless Electric Co.). Motors are still marketed with trade name Peerless.

Reliance Electric. Makes a line of gear reducers. (See Chap. 17, Art. 17-23.)

Rotating Components Division, Aeroflex Laboratories, Inc. Motors are built to military specifications. Line includes servomotors, motor generators, drag-cup motors, inverter-driven ac motors (brushless dc), and hysteresis synchronous and polarized synchronous motors.

UMC Electronics, Inc. Makes "inside-out" motors, as well as a line of brushless motors.

MS = multispeed hysteresis synchronous

SP = single-purpose hysteresis synchronous

VS, VT = ac torque motor

Welco Industries, Inc. Makes a line of speed-control systems.

Motors are often designed to drive a specific application or for a specific purpose. The nameplate of Fig. 1-1 indicates a motor that is designed to drive a fan or blower, such as a furnace blower.

1-6 Model Number. The model number shown on the nameplate of Fig. 1-1 after the abbreviation "MOD." uniquely identifies the motor to the manufacturer. Nearly all motors carry some form of identification unique to the manufacturer. It may be a style number, model number, identification number, or instruction sheet. Many manufacturers use a coded serial number that identifies the approximate date of manufacture.

1-7 Horsepower. Electric motors, like any other motors, are usually rated in *horsepower*, abbreviated "HP" on the nameplate in Fig. 1-1.

Originally, a motor was said to be developing 1 hp when it was doing work at the rate of 33,000 ft-lb/min. This figure is said to have been arrived at by determining the average rate at which London dray horses could do work over an 8-h period. The watt (W) is also a unit of power. In the United States, a *horsepower* is now defined as 746 W exactly.

Millihorsepower (abbreviated mhp) is often used for rating subfractional horsepower motors. A millihorsepower is 0.001 hp or 0.746 W.

In Europe, and elsewhere that the metric system is used, rated power output may be expressed in chevaux-vapeur, abbrevated cv; 1 cv is equivalent to 735.5 W, which is close enough to the 746-W horsepower for most practical purposes. Often the rated power output is expressed in kilowatts or watts by European manufacturers.

The horsepower figure stamped on the nameplate is the horsepower the motor is rated to develop when connected to a circuit of the voltage, frequency, and number of phases specified on the motor nameplate. Some motors will develop their rated horsepower continuously, and others are designed for use for intermittent service and are rated to develop their horsepower for only a limited period. This length of time is stamped on the nameplate in some fashion and signifies the time the motor will carry its rated load without exceeding the temperature rating, which is also stamped on the nameplate. This time may be indicated by the abbreviation cont. for continuous or by a definite time interval, such as 30 min. (See Art. 1-18.)

NEMA defines horsepower on the basis of the breakdown torque of the motor. Commonly used horsepower ratings are shown in Table 19-1.

1-8 RPM. It hardly needs saying that rpm, also written r/min, stands for revolutions per minute. This figure represents the approximate speed at which the motor will run when properly connected and delivering its rated output. In the case of synchronous motors, this figure is exact.

The full-load speed of a normal *induction* motor is determined primarily by the frequency and number of poles.* For fractional horsepower induction motors, except permanent-split and shaded-pole types, approximate full-load speeds are given in Table 19-1. Full-load speeds for shaded-pole and permanent-split capacitor motors are given in Table 19-5. Conversely, these tables can be used to determine the number of poles for which the motor is wound.

Occasionally, two or more speeds are stamped on the nameplate. If

*For the significance of the term number of poles, see Art. 2-15.

the speeds are close together, as in 1725/1425 rpm, the motor is probably a dual-frequency motor; but this point can be checked definitely by referring to the frequency stamping. Such a motor is, however, a single-speed motor. Dual-frequency stamping is more common in the case of polyphase motors than it is in the case of single-phase motors. If the two speeds are in the ratio of approximately 3:2 or 2:1, the motor is probably a *multispeed motor.* The changing in speeds may be effected by changing the number of poles in the winding, in which case definite speeds that are substantially independent of the load are obtained; or the change in speed may be effected by changing the slip of the motor, in which case the speed, particularly on the low-speed connection, depends to a large extent upon the load. Pole-changing motors are discussed in Arts. 4-23 to 4-26, 4-28, and 5-19; slip-changing motors are treated in Chap. 7.

If a motor is designed for more than two speeds of operation, often only the highest rpm is marked on the nameplate.

Constant-speed shunt or compound-wound dc motors ordinarily are rated to operate at one of the speeds in Table 19-1, but the speed bears no relation to the number of poles. DC series motors may operate at almost any speed.

Universal motors may operate at almost any speed independent of the number of poles or frequency.

1-9 Volts. The *volt* is the unit of electrical pressure. Just as hydraulic pressure is necessary to force water, or any other fluid, through pipes, so is it necessary to have electrical pressure to drive current through a circuit, and this electrical pressure is measured in terms of volts. A corresponding unit of water pressure would be pounds per square inch, or kilograms per square centimeter.

The voltage figure given on the motor nameplate refers to the voltage of the supply circuit to which the motor should be connected. Sometimes two voltage figures are given, such as 115/230. In this case, the motor is intended for use on either a 115- or a 230-V circuit, and special instructions are furnished as to how to connect the motor for each of the different voltages. These instructions may appear on the nameplate itself (not in the instance shown in Fig. 1-1), or on a separate plate or label affixed to the outside of the motor or to the inside of the cover plate for the conduit box, or as an instruction tag. Standard voltage ratings are given in Art. 19-4.

The discussion in Arts. 1-10 and 1-13 as to the nature of an alternating current applies equally well to an ac voltage. Thus, the relationship between effective volts and peak volts is the same as the relationship between effective amperes and peak amperes, which is explained in

Art. 1-10. In practice, unless otherwise specified, "115 volts" means "115 volts effective."

1-10 Amperes. The *ampere* (A) is the unit of measurement of rate of current flow. The unit of quantity of electricity is the *coulomb* (C), which corresponds to gallons, or liters. The rate of flow of water could be expressed in gallons per second, or liters per second, and the rate of flow of electric current could be expressed in coulombs per second. Instead, the shorter term ampere is applied. (One dc ampere is actually 1 C/s.)

The amperes figure given on the motor nameplate represents the approximate current drawn by the motor when developing rated horsepower on a circuit of the voltage and frequency specified on the nameplate. In Fig. 1-1, the motor is a single-voltage motor, and one ampere figure is given, 5.5. On *adjustable varying-speed* fan motors (Chap. 7), it is customary to give only the current at the top speed. Dual-frequency, 50/60-hertz (Hz) motors usually have two stampings of full-load amperes; the 50-Hz current is approximately 10 to 15 percent higher than the 60-Hz current.

The nature of alternating current is described in Art. 1-13 and illustrated in Fig. 1-2. (Figure 1-2 describes the pulsating nature of an ac

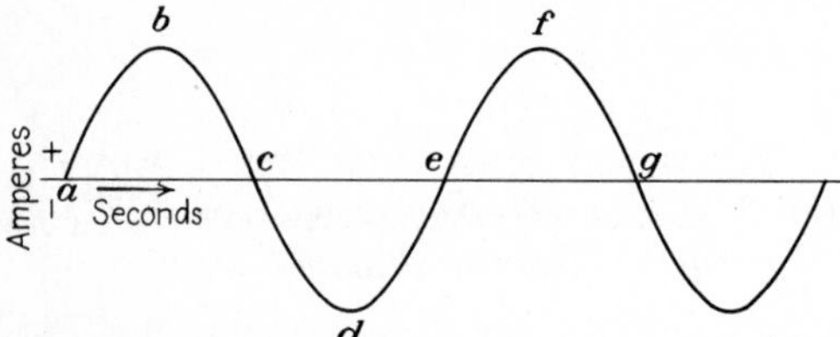

Fig. 1-2 Current flow in an ac circuit.

voltage equally well.) In a dc circuit, the current is flowing continuously in one direction. In an ac circuit, the direction of current flow periodically reverses many times per second, so that the total net flow, as measured in coulombs per second over any whole number of cycles, is zero. In direct current, the ampere is easy to define: it is simply a current flow of 1 C/s. What is an ac ampere? The *ac ampere* is defined as that amount of alternating current that will produce the same heating effect in an ohmic resistance that 1 A of direct current would produce if it were flowing through the same ohmic resistance. Since the heating effect is proportional to the square of the current, the *effective value* of an alternating current is the square root of the average (mean value) of the squares of the instantaneous currents, these instantaneous currents being taken at a large number of uniform time intervals throughout one cycle. Often the effective value is referred to as the *root-mean-square (rms)* value, because of the manner in which it is found. The *maximum* or *peak* value is the current at the instant *b, d,* or *f* in Fig. 1-2. If the wave is a sine wave, as is usually the case, the following relationship holds:

$$\text{Effective value} = \text{rms value} = 0.7071 \times (\text{maximum or peak value})$$

In practice, the ac ampere is always taken as the effective or rms value, or the equivalent value of dc current necessary to produce the same heating effect. One ac volt impressed across a noninductive resistance of one ohm (Ω) will produce a current of one ac ampere.

1-11 Frame. Frame is abbreviated "FR." on the nameplate in Fig. 1-1. Frame size refers principally to the physical size of the machine, as well as to certain constructional features. Standard NEMA frame sizes for fractional horsepower motors are shown in Fig. 19-1, together with their principal mounting dimensions.

Basic frame number is the D dimension in inches, multiplied by 16. Dimensions BA, E, and H are constant for any given D dimension. (These dimensions are illustrated in Fig. 19-1.)

Suffix letters may be added to a basic frame number to denote variations, as follows:

C = face mounting

G = gasoline pump motors (Fig. 19-3)

H = a frame having an F dimension larger than that of the same frame diameter without the H (Fig. 19-1)

J = jet pump motors (Fig. 19-6*b*)

K = cellar drainer and sump pump motors (Fig. 19-2)

L = motors for home-laundry equipment

M and N = oil burner motors (Fig. 19-4)

Y = special mounting dimensions (which must be obtained from the motor manufacturer)

Z = all mounting dimensions standard except the shaft extension

Prefixes are added to NEMA frame numbers by some manufacturers. There is no standardization of such prefixes, and each manufacturer uses an individual system.

1-12 Hertz (Frequency). "HZ" on the nameplate of Fig. 1-1 is the abbreviation for hertz; 1 Hz is one cycle per second. As pointed out in

Art. 1-13, a cycle consists of two reversals. *Frequency* refers to the rate of these reversals. Frequency had long been expressed in cycles per second or, loosely, cycles; however, common practice today is to use hertz.

Mexico converted from 50 to 60 Hz in 1975, so that all of North America is now using 60 Hz as a standard. Many countries, especially in Europe, use 50 Hz.

1-13 AC. The universal abbreviation for alternating current, as distinguished from direct current (or continuous current, as it is sometimes called), is ac. Although the nameplate of Fig. 1-1 does not have this marking, the presence of the markings "HZ" and "PH" signifies that the motor is intended for use on a circuit of alternating current.

In an ac circuit, the current repeatedly changes its direction of flow through the circuit many times every second. In any dc circuit, the current flows in one direction continuously; although the *magnitude* of the current flowing into a dc motor varies with the amount of the load imposed, the *direction* of flow in the circuit does not change. For this reason, the flow of direct current in a wire is usually likened to the flow of water in a pipe.

The nature of alternating current can be seen by referring to Fig. 1-2, which is a graph of instantaneous current flowing in an ac circuit plotted against time. Points *a*, *c*, *e*, and *g* represent instants when the current is zero. From *a*, the current is seen to start at zero, increase to a maximum value at *b*, and decrease to zero at *c*; the current then becomes negative, which signifies that it reverses its direction of flow in the circuit. Flowing in this opposite direction, the current again builds up to a maximum value at *d* and decreases to zero again at *e*, and one cycle is complete.

In other words, the current is reversing its direction of flow continually; hence the name "alternating current." One reversal, as from *b* to *d*, is called an *alternation;* two alternations, as from *a* to *e*, make one *cycle.*

1-14 Phases. Right after "HZ" is found the inscription "PH. 1." AC systems may have one, two, or three phases. Single-phase and three-phase systems are the most common.

(1) Single-phase system. The single-phase system is the simplest form of ac system. A typical system is shown in Fig. 1-3. Residences, and most other places where fractional horsepower motors are used, usually are wired only for single-phase alternating current. For this reason, a great majority of fractional horsepower ac motors are designed for operation on a single-phase system.

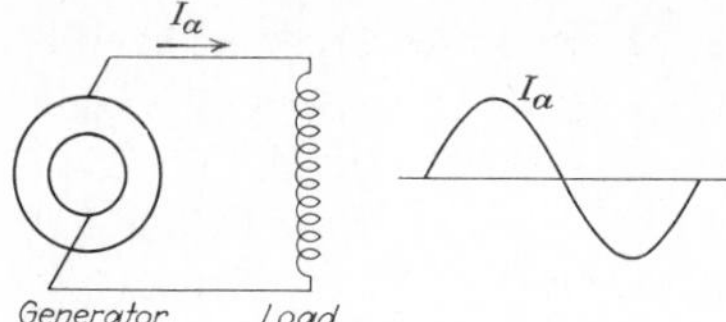

Fig. 1-3 A simple single-phase ac system.

(2) Two-phase systems. A two-phase system (see Fig. 1-4) can be thought of as two electrically distinct single-phase systems, as illustrated. There, the two ac generators are shown mechanically coupled, and the loads are electrically distinct, one from another. The currents in both phases alternate through their respective cycles, as shown in Fig. 1-4*b*, but they do not reverse simultaneously. The current in phase

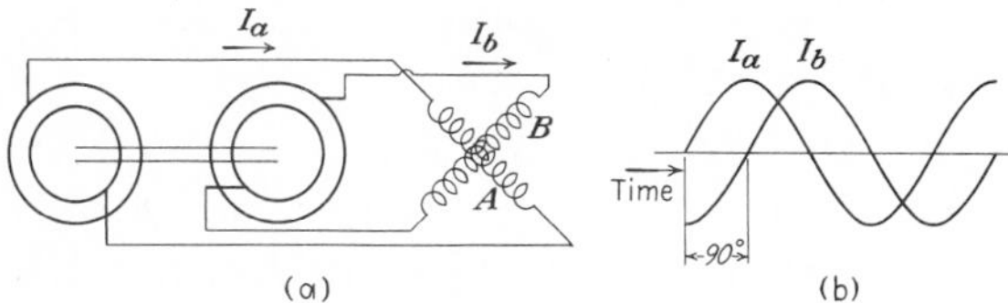

Fig. 1-4 A two-phase system with two single-phase generators. (*a*) Mechanically coupled generators with electrically distinct loads; (*b*) relation of currents in the two phases.

B reaches its maximum value one-quarter cycle behind the current in phase A; in fact, the current in phase B always reaches any particular point in its own cycle 90°, or one-quarter cycle, behind the time when the current in phase A reaches the same point in its cycle; hence the two-phase system is sometimes designated as a quarter-phase system.

In practice, however, only a single ac generator is used, instead of the two shown in Fig. 1-4. This single machine has two electrically distinct windings, or phases, which are displaced 90 electrical degrees in space so that the voltages induced in them are 90° out of phase. A two-phase system may be four-wire or three-wire, as shown in Fig. 1-5. Two-phase systems are rarely used today.

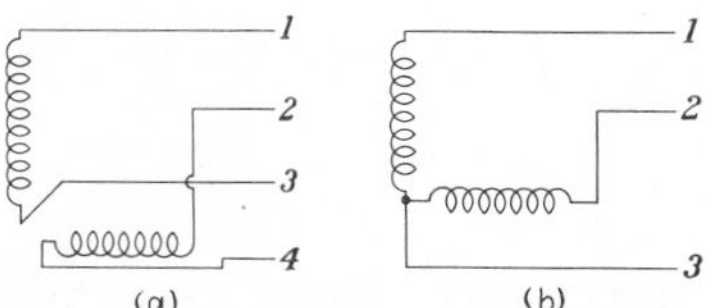

Fig. 1-5 Practical two-phase systems. (*a*) Four-wire system; (*b*) three-wire system.

(3) Three-phase systems. For simplicity's sake, it may be well to think of a three-phase system as three electrically distinct single-phase systems with three distinct sources of voltage, and two wires for each phase, making six wires in all, as shown in Fig. 1-6*a*. In this figure, each source of voltage is represented by a single coil as A, B, or C. This system is arranged so that the currents in the respective three phases reach their maximum value in time at three different instants, differing by one-third cycle, i.e., by 120°, as shown in Fig. 1-6*b*. In practice, however, the three phases are not usually electrically distinct as they generally are in two-phase systems. The reason, of course, is the unnecessary complication and expense of using six wires, necessitating the use of other systems, described in the following paragraphs.

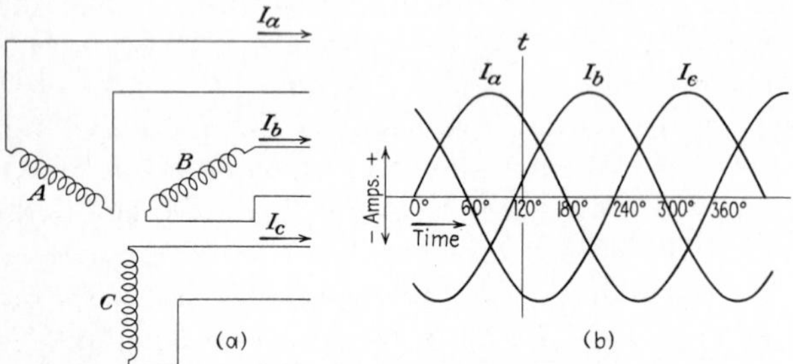

Fig. 1-6 A three-phase six-wire system. (*a*) Three electrically distinct single-phase windings displaced 120° from each other; (*b*) relation of currents in the three phases.

Figure 1-7*a* represents the system shown in Fig. 1-6, except that a common return wire has been used for all three phases. (The three phases are now no longer electrically distinct, for they are using a common return wire.) This type of system is often used in practice. Sometimes the common return, or neutral, is grounded; sometimes the earth itself is made to act as the common return circuit. If the load is balanced, the current in this return wire is zero at all instants; and since

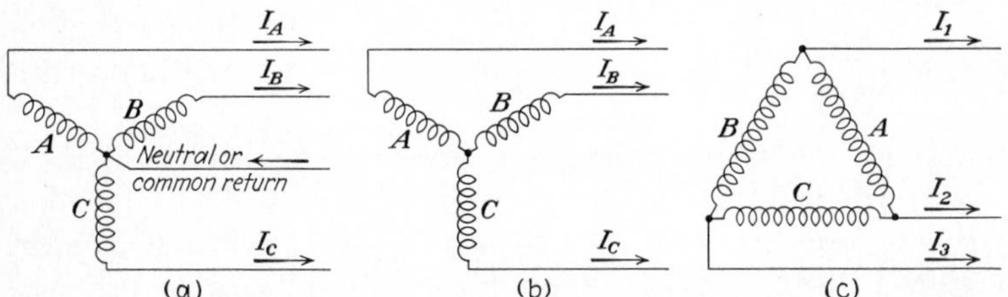

Fig. 1-7 Three-phase Y- and delta-connected systems. (*a*) Y-connected system with neutral (four-wire system); (*b*) Y-connected system without neutral (three wire system); (*c*) delta-connected system.

the wire carries no current, it can be omitted, as shown in Fig. 1-7*b*. The reader may check this point by drawing any vertical line, as *t*, in Fig. 1-6*b*, and adding up the instantaneous values of the three currents; the sum of the three currents will be zero. In like manner, if the line *t* is drawn at any other point, it will also be found that the sum of the three currents is zero. Therefore, the fourth wire is unnecessary in a balanced three-phase system and often is omitted. Both of these connections are *Y connections,* also called *wye connections* or *star connections.* The line-to-line voltage = $\sqrt{3}$ × line-to-neutral voltage, and line-to-neutral voltage = 0.577 × line-to-line voltage. Windings of three-phase fractional horsepower induction motors are usually star-connected.

Another form of connection is the *delta connection,* shown in Fig. 1-7*c*. This connection undoubtedly derives its name from its similarity of appearance to the Greek capital delta. With this connection and a balanced three-phase load, the line currents are all equal to $\sqrt{3}$ multiplied by the current in the individual phases or coils.

All the connections and systems described are applicable to motors as well as to generators.

1-15 KVA Code. The code letter on the nameplate indicates the locked-rotor kilovolt-amperes per horsepower, according to Table 1-2. By referring to the table, it can be seen that the N means that this motor has a maximum locked rotor kilovolt-amperes per horsepower of 12.49. Since the motor is rated ⅓ hp, the maximum locked-rotor kilovolt-amperes is, therefore, 4.163. The maximum locked-rotor amperes at 115 V will therefore be 4163/115 = 36.2.

For a three-phase motor, the code letter is determined by the locked-rotor kilovolt-amperes per horsepower per phase.

Table A-6 in the Appendix is a tabulation of locked-rotor currents and corresponding code letters for various horsepower ratings.

1-16 Insulation Class. The insulation class, abbreviated "INS. CL." in Fig. 1-1, refers to the class of insulation used in the motor; most commonly used are Classes A (105°C) and B (130°C). For a description of insulation class, see the Glossary.

1-17 Maximum Ambient. Maximum ambient, abbrevated "MAX. AMB." in Fig. 1-1, refers to the maximum ambient temperature in which the motor is capable of operating. Operation of the motor in a higher ambient may result in shortened life and reduced torques.

The figure of maximum ambient, expressed in degrees Centigrade, should not be confused with the figure formerly used to indicate rated

TABLE 1-2 Locked-Rotor-Indicating Code Letters

Code letters	Kilovolt-amperes per horsepower with locked rotor
A	0–3.14
B	3.15–3.54
C	3.55–3.99
D	4.00–4.49
E	4.5–4.99
F	5.0–5.59
G	5.6–6.29
H	6.3–7.09
J	7.1–7.99
K	8.0–8.99
L	9.0–9.99
M	10.0–11.19
N	11.2–12.49
P	12.5–13.99
R	14.0–15.99
S	16.0–17.99
T	18.0–19.99
U	20.0–22.39
V	22.4 and up

Source: From *National Electrical Code, 1984,* NFPA no. 70, National Fire Protection Association, Quincy, Mass., 1984; *NEMA Motor and Generator Standards,* NEMA Publ. no. MG1-1978, National Electrical Manufacturers Association, Washington, 1978.

temperature rise on nameplates and elsewhere. The rated temperature rise was based on temperatures taken by thermometer, a method so inaccurate it has long been abandoned. However, the rated rise figure in degrees Centigrade may still be found on old motors and in old literature.

1-18 Duty. The duty, or time rating, denotes the length of time the motor is expected to be able to carry rated load under usual service conditions. In the case illustrated, it is "CONT.," which means continuously. Intermittently rated motors carry a time rating, generally 5, 15, 30, or 60 min; this means that the motor will carry rated load for only the time given, and the motor must be allowed to cool down, generally to within 5°C of room temperature, before being restarted.

1-19 Service Factor. Service factor, abbreviated "S.F." in Fig. 1-1, is a term usually applicable only to general-purpose ac motors, although it is also used for certain definite-purpose motors. It is defined by NEMA as follows: "The service factor of a general-purpose alternating-current motor is a multiplier which, when applied to the rated horsepower, indicates a permissible horsepower loading which may be carried under the conditions specified for the service factor." Service factors for general-purpose fractional horsepower ac motors are given in Table 1-3.

TABLE 1-3 Service Factors for General-Purpose AC Motors

Hp ratings	Service factor
1/20	1.4
1/12	1.4
1/8	1.4
1/6	1.35
1/4	1.35
1/3	1.35
1/2	1.25
3/4	1.25
1 (2-pole)	1.25

Source: From *NEMA Motor and Generator Standards*, NEMA Publ. no. MG1-1978, National Electrical Manufacturers Association, Washington, 1978.

1-20 Service Factor Amperes. Abbreviated "S.F.A." in Fig. 1-1, this value of current given on the motor nameplate corresponds to the approximate current drawn by the motor when it is loaded to its rated service factor horsepower and when connected to a circuit of the voltage and frequency specified on the nameplate.

1-21 Bearing. Bearing is abbreviated "BRG." in Fig. 1-1. Fractional and subfractional horsepower motors normally use one of two types of bearings, sleeve or ball. General constructional features of bearings are given in Chap. 17. The nameplate of Fig. 1-1 indicates that this motor has sleeve bearings.

1-22 Motor Reference. Abbreviated "MTR REF" in Fig. 1-1, this reference number is part of the manufacturer's identification and is used, along with the model number described in Art. 1-6, to identify the mechanical and electrical features of the motor.

1-23 Date Code. Most manufacturers employ a series of numbers to designate when the motor was manufactured. The series is usually in code and in some cases may tell the day of the year and even the shift.

1-24 Other Information on the Nameplate. A variety of additional information may appear on the nameplate. This may include instructions for connecting the motor to a source of supply, reversing the direction of rotation, lubricating the motor, or operating it safely.

1-25 Mirror-Image UR. The nameplate shown in Fig. 1-8 has information similar to that in Fig. 1-1. However, one additional item is the mirror-image UR. This marking indicates that the motor is a UL recognized component and meets the requirements in UL 1004.[6] These requirements include materials, spacings, and construction, but not performance.

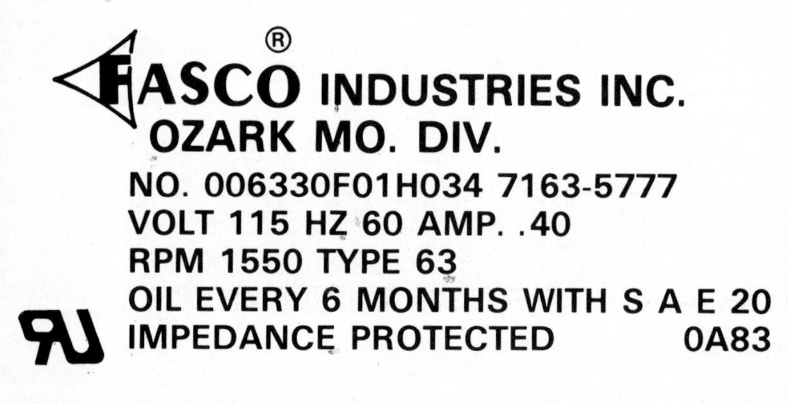

Fig. 1-8 A motor nameplate showing the mirror-image UR symbol.

1-26 Impedance-Protected. When a motor has inherent impedance sufficiently high to limit the input current under any load, including locked-rotor, to a value that does not cause dangerous overheating, the motor is said to be impedance-protected. The nameplate in Fig. 1-8 has the impedance-protected marking, indicating the motor has such protection. Impedance protection is normally limited to small shaded-pole and permanent-split capacitor motors. Some subfractional horsepower motors may have impedance-protected abbreviated as ZP. The requirements for impedance protection are given in UL 519.[7]

1-27 Nameplate Materials. Many manufacturers of motors today use some form of adhesive-backed material on which to convey the necessary information. The nameplates shown in Figs. 1-1 and 1-8 are examples of the use of such materials. The information is printed on adhesive-backed metallized Mylar, and a clear adhesive Mylar film is placed over the entire nameplate for protection. Other manufacturers use adhesive-backed polyester on which to print the information. Metal nameplates, which for years were used, have gradually been replaced by adhesive nameplates and ink stamping.

References

1. *Motors and Generators,* CSA Standard C22.2 no. 100-1983, Canadian Standards Association, 178 Rexdale Blvd. Rexdale, Ontario, Canada, M9W 1R3, 1983.
2. *National Electrical Code, 1984,* NFPA no. 70, National Fire Protection Association, Batterymarch Park Quincy, MA 02269.
3. *NEMA Motor and Generator Standards,* NEMA Publ. no. MG1-1978, National Electrical Manufacturers Association, 2101 L Street N.W., Washington, DC 20037, 1978.
4. *Thermal Protectors for Motors,* UL 547, Underwriters Laboratories, Inc., 333 Pfingsten Rd., Northbrook, IL 60062, 1980.
5. *Motors with Inherent Overheating Protection,* CSA Standard C22.2 no. 77-1976, Canadian Standards Association, 178 Rexdale Blvd., Rexdale, Ontario, Canada, M9W 1R3, 1976.
6. *Electric Motors,* UL 1004, Underwriters Laboratories Inc., 333 Pfingsten Rd., Northbrook, IL 60062, 1984.
7. *Impedance Protected Motors,* UL 519, Underwriters Laboratories Inc., 333 Pfingsten Rd., Northbrook, IL 60062, 1982.

Bibliography

American National Standards Institute: *Voltage Ratings for Electric Power Systems and Equipment (60 Hz),* ANSI C89.1, ANSI, 1430 Broadway, New York, NY 10018, 1982.

Staff editors: "How to Read Electric Motor Nameplates," *Electr. Appar.,* September 1982, pp. 18–24.

CHAPTER
2
WHAT MAKES AN INDUCTION MOTOR RUN?

Of all the various types of ac motors, the induction type is the most popular, whether for use on single-phase or polyphase circuits. This statement is equally true for fractional horsepower motors, the majority of which are operated on single-phase circuits. The enormous popularity of the induction motor is principally due to its simple construction, ruggedness and reliability, and constant-speed characteristics; i.e., the speed is substantially independent of the load within the normal working range. Constant-speed motors are required for the majority of applications. The polyphase induction motor is the essence of simplicity. Current is conducted into the primary windings (usually on the stator) but *induced* in the secondary windings (usually on the rotor) by electromagnetic action: hence the name *induction motor.* The secondary windings are short-circuited upon themselves, either directly or through an external resistance, and are not connected to the power supply.

2-1 Single-Phase and Polyphase Motors. Because the secondary current is entirely an induced current, the secondary winding usually is put on the rotating member, for, with this arrangement, no brushes, collector rings, or commutator is required. In the squirrel-cage form of construction of secondary windings, the "winding" consists of aluminum conductors which are molded into rotor slots and which are integral with a short-circuiting end ring of the same material at each end of the bars. To obtain special characteristics (principally high rotor resistance), sometimes the conductors and end rings are cast of aluminum or magnesium alloys. In other cases, individual bars of copper or brass, bonded to end rings, may be used. End rings are often referred to as *resistance rings* (see Fig. 5-2*g* in Chap. 5). This type of rotor is known as a squirrel-cage rotor because of the resemblance of the current-carrying conductors and end rings to the cylindrical cages originally made to exercise pet squirrels.

The squirrel cage is by far the most popular but not the only form of secondary winding used in induction motors. Polyphase induction motors may have a secondary winding generally resembling the primary winding, in which case collector rings and brushes are necessary. Such motors are known as *wound-rotor induction motors* and may be used with external resistors for applications requiring adjustable varying-speed motors. Single-phase repulsion-start induction motors use a secondary winding that is similar to the armature winding of a dc motor.

Single-phase induction motors have no inherent locked-rotor torque and must be started by other means. There are a number of different methods employed for starting single-phase induction motors, giving rise to a number of different types of motors, each type being named after the starting method.

Full and complete treatment of the theory of polyphase and single-phase induction motors would fill a book in itself.[1,*] However, for those who wish a short explanation of the basic principles of induction motors, a brief discussion is given here. It is believed that this short discussion will help the reader understand and remember some of the simpler aspects of induction motors such as why they run at all; why collector rings and brushes are seldom necessary; and why polyphase motors are inherently self-starting, whereas single-phase motors require some special starting arrangement. Such an understanding of the principles of what actually makes the motor run is of almost invaluable assistance in diagnosing and remedying faults in motors. To the student, it may well serve as an introduction to the study of fractional horsepower motors, a study that may be pursued further in the references and bibliography given at the end of this chapter.

In the fractional horsepower sizes, as previously mentioned, the sin-

*For numbered references, see References at end of this chapter.

gle-phase induction motor is far more commonly used than the polyphase induction motor. However, the principle of the operation of the polyphase induction motor is a little easier to grasp, and this understanding materially helps in understanding single-phase induction motors. Therefore, we shall be concerned first with the polyphase induction motor.

2-2 Rotation of a Copper Disk Produced by a Rotating Magnet. The basic principle of induction-motor action is illustrated by the horseshoe magnet and the copper disk shown in Fig. 2-1. Here the magnet is

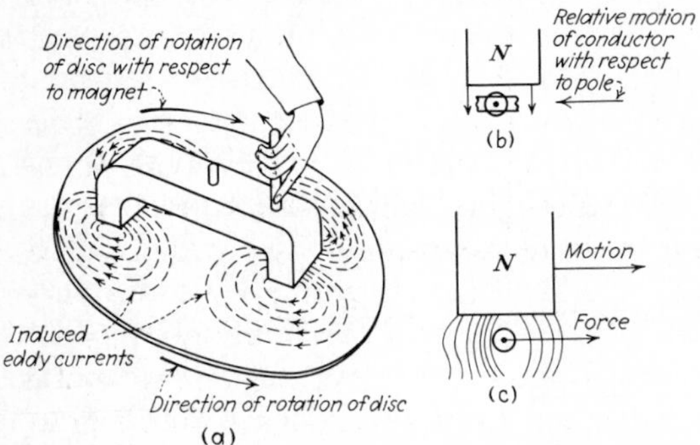

Fig. 2-1 Rotation of a copper disk produced by the rotation of a permanent magnet. (*a*) Induced eddy currents in a disk due to rotation of a magnetic field; (*b*) induced current in a conductor cutting magnetic lines of force causes magnetic flux to flow around conductor in a counterclockwise direction; (*c*) resultant increase in flux to the left of and decrease in flux to the right of the conductor cause a force to act on the conductor to move it to the right. *(From Chester L. Dawes, A Course in Electrical Engineering, vol. II, 4th ed., McGraw-Hill Book Company, New York, 1947.)*

being rotated by hand, causing the disk to rotate likewise, though at a slower speed. This disk follows the magnet, not because of any magnetic attraction between the two, for the former is of copper, a nonmagnetic material, but because of the action which is due to the eddy currents induced in the disk and to their reaction against the revolving magnetic field. As the magnet and its lines of force are caused to revolve, the lines of force cut the disk, which is a metallic conductor of low resistance. Eddy currents are, therefore, induced in the disk. The flow of these currents is in such a direction as to tend to oppose the motion of the magnetic field, i.e., to try to stop the magnet from rotating. (It is a fundamental principle of electricity and magnetism that the current flow in an electric circuit, induced by a change of magnetic flux linking the circuit, is always in a direction tending to oppose the change of flux.) The result is a retarding drag on the magnet, but, since action is always equal to reaction, there is an equal and opposite force exerted on the disk, which, if free to turn, will revolve.

The disk can never rotate as fast as the magnet (assuming that the magnet is being rotated at a uniform speed). For, if the disk were to rotate at the same speed as the magnet, there would be no cutting of lines of force by the conductor, hence no voltage induced and no current set up to produce torque. If no torque were produced, the disk would slow down until sufficient current were induced to develop enough torque to rotate the disk. Therefore, the disk must rotate more slowly than the magnet.

Here is illustrated the principle of the induction motor. In induction motors, a rotating field is set up by the stator, and the rotor can never revolve as rapidly as this field. The difference between the speed of the rotating magnetic field and the speed of the disk, or the rotor, is called the *slip*. The speed of the rotating field is known as the *synchronous speed*.

But the device shown in Fig. 2-1 does not resemble a commercial induction motor very closely in actual appearance. That it does represent the *principle* is developed in the following paragraphs.

2-3 Rotation of a Squirrel-Cage Rotor Produced by a Rotating Field Magnet. A dc motor has a field structure generally resembling the one shown in Fig. 2-2. Here the pole pieces and the yoke are shown as a single piece, and only the cross section of the field coils is shown. The rotating member is an ordinary cage rotor. A squirrel-cage rotor, as previously discussed in part, consists of steel punchings to carry the magnetic flux and to support the individual rotor conductors, which are perpendicular, or approximately perpendicular, to the plane of the

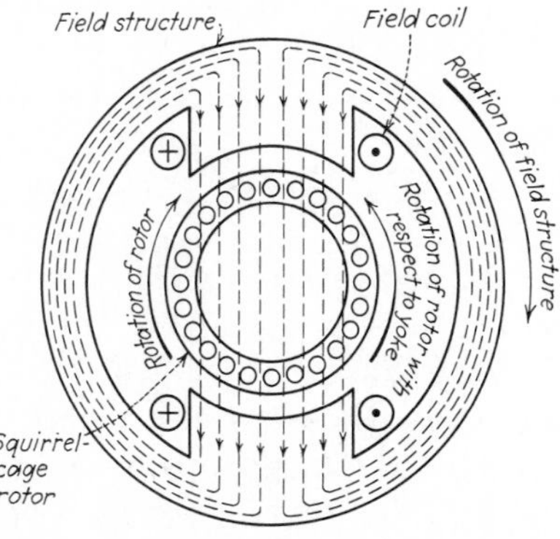

Fig. 2-2 Rotation of a squirrel-cage rotor produced by a rotating field magnet.

paper; all these conductors are mutually short-circuited at each end by means of a resistance ring.

Suppose that the field coils were excited with direct current. Flux lines would be set up as shown by the dotted lines. Assume now that the field structure is arranged so that it can be rotated (collector rings, of course, would be necessary to introduce the direct current into the field coils) and that it is revolved mechanically. This rotation of the magnetic field structure causes the flux to revolve with it. As this flux revolves, it will tend to make the rotor follow its rotation for the same reasons that the magnet of Fig. 2-1 makes the metal disk follow it. To repeat, flux lines that emanate from the field structure are caused to revolve around the rotor by the mechanical rotation of the field structure, thereby cutting the conductors in the squirrel-cage rotor and inducing currents in them. These induced currents react with the revolving magnetic field, tending to make the squirrel-cage rotor follow this field, but at a speed slightly lower than that of the field structure.

The structure shown in Fig. 2-2 is a rather hypothetical motor, and it is doubtful that any practical motors were ever built like this, although electric clutches embodying the principle are made. It does, however, resemble an induction motor in appearance more than does the structure of Fig. 2-1. The rotor is of the familiar squirrel-cage construction actually used in induction motors. The stator member, however, differs from the primary member of an induction motor in two respects: (1) The primary field structure revolves mechanically, whereas the stator structure of an induction motor is stationary and (2) the stator windings of an induction motor are energized with alternating current instead of with direct current.

It will be the purpose of subsequent paragraphs to show that the alternating currents flowing in the windings of the stationary member of a two-phase induction motor actually do set up a rotating field which has the same effect on the rotor that the revolving field magnet shown in Fig. 2-2 has on its rotor.

PRODUCTION OF A ROTATING FIELD IN A TWO-PHASE STATOR

2-4 A Simple Two-Phase Motor-and-Generator System. Although the polyphase motor (in fractional horsepower sizes) is used less than the single-phase motor, it is being discussed here first because it is the simpler to understand. For this reason and because the two-phase motor is simpler to explain than the three-phase motor, the two-phase motor has been chosen for purposes of illustration, not because it is more important commercially. In addition a discussion of the single-phase motor follows more directly from it than from a discussion of the three-phase motor.

A two-phase generator and motor are represented in Fig. 2-3*a*. Here the generator is shown diagrammatically as having two electrically distinct circuits or phases, designated as phase A and phase B. These

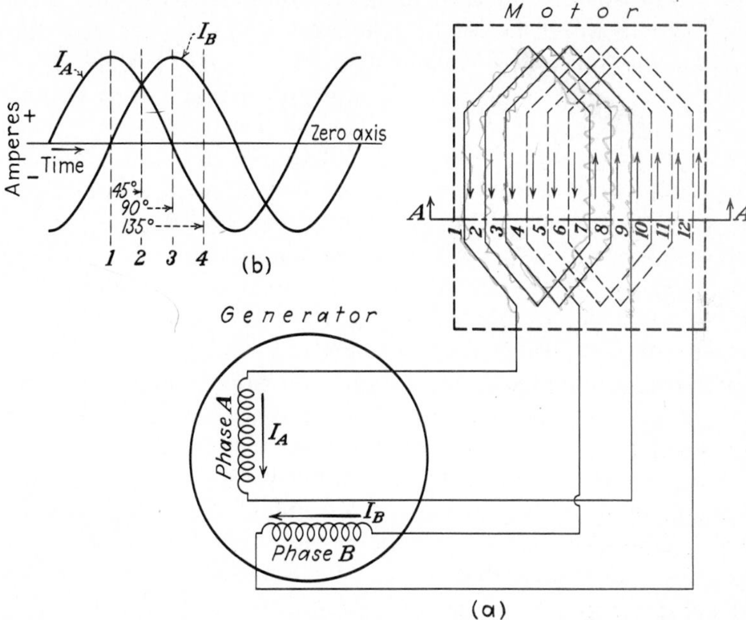

Fig. 2-3 A two-phase motor-and-generator system. (*a*) Direction of instantaneous current flow in windings; (*b*) relation of currents in the two phases.

phases are electrically distinct circuits, and in them are generated voltages which are equal in magnitude and 90° out of phase. If a balanced two-phase load is connected to the generator, the currents in the two phases will be balanced also; these are represented in Fig. 2-3*b* and the instantaneous values of currents I_A and I_B are shown. In both the generator and motor diagrams of Fig. 2-3*a*, arrows indicate the directions of the currents I_A and I_B. These directions of current correspond to positive values of current, i.e., to values of current indicated above the zero axis in Fig. 2-3*b*. Negative currents in Fig. 2-3*b*, i.e., currents represented below the zero axis, flow in a direction opposite to that indicated by the arrows in Fig. 2-3*a*.

2-5 Windings of a Two-Phase Motor. The motor winding represented in Fig. 2-3 is a simple type of single-layer diamond-coil winding. In this figure, the motor is represented as though the stator had been rolled out flat. The windings in the circuit of phase A, including the active conductors (that portion of the conductors embedded in the stator iron) and end windings, are shown with a solid black line. The windings comprising the circuit of phase B are shown in dotted lines. The directions of the currents through the active conductors are represented by arrows drawn parallel to the latter. As in the case of the generator, these arrows represent the positive directions of current, i.e., the actual direction of flow at the instants when the latter is positive in Fig. 2-3*b*; at the instants when the current is negative, the actual direction of current flow in the active conductors is opposite to that shown by the arrows.

Now, assume the motor to be rolled back into its normal cylindrical form. Further, assume an imaginary plane parallel to the punchings (i.e., perpendicular to the shaft) as shown at *A-A* in Fig. 2-3*a*. Four cross-sectional views of the motor taken at this plane are given in Fig. 2-4.

2-6 How the Rotating Field Is Set Up. The four views shown in Fig. 2-4 represent the instantaneous currents in the various conductors and the corresponding fields set up in the motor at each of four different instants of time; the instantaneous fields are set up by the currents flowing in both windings *at that particular instant.* Each of the views 1, 2, 3, and 4 corresponds to the current and flux conditions in the motor at the instants 1, 2, 3, and 4, respectively, in Fig. 2-3*b*. These instants of time will be discussed separately.

Time 1. At this instant, the current in phase A is at its maximum value *in a positive direction,* which means that the currents are flowing through the conductors of phase A in the directions shown by the arrows in Fig. 2-3*a*. In view 1 of Fig. 2-4, a single dot, or point, represents the head of an arrow and indicates that current is flowing directly toward the observer; a cross represents the tail of an arrow and indicates that the current is flowing directly away from the observer. In this first view, the slots are numbered 1 to 12, inclusive, and these numbers correspond to the numbering of the slots in Fig. 2-3*a*. At this instant, there is no current flowing in phase B. Therefore, a field will be set up as shown by the dotted lines and in the directions indicated by the arrows. The direction of this field flux can be determined by application of the right-hand rule.* The field flux set up, at least so far

*The right-hand rule is used to determine the direction of the magnetic flux which is produced by current flowing in a coil of wire. It is stated as follows: If an electromagnet be taken in the right hand with the fingers coiling in the direction of the flow of current, the thumb, extending along the magnet away from the hand, will point in the direction of flow of magnetic flux, i.e., to the north pole of the magnet.

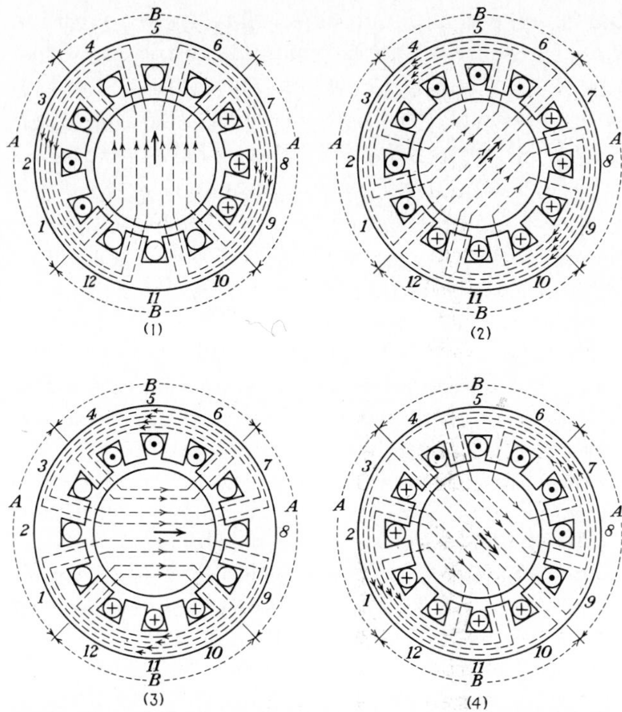

Fig. 2-4 The rotating field set up by a two-phase stator. (1) Current in phase A is positive, and current in phase B is zero; (2) currents in both phases positive; (3) current in phase A is zero, and current in phase B is positive; (4) current in phase A is negative, and current in phase B is positive.

as any action on the rotor is concerned, is practically identical with the field flux as shown in Fig. 2-2.

Time 2, 45° later in time. At this time, as can be seen by reference to Fig. 2-3*b*, the currents in both phases are positive, although neither is at its maximum value; however, the currents in both of the phases are equal. The currents in the individual conductors are shown in view 2. Here, it is seen that the resultant field, or flux, as shown by the dotted lines, is similar in character to that produced at position 1, *except that the field is rotated clockwise 45° in space.*

Time 3, 90° in time later than time 1. The current in phase A is zero, and the current in phase B is positive and at its maximum value. The currents in the individual conductors and the field flux are as shown in view 3. Note that the field has again shifted 45° in space.

Time 4, 135° later than time 1. The currents in phases A and B are again equal, but the latter current is positive, whereas the former current is negative. That is, the current in phase A is flowing in opposition to the arrows shown in Fig. 2-3*a*, and the currents in the individual conductors flow as shown in Fig. 2-4 in view 4. These currents set up a stator field as shown in view 4. Note that this field again has moved 45° in space from position 3, or 135° in space from position 1.

It seems unnecessary to prolong the discussion further to show that the field will make one complete revolution in one cycle of current, since the field has been shown to revolve 45° in space for each 45° of lapsed time. It will be observed that there is set up an air-gap flux, or field, that actually rotates in space, just as the field of Fig. 2-2 was caused to rotate in space by actual rotation of the field-magnet structure; in the case of Fig. 2-4, however, there was no actual rotation of the field structure itself, although the magnetic field set up by the stationary field structure did itself rotate in space.

Thus it has been shown how alternating current, flowing in the windings of a stationary primary member of a two-phase induction motor, can actually set up a rotating field that is practically the same, so far as any action on the rotor is concerned, as the rotating field that is set up by a mechanically revolving field magnet similar to the one represented in Fig. 2-2.

2-7 Conditions Necessary To Set Up a Rotating Field in a Two-Phase Motor. The two conditions necessary to set up a rotating field in a motor excited from a two-phase source are, as we have seen, the following:

1. Two separate phase windings in the motor located 90° apart in space. (The windings should be separated by 90 electrical degrees and not necessarily by 90 mechanical degrees. The distinction between electrical and mechanical degrees is discussed in Art. 2-16.)

2. Two phases, or sources of ac voltages, equal in magnitude and displaced by 90° in time phase.

In the two-phase motor, there are, in effect, two distinct magnetic fields, 90° apart in time and 90 electrical degrees apart in space. One of these fields is set up by the winding of phase A, and the second is set up by phase B. Actually, both fields do not exist simultaneously in the motor; at any particular instant, only the resultant exists. This resultant is a uniformly rotating magnetic field, practically constant in magnitude, as shown by the four pictures in Fig. 2-4.

It can be shown that, in three-phase induction motors, a rotating field is produced in a somewhat similar manner.

PRODUCTION OF A ROTATING FIELD IN A SINGLE-PHASE INDUCTION MOTOR

In the four articles that follow, the single-phase motor is discussed from the point of view of the cross-field theory.

2-8 The Single-Phase Motor: A Special Case of the Two-Phase Motor. Suppose the winding of phase B in the two-phase motor just discussed is omitted; we would then have a single-phase induction motor such as is represented in Fig. 2-5*a*. In the two-phase motor, as just learned, there are two component stationary fields which combine to form a single resultant field which rotates. One of these component fields is set up by one phase, and the other component field by the second phase, *but both components of the rotating field are set up by stator windings.* However, in the single-phase motor, there is only one winding on the stator, and this one winding can set up only one of the two components required to produce a rotating field. How a second component is actually set up by the rotor when the latter is turning is developed in Art. 2-10. But first it is necessary to examine the field set up by the stator winding.

2-9 Field Set Up by Stator Winding Alone. Assume that the stator winding of the motor in Fig. 2-5*a* is excited with alternating current. At any particular instant when the current is positive, i.e., flowing in the directions indicated by the dots and crosses in the stator slots, a magnetic field will be set up as shown. Neglecting the effect of the rotor, this field will be stationary in space but will pulsate in magnitude; it will be at a maximum value when the current is maximum and zero when the current is zero—the significant point, however, being that the field is stationary in space and does not revolve, as in the case of the two-phase motor.

Since the field set up by the stator winding does not revolve, there is no tendency for the rotor to turn; hence, there is no inherent locked-rotor torque. Once the motor is started and running, the single-phase motor will develop torque because of the action of the cross field set up by the rotor. How this field is set up is the subject of the following article.

2-10 Field Set Up by the Rotor. Assume that the squirrel-cage rotor is revolving, having been started by some means or other, for it was not inherently self-starting. Voltages will be induced in each of the individual conductors in the squirrel cage because they are cutting magnetic

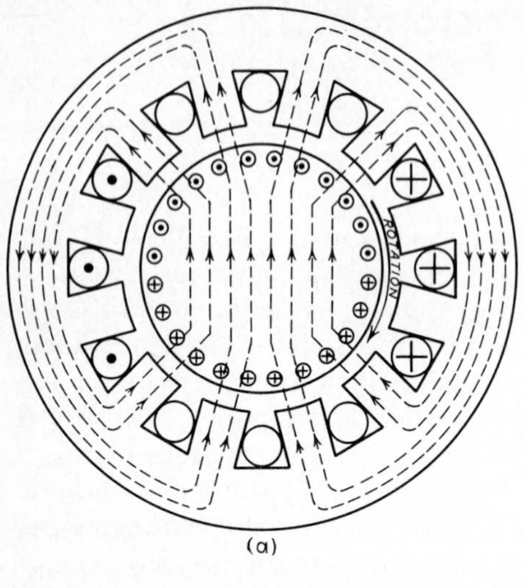

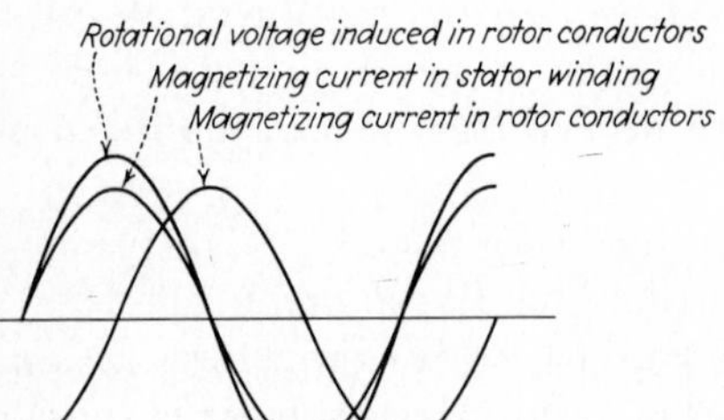

Fig. 2-5 The fields of a single-phase induction motor. (*a*) The stator field and the rotational voltages induced in the rotor by it; (*b*) magnetic field set up by the currents flowing in the rotor bars (cross field); (*c*) instantaneous values of stator current, rotor voltage, and current in the cross-field axis.

lines of force. By applying Fleming's right-hand rule,* it can be ascertained that all the currents above the center of the rotor would tend to flow toward the observer and all the currents below the center of the rotor would tend to flow away from the observer. This phenomenon is indicated by the use of dots and crosses in the rotor bars.

It should be noted that the *rotational voltages induced in the rotor conductors* are in phase with the stator field and stator magnetizing current, as shown in Fig. 2-5*c*. Now, since voltages are induced in the rotor conductors, currents will be caused to flow. The impedance to the currents in the rotor bars is almost entirely reactive, so that these currents will lag the voltage by nearly 90°, as shown in Fig. 2-5*c*. (For a discussion on the phase relation between the voltage and current in an ac circuit containing principally inductance, the reader is referred to any elementary text on alternating currents.)

These rotor currents will set up a magnetic field, known as the *cross field*, as shown in Fig. 2-5*b*. This field is displaced 90° in space from the field shown in Fig. 2-5*a*, i.e., 90° in space from the field set up by the stator winding. Moreover, this field set up by the rotor is not displaced only 90° in space from the stator field, but it also lags the latter by approximately 90° in time. In other words, the effect of the magnetizing current flowing in the rotor is somewhat similar to the effect of the current flowing in phase B in Fig. 2-4(3).

2-11 Rotating Field. We have just seen that the stator will set up a pulsating field and that the rotor, *if revolving*, will set up a second pulsating field 90° behind the stator field in time and in space. If the rotor is revolving at synchronous speed, this field set up by the rotor will be substantially equal in magnitude to the stator field. Therefore, a rotating field will be set up in the single-phase motor just as it was in the two-phase motor; in general, this field will be similar in character to the rotating field set up in Fig. 2-2, except as noted in the following paragraphs.

*Fleming's right-hand rule can be applied as follows: Extend the thumb and the first and second fingers of the right hand so that all three are mutually perpendicular and in the most natural and comfortable position; this position is with the thumb and index finger both lying in the plane of the hand with the second finger perpendicular to the plane of the hand. With the hand held in this position, the thumb points in the direction of motion of the conductor, the first finger in the direction of the field, and the second finger in the direction of induced voltage. If this rule is applied in the case in point, it will be seen that the currents and voltages in the rotor bars will be in the directions shown in Fig. 2-5*a*. As an aid to remembering this rule, some students memorize the expression "my fine clothes" to stand for motion, field, current, as being represented by the thumb, first finger, and second finger, respectively.

COMPARISONS BETWEEN SINGLE-PHASE AND POLYPHASE INDUCTION MOTORS

2-12 Differences in the Rotating Field. In the well-designed polyphase induction motor, the strength of the rotating magnetic field does not vary appreciably as it rotates. This condition can be described by the use of a phasor, as shown in Fig. 2-6*a*; the angular position of the field

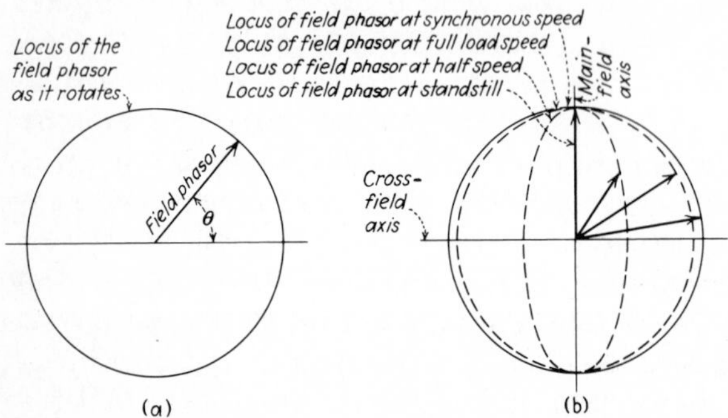

Fig. 2-6 Rotating fields of single-phase and polyphase induction motors compared. (*a*) Polyphase motor; (*b*) single-phase motor.

is represented by the angular position θ of the phasor, and the strength of the field is represented by the length of the phasor. In the case of the polyphase motor, the locus (the imaginary line that is traced by the extremity of the arrow) of this field phasor is a circle. In the case of the single-phase motor, the locus of this field phasor is an ellipse, as shown in Fig. 2-6*b*. It is because of this method of representation of the field that one sometimes hears the statement that a polyphase motor has a circular field, whereas a single-phase motor has an elliptical field. The words "circular" and "elliptical" refer to the shape of the locus of the phasor describing the field, and not to any characteristic of the field itself.

Why is the locus of the field phasor of a single-phase motor an ellipse? The strength of the cross field depends upon the speed of the motor, for the rotational voltage induced in the rotor conductors must depend upon the speed. Therefore, since the strength of the field in the cross-field axis decreases with a reduction in speed, the shape of the ellipse must change with speed. At synchronous speed, the cross field is practically equal to the main field, and the field locus is a circle, as in a polyphase motor. As the speed decreases, the ellipse becomes flatter and flatter in shape, until finally, at standstill, it collapses to a straight line; i.e., when the rotor is at rest, the field pulsates in value along the stator axis but does not rotate.

2-13 Losses, Breakdown Torque, and Torque Pulsations. The rotor of a polyphase motor carries no magnetizing current, unless the supply voltages are unbalanced, and the rotor I^2R losses at no load are negligible. However, the rotor of a single-phase motor carries the magnetizing current to set up the cross field; hence, there are appreciable rotor I^2R losses even at no load.

It is quite commonly known that the breakdown torque of a polyphase induction motor is not affected by the rotor resistance. However, such is not the case with a single-phase induction motor, because rotor resistance limits, to some extent, the amount of current that will flow in the rotor conductors to set up the cross field; hence, increasing the rotor resistance means weakening the cross field and thereby reducing the maximum torque that can be developed. Thus, if the rotor resistance rings of a single-phase motor are machined to a smaller cross section, the rotor resistance is increased and the breakdown torque decreased. Further, it is an interesting fact that the breakdown torque of a single-phase motor always occurs at some speed above two-thirds synchronous, whereas the breakdown torque of a polyphase motor can occur at any speed, even including negative speeds.

The torque of a single-phase motor pulsates between wide limits at twice power frequency. This point is discussed and explained more fully in Art. 17-13. Polyphase motors have no such torque pulsations.

2-14 Necessity for Starting Arrangement for Single-Phase Induction Motors. It has been brought out in the foregoing discussion that, unlike a polyphase induction motor, a single-phase induction motor has no revolving field at standstill and no inherent locked-rotor torque. It is, therefore, necessary to employ a starting device of some sort for any single-phase induction motor. The means employed for starting are many, and the different types of single-phase induction motors are named after the method employed for starting them. For example, there are the split-phase induction motor, the capacitor-start motor, the repulsion-start motor, and the reactor-start motor. A single-phase induction motor will continue to run in the direction in which it is started.

MISCELLANEOUS

2-15 Number of Poles. In dc motors, there are definite pole pieces which can be seen and counted, so that there is no doubt as to the number of poles. In an induction motor, there are no such salient poles, but the windings are arranged so that a revolving magnetic field is set up just as though it were produced by a rotating dc field structure, as has been explained previously. A motor is said to be wound for four poles if the winding sets up a magnetic field with four poles.

2-16 Electrical Degrees. The term electrical degrees was used in the preceding articles and is frequently heard in connection with motors. It is defined as follows:

$$\text{Electrical degrees} = (\text{mechanical degrees}) \times (\text{number of pairs of poles})$$

Why is the term "electrical degrees" used? It was shown in Art. 2-6 that the magnetic field of the induction motor of Fig. 2-4 moved 45° in space for each 45° of lapsed time. In 360° of time, the field moves a distance of two poles. If there are only two poles, as in Fig. 2-4, this distance is a complete revolution. If the machine is wound for four poles, 720° of time is necessary for a complete revolution. In other words, one pair of poles always represents 360 electrical degrees, regardless of the total number of poles. Thus it happens that one mechanical degree represents as many electrical degrees as there are pairs of poles. In dealing with electric motors, the meaning of the term electrical degrees should be borne in mind continuously, because this term is usually far more significant than mechanical degrees.

2-17 Synchronous Speed. The speed of the rotating field of an induction motor is known as the *synchronous speed*. The synchronous speed can always be figured from the line frequency and the number of poles for which the motor is wound. Since the field moves one pair of poles per cycle

$$\text{Synchronous rpm} = \frac{60 \text{ seconds per minute} \times \text{cycles per second}}{\text{pairs of poles}}$$

or

$$\text{Synchronous rpm} = \frac{120 \times \text{frequency}}{\text{number of poles}} \tag{2-1}$$

The actual operating speed, under full-load conditions, is about 5 percent less than synchronous speed, as can be seen by referring to Table 19-1 in Chap. 19.

2-18 References for Further Study of Induction-Motor Theory. It was pointed out in the first part of this chapter that full treatment of the theory of induction motors was beyond the scope of this book. For the eager student who wishes to pursue this fascinating subject further, the literature is full of help. Every elementary textbook on ac machinery carries an explanation of the theory of operation of the polyphase induction motor. A classic, nonmathematical, and comprehensive explanation of polyphase motor theory has been given by Lamme.[2] A procedure for calculating the running performance of polyphase induction motors from constants of the motor, suitable for routine design calculations, is given by Veinott.[3] Fortunately, polyphase motors are uniformly treated from the viewpoint of a single theory: the revolving-field theory. Unfortunately the single-phase motor is explained by two theories: the cross-field theory and the revolving-field theory, each of which has its staunch advocates.

For purposes of explanation, the cross-field theory was used in this chapter. This theory was used by Branson in his classic paper in 1912 wherein he developed a circle diagram for the single-phase motor.[4] This theory has found its way into some textbooks[5,6] and was used by West[7] to develop mathematical equations for calculating the performance of single-phase induction motors; these equations form the basis of a modified calculating method developed by Veinott for routine design calculations on single-phase induction motors.[3] Later, Beach published a comprehensive nonmathematical explosion of the cross-field theory, which presents a clear interpretation of how the stator and rotor fluxes and currents mutually interact to develop torque and maintain rotation; his treatise is recommended as an introduction to the mathematical treatments.[8]

The second point of view is quite different from the cross-field theory and is known as the revolving-field theory. This theory essentially holds that the resultant field in a single-phase induction motor is made up of two component *revolving* fields rotating in opposite directions. This point of view is explained very fully and exceptionally clearly in a nonmathematical manner by Lamme.[9] It is used by Morrill to develop a calculation procedure for single-phase motors[10] as well as for his classic paper on capacitor-motor theory.[11] It is explained in some textbooks[5,6] and has been used in a number of American Institute of Electrical Engineers (AIEE) and Institute of Electrical and Electronics Engineers (IEEE) papers.

In another work, the single-phase motor is explained by both cross-field and revolving-field theories, both qualitatively and quantitatively.[1] Analytical methods for computing the running performance by each method were developed and illustrated using practical examples.

More recently, a third approach to the theory of electric machinery has been developed: the dynamic-circuit or coupled-circuit approach. This approach views the machine as having a number of circuits, dynamically coupled, and performance is determined by setting up the equations for these circuits and solving them by means of matrices, or tensors, usually by means of various transformations of variables. Although there have been many papers that have used this general approach, the one by Messerle[12] is representative. The approach is exemplified in a book by Adkins[13] and by a good many textbooks written since then. Many more references on machinery theory are given in three bibliographies prepared by AIEE and IEEE.[14–16]

References

1. Veinott, C. G.: *Theory and Design of Small Induction Motors*, McGraw-Hill Book Company, New York, 1959. (Out of print, but available from University Microfilms, P.O. Box 1647, Ann Arbor, Mich.)
2. Lamme, B. G.: *Electrical Papers*, Westinghouse Electric Corporation, East Pittsburgh, 1919, pp. 5–39.
3. Veinott, C. G.: "Performance Calculations on Induction Motors," *AIEE Trans.*, vol. 51, 1932, pp. 743–752.
4. Branson, W. J.: "Single-Phase Induction Motors," *AIEE Trans.*, June 1912, p. 1749.
5. Lawrence, R. R.: *Principles of Alternating-Current Machinery*, 3d ed., McGraw-Hill Book Company, New York, 1940.
6. Puchstein, A. F., and T. C. Lloyd: *Alternating-Current Machines*, John Wiley & Sons, Inc., New York, 1951.
7. West, H. R.: "The Cross-Field Theory of AC Machines," *AIEE Trans.*, February 1926, p. 466.
8. Beach, Robin: "A Physical Conception of Single-Phase Motor Operation," *Electr. Eng.*, July 1944, pp. 254–263.
9. Lamme, B. G.: "A Physical Conception of the Operation of the Single-Phase Induction Motor," *AIEE Trans.*, April 1918, p. 627.
10. Morrill, W. J.: "The Apparent-Impedance Method of Calculating Single-Phase Motor Performance," *AIEE Trans.*, vol. 60, 1941, pp. 1037–1041.
11. ———: "The Revolving-Field Theory of the Capacitor Motor," *AIEE Trans.*, vol. 48, 1929, pp. 614–629.
12. Messerle, H. K.: "Dynamic Circuit Theory," *AIEE Power Apparatus and Systems*, April 1960, pp. 1–12.
13. Adkins, Bernard: *The General Theory of Electrical Machines*, Chapman & Hall, Ltd., London, 1959.
14. Rotating Machinery Committee: *Bibliography of Rotating Electric Machinery 1886–1947*, AIEE Publ. S-32, Jaunary 1950.
15. ———: *Bibliography of Rotating Electric Machinery for 1948–1961*, IEEE Trans. Paper no. 64-1, 1964.
16. ———: *Bibliography of Rotating Electric Machinery for 1962–1965*, IEEE Trans. Paper no. 31 TP 67-477, 1967.

Bibliography

Button, C. T.: "Single-Phase Motor Theory—Correlation of Cross-Field and Revolving-Field Theory," *AIEE Trans.*, vol. 60, 1941, pp. 664–665.

Kimball, A. L., and P. L. Alger: "Single-Phase Motor Torque Pulsations," *AIEE Trans.*, June 1924, p. 730.

Slemon, Gordon R.: *Magnetoelectric Devices, Transducers, Transformers, and Machines*, John Wiley & Sons. Inc., New York, 1966.

CHAPTER 3

SINGLE-PHASE INDUCTION MOTOR WINDINGS AND CONNECTIONS

Before proceeding to a description and a discussion of specific types of single-phase induction motors, it seems advisable to give a general description of the more common types of windings used. To illustrate the different classifications, a simple winding is presented in four different common forms (see Figs. 3-1, 3-2, 3-8, and 3-9). Also, in this chapter the various types of diagrams that are used later in this book are described and defined, and an explanation of how to interpret these diagrams is provided. It is not intended, however, to cover all phases of repair-shop technique; information of this sort can be found, if desired, in books on the subject and in current magazines.

3-1 Concentric Winding. Single-phase induction motors are generally wound with concentric coils. For this reason, it may be well to give a rather general description of the concentric winding, comparing it with the more familiar diamond-coil lap winding such as is used both in dc armatures and in polyphase motors. A diamond-coil lap winding was used in Fig. 2-3 to explain the production of a rotating field by polyphase alternating current. Another form of this winding, but with eight slots per pole instead of six, is shown in Fig. 3-1. An exactly equivalent concentric type of winding is shown in Fig. 3-2. To understand why the windings of Figs. 3-1 and 3-2 are exactly equivalent, compare the two winding arrangements: in both, slots 15, 16, 1, and 2 carry A phase current upward; slots 3, 4, 5, and 6 carry B phase current upward; slots 7, 8, 9, and 10 carry A phase current downward; and slots 11, 12, 13,

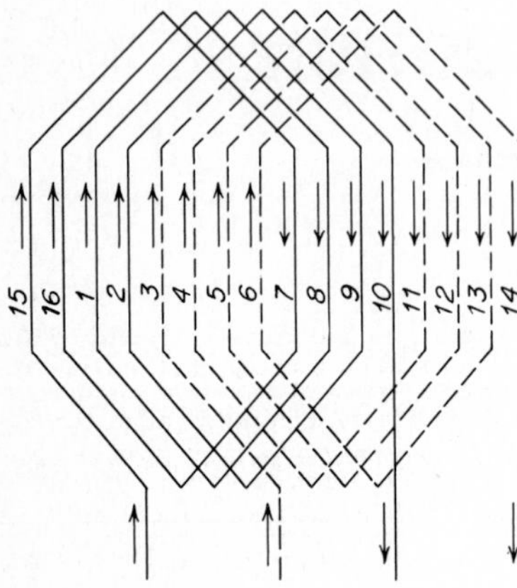

Fig. 3-1 A two-phase progressive lap winding with diamond coils and eight slots per pole. This winding is shown here merely to illustrate one of the various possible types of windings and should not be considered as a recommended arrangement for two-phase motors because it is a single-layer winding. A two-layer winding is normally used for two-phase motors.

and 14 carry B phase current downward. In other words, the direction of current flow through the active conductors of every slot of the winding of Fig. 3-2 is identical with the direction of current flow in every slot of the winding of Fig. 3-1; the difference between these two winding arrangements is only in the end connections. The strength, location, and distribution of the useful magnetic fields in the motor depend wholly upon the total number and arrangement of ampere-conductors in the slots; in no respect do these factors depend upon how the end connections are made between these conductors. Therefore, one wind-

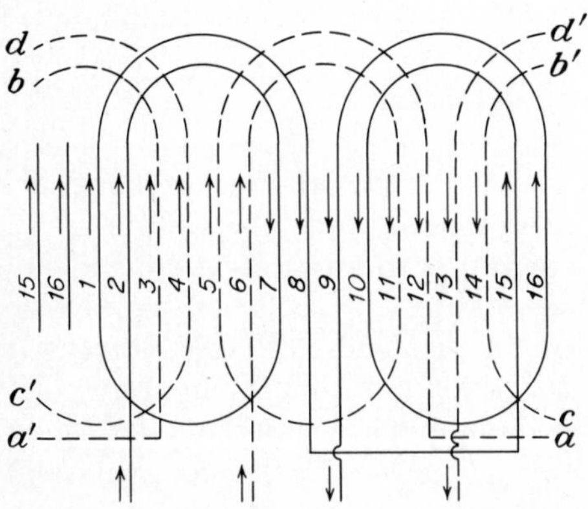

Fig. 3-2 A two-phase concentric winding. This winding is exactly equivalent to the winding of Fig. 3-1.

ing is said to be *exactly equivalent* to a second winding if, with the same current flowing in the external leads of each winding, there are exactly the same number of ampere-conductors in every slot of the first winding as in every slot of the second winding. (The number of ampere-conductors in a slot is the product of the number of conductors multiplied by the amperes flowing in each conductor.)

It is always possible to find a concentric winding that is exactly equivalent to a progressive, or diamond-coil, winding as has just been done. The reverse of this statement, however, is not always true, for a concentric winding is the more flexible arrangement. Concentric windings are used in single-phase induction motors because of this greater flexibility. With this type of winding, it is possible and practicable to use more copper in the main winding than in the auxiliary or starting winding, and, moreover, the number of turns does not have to be the same in both windings. In most single-phase induction motors, it is desirable to use more copper in the main winding than in the auxiliary winding and to employ a different number of turns. Split-phase motors, in particular, start on the principle that the ratio of resistance to reactance of the starting winding is higher than the ratio of resistance to reactance of the main winding.

In the diamond-coil or polyphase type of winding, all the coils have the same pitch or "throw," and the center of each coil is displaced—usually by one slot—from the center of the neighboring coil. In the concentric type of winding, every coil in any given pole group has a different throw. But the centers of all the coils in a single pole group coincide; i.e., the coils are concentric with one another.

3-2 Necessity for Different Types of Connection Diagrams. The complete specification of all the details of a winding could be shown on a single drawing. Such a drawing would indicate size of wire; number of turns or conductors in each and every slot; end connections of all coils to form the groups; connections of the coil groups to one another to form both the main and auxiliary windings, respectively; and the connections of the main and auxiliary windings to the line and to the starting switch, capacitor, reactor, or any other piece of apparatus essential to the operation of the motor. All this information is absolutely necessary in order to wind and connect a motor. Yet if it were all on a single drawing, such a drawing would be necessary for every type, design, and rating; the number of diagrams would be literally without limit; and fundamental principles would be lost in a hopeless multiplicity of diagrams. By breaking this information down into a few different *kinds* of diagrams, the total number necessary can be reduced to a practical number, and fundamental principles are not overlooked.

There are at least three different independent operations incident to the winding and connecting of a motor:

1. Winding and connecting the stator coils in groups
2. Connecting these groups
3. Assembling the motor, which involves connecting the windings to the switch or capacitor, if either is used, and to the external or line leads

For each of these operations, there is a specific type of diagram. For connecting the motor to the line or to auxiliary apparatus, such as a controller or capacitor, a fourth type of diagram is essential. What these diagrams are and how to interpret them are the basis of the following articles.

3-3 Distribution Chart. A distribution chart gives the number of turns or conductors in each slot for all the windings. The winding distribution chart for Fig. 3-2 is Fig. 3-3. Since Fig. 3-2 is a very simple winding,

Slot number	1	2	3	4	5	6	7	8	9	10	11	12	13	14	15	16
Main winding	1	1	x	x	x	x	1	1	1	1	x	x	x	x	1	1
Auxiliary or starting winding	x	x	1	1	1	1	x	x	x	x	1	1	1	1	x	x

Fig. 3-3 A winding distribution chart for Fig. 3-2.

consisting of only one strand of wire throughout, the corresponding distribution chart is simple. A typical winding distribution chart may take the form of Fig. 3-4, which specifies the size of wire used in each winding, the insulation covering on the wire, the number of strands per conductor, the total number of strands in each and every slot, and the end connections of the coils in a group. In short, the distribution chart of Fig. 3-4 gives all the information necessary for *winding the stator coils* and connecting them in groups, except for the coil perimeters and the winding method. The winding method is more or less optional but can be given on this chart if desired.

3-4 A Stator (or Rotor) Connection Diagram. After the stator coil groups are wound, the next step is to connect the groups together. A *group* is simply the combination of all the coils in a given winding or phase under one pole. For example, in Fig. 3-4, both windings have two coils per group; each main-winding group consists of two coils of 36 turns (72 strands) each. One of these coils has a throw of 2 and 7, the other a throw of 1 and 8.

Slot number		1	2	3	4	5	6	7	8	9	10	11	12	13	14	15	16	Wire size (gauge)	Insulation and material	Strands per conductor
Main winding	Conductors	72	72	x	x	x	x	72	72	72	72	x	x	x	x	72	72	21	Heavy; copper	2
Auxiliary or starting winding		x	x	43	43	43	43	x	x	x	x	43	43	43	43	x	x	19	Heavy; copper	1

Fig. 3-4 A more detailed winding distribution chart.

How to connect the groups together is shown in Fig. 3-5. In this diagram, each group is represented by a single, solid, black, elongated rectangle. The connections of the ends of these groups are shown. In the diagram of Fig. 3-5, it will be noted that one of the rectangles representing a group has to be drawn in two sections: in the figure, *b* should

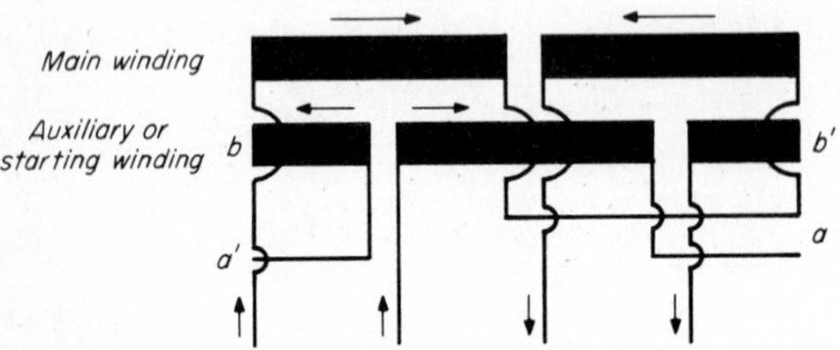

Fig. 3-5 Group connections for Fig. 3-3.

be joined to *b′*. To avoid this difficulty, such diagrams are usually drawn in the form of a circle; a diagram of this kind is a *stator connection diagram.* A number of stator connection diagrams drawn in circular form are given throughout this book; the one that corresponds to the connection diagram just discussed is Fig. 4-8 in Chap. 4. A diagram such as the one just discussed would be a *rotor connection diagram* if it were used to connect a rotor winding.

Thus, we have shown how a winding, such as that represented in Fig. 3-2, is represented by two diagrams, such as Figs. 3-3 and 4-8. It is more convenient to use two diagrams than one, for one stator connection diagram can be used with a large number of distribution charts. Moreover, the distribution chart could be applicable to a motor having two, four, six or more poles, so long as the motor has eight slots per pole.

3-5 Wiring Diagram. A wiring diagram shows the internal connections of the various windings to one another and to any auxiliary devices such as a starting switch, capacitor, reactor, or thermal protective device. Examples of this type of diagram are Figs. 4-16, 4-17, and 5-6. A wiring diagram may or may not be necessary to supplement the stator connection diagram.

3-6 Line Connection Diagram. A line connection diagram shows the permissible connections of the motor leads to the line. Representative examples are Figs. 4-16, 4-17, and 5-6. If there are only two line leads brought out of the winding, a line connection diagram is generally not necessary.

3-7 Magnetic Polarity. It would be impossible in any book of this size to give all possible connection diagrams. Some knowledge of how to check the correctness of diagrams, and also of how to make new ones, is absolutely essential to one who has to rewind or to reconnect motors.

One fundamental principle of winding is that adjacent poles must be wound to give opposite *magnetic polarity* (except in the case of consequent-pole windings); i.e., if one pole is wound in a clockwise direction, the next adjacent pole must be wound in a counterclockwise direction, as has been done in Fig. 3-2. This statement does not mean that the coils actually have to be wound in this direction before or while they are being put into the stator, but it does mean that the winding must be connected so that the current proceeds through one pole in a clockwise direction and through the next adjacent pole in a counterclockwise direction.

This principle—namely, that the magnetic polarity must alternate from pole to pole as we progress about the winding—is used to determine the correctness of connection diagrams. It is essential to understand this principle in order to check these diagrams, or the winding itself, particularly if it is necessary to draw up a new connection diagram. Taking a representative diagram, such as Fig. 4-9, and tracing through the main winding, we find that we go from pole to pole, but we go through one pole in one direction and through the next adjacent pole in the opposite direction; i.e., if we enter at lead T4, we proceed clockwise around the diagram through the first group, counterclockwise through the second group, and so on through the rest of the groups until we come out at T1. This tells us that all adjacent poles are of opposite magnetic polarity, as they should be.

3-8 Gun Windings. Formerly coils were often wound directly in the stator punchings by means of a hand-operated gun, one turn at a time. For example, assume that a motor is to have the winding of Fig. 3-2 put in by the gun method. Starting in slot 2, the operator first winds the innermost coil (2 and 7), and then, in the same direction, the next coil (1 and 8) in the two adjacent outside slots. The operator carries the wire over to slot 15 and winds all the turns in the innermost coil (15 and 10), then all the turns in the next adjacent coil (16 and 9).

Since the sharp edges of the slot openings may cut or scrape the insulation off the wire, it is the usual factory practice to employ winding guides. These guides are arranged so as to cover the sharp corners of the teeth to protect the wire so that it can be wound into the slots more rapidly.

Of importance to service departments is that winding with a gun gives a very short wire connecting adjacent pole groups; this wire is usually tight against the iron, so that it may be very difficult to reconnect such a winding in parallel.

3-9 Machine Windings. Most fractional horsepower motors are now wound in the factory by machines. Some machines wind the coils directly in the slots in much the same manner as a human operator would do it with a gun, as described in the preceding article. The machine may use one or more needles (guns), winding a single coil, or a number of coils simultaneously, at a speed in excess of 300 strokes per minute! Thus, the machine winds more rapidly and uniformly than any human can.

A more modern method is first to wind the coils on forms and then insert the whole groups by machine. Such a machine often inserts the slot wedges at the same time. If the total winding is not too bulky, both the main and auxiliary windings are inserted simultaneously. If the main winding is bulky, it is inserted in one operation and the auxiliary winding in a second operation. When the windings are inserted separately, the main winding must be blocked or pressed into the slots as far as possible. A method called *Electro-press*, developed by the General Electric Company, accomplishes this pressback without damage to the wire. A bank of capacitors, discharged into the winding, causes a current to flow in the coil sides with resultant magnetic field induced around each coil side. This magnetic field, being stronger near the bore than at the yoke, causes the individual strands of wire to move from the bore and against the bottom of the slot.

The winding machine shown in Fig. 3-6 is an example of a combination winding-insertion machine. The stator winding is programmed into the winder, which winds the pole groups and places them over the vertical fingers which represent the stator teeth. After all pole groups are wound and placed over the fingers, the horizontal table is indexed 180°. The stator core is lowered over the vertical member and clamped in place, and then the winding is transferred into the stator slots. The large vertical disk near the bottom of the machine supplies slot wedge material, which is cut to length and inserted with the winding.

3-10 Mold Windings. Mold windings are practical for repair shops. Coils are usually of the concentric type, though progressive-type coils could be used. Typical coils are illustrated in Fig. 3-7.

(1) Concentric coils. Concentric coils are wound on a stepped block, or mold, provided with as many steps as there are parts to the coil. With this method, all the coils for a single group are wound in succession with a single strand of wire, or with two strands of wire, if there are two strands per conductor.

For repair shops, a few simple adjustable molds can be made out of wood for permanent shop equipment. These molds should be arranged so that they can be padded on any or all of the sections. If only a few

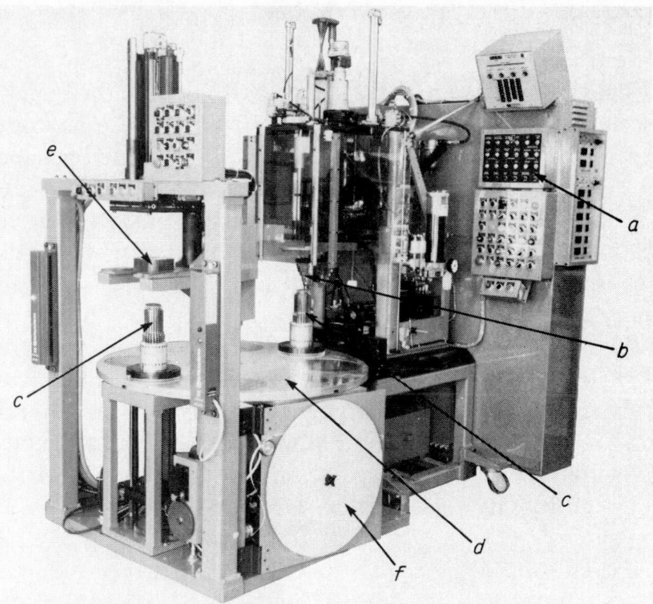

Fig. 3-6 A modern winding machine for small motors. *a*, Computer for programming and counting turns in each coil group; *b*, coil forms on which pole groups are wound; *c*, mandrel with fingers for receiving formed coil groups; *d*, indexing table; *e*, insulated stator in position to be placed over finger mandrel; *f*, reel containing slot wedge material. *(Industra Electric Products.)*

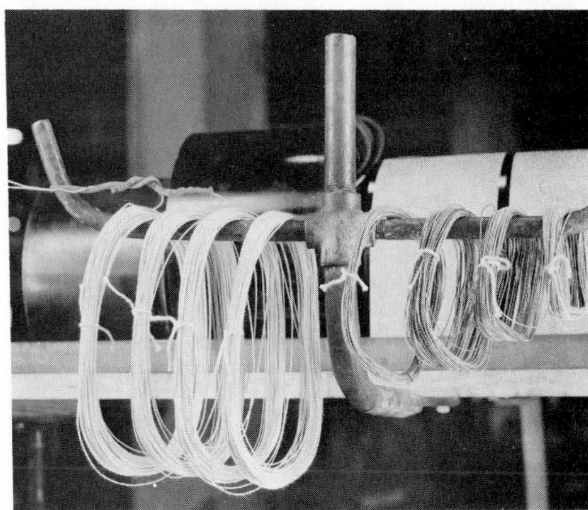

Fig. 3-7 Typical mold-wound coils. The ones at the right comprise one pole group for a concentric winding. The four coils at the left all have the same perimeter and are suitable for a polyphase winding of a single-phase progressive winding. *(Westinghouse Electric Corporation.)*

jobs of this type are done, wire finish nails can be driven into the bench, and the coils wound around these nails. The perimeter of the molds is best determined by a trial with a single strand of wire looped around the stator in the same position that the finished coil is to occupy.

Mold winding has a further advantage over hand or gun winding in that the insulation on the wire is less likely to become damaged during the winding process.

(2) Progressive coils. In the progressive type of winding, illustrated in Fig. 3-8, all the coils in a pole group have the same throw (but not necessarily the same number of turns), but the center of each coil is displaced from the center of the preceding coil. These coils are, therefore, a little simpler to wind on an external mold because all the coils have the same perimeter. The progressive winding of Fig. 3-8 is exactly equivalent to the windings of Figs. 3-1 and 3-2 and is similar to a polyphase winding.

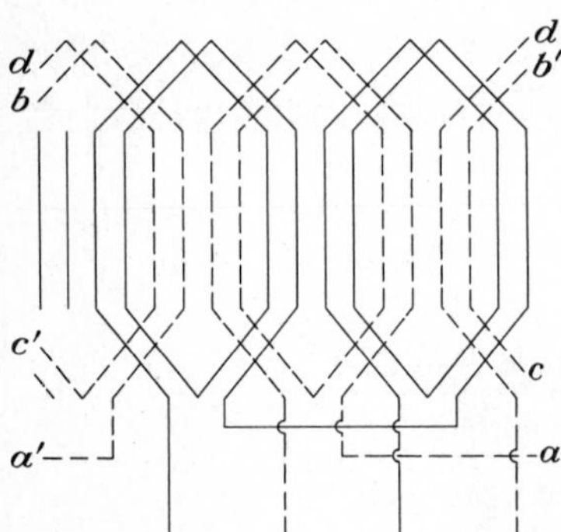

Fig. 3-8 A single-phase progressive winding with diamond coils and eight slots per pole. This winding is an exact equivalent of Fig. 3-2.

3-11 Skein Windings. Skein winding is a popular method of winding, although it is probably less popular than formerly, at least for main windings. This method is somewhat difficult to handle if the weight of the wire is more than 2 or 3 oz (60 or 85 g) per skein, or if the wire is larger than no. 20 or 21. Of the various winding methods just discussed, this one is probably the most complicated and most difficult to explain, although it is not at all difficult to execute, once it is understood and mastered. It is not particularly flexible from a design standpoint, for the choice of distributions is limited.

Successive steps required to produce a skein winding which is exactly equivalent to the concentric winding shown in Fig. 3-2 are given in Fig. 3-9. The operation consists of three steps as shown. Additional information on how to use skein windings is given in an earlier work.[1,*]

*For numbered references, see References at end of this chapter.

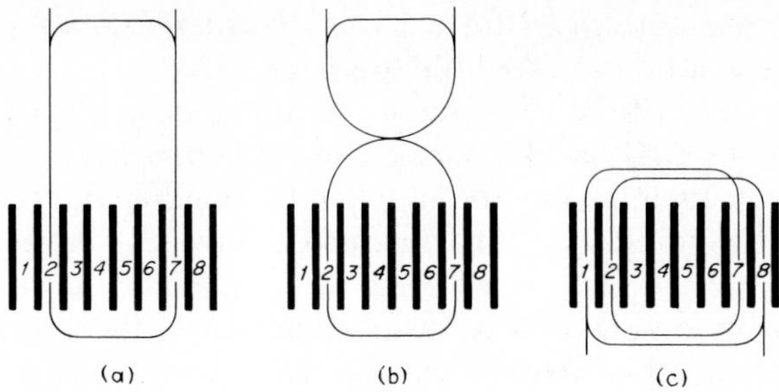

Fig. 3-9 Procedure for winding a skein winding that is an exact equivalent of the concentric winding of Fig. 3-2. (*a*) Coil of wire positioned in slots 2 and 7; (*b*) portion of coil twisted one-half turn; (*c*) twisted portion folded back and positioned in slots 1 and 8.

The best way to determine the length of the skein is to try first a single wire, looping it in the same manner that the final skein will be looped. Then a trial skein can be made up to this perimeter and tried in the stator slots.

3-12 Windings with Dissimilar Coils and Consequent-Pole Windings. Windings with dissimilar coils on adjacent poles are often used in single-phase motors to avoid the use of split coils. Figure 3-10*a* shows a conventional balanced-coil arrangement requiring two coil sides in slots 1, 7, 13, and 19. However, by the use of unequal coils, as shown in Fig. 3-10*b*, the split coils can be avoided. A logical extension of this same idea is the arrangement in Fig. 3-10*c*, which uses only one coil per pair

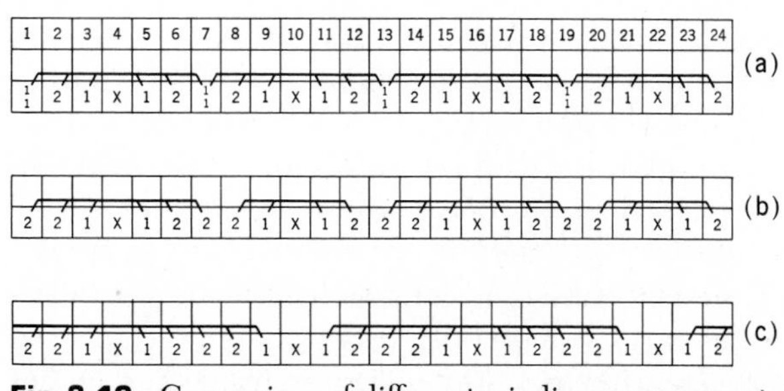

Fig. 3-10 Comparison of different winding arrangements. All three arrangements are exactly equivalent—within the scope of the definition in Art. 3-1—but all three are not equally desirable. (*a*) Conventional winding with identical coil groups on all poles and adjacent poles connected for opposite magnetic polarities; (*b*) dissimilar coil groups on adjacent poles with no split coils and adjacent poles connected for opposite magnetic polarities; (*c*) consequent-pole winding with one coil group per pair of poles and all coil groups connected for the same magnetic polarity.

of poles, and which connects all coils for the same magnetic polarity; this is a consequent-pole winding. All three arrangements are exactly equivalent, within the scope of the meaning of that term, as defined in Art. 3-1, but there are differences to be noted. Use of identical coils on all poles is the best arrangement from an engineering standpoint; end extensions are uniform around the stator; greatest flexibility is permitted in parallel connections because any pole can be paralleled with any other one. Dissimilar coils are perhaps easier to wind because there are no split coils, which are usually tighter in the slots because of the additional insulation required; however, care must be taken in parallel connections that a small coil is never paralleled with a large one; for example, if large and small coils alternate, the parallel diagram of Fig. 4-10 can be used, but the one of Fig. 4-11 cannot. Consequent-pole windings, such as shown in Fig. 3-10*c*, generally have little to recommend them; they result in bulky and nonuniform end extensions and may produce somewhat lower torque because of the slightly higher reactance. The distribution factor is the same for all three arrangements in Fig. 3-10.

3-13 Some Practical Hints on Winding. Detailed instructions on the technique of winding and the tools and equipment necessary for rewinding small motors are given in other books, but a few points are noted here.

If the motor is to be rewound with Class A insulation materials, the slots should be insulated with a U-shaped cell of Mylar, 0.0075 to 0.010 in (0.19 to 0.25 mm) thick. For Class B insulation, cells made of Dacron-Mylar-Dacron or Nomex having a minimum thickness of 0.010 in are recommended, although many manufacturers use Mylar for both Class A and B. For temperatures above Class B, not less than 0.010-in thick Nomex is recommended. Some motors have the stator slots lined with a coating of epoxy or polyester material molded in place in lieu of slot cells. If a winding is burned out in such a motor, it is recommended that the slot cells just described be used when the motor is rewound.

Slot cells should be about ⅜ in (9.5 mm) longer than the length of the core and should preferably be cuffed on both ends. Cuffing reduces the likelihood of the slot cells' tearing and prevents them from slipping out of place axially. When winding and shaping the coils, care should be taken not to split the slot cells, thus allowing the wire to scrape the iron core. The coils should be started in the proper place so that the finished coils avoid any through bolts or end-shield screws. Usually the main winding has the greater amount of copper in it and is the bulkier winding. Therefore, it is advisable to wind the centers of the main poles halfway between the through-bolt holes. If the stator has various slot sizes, the windings must be placed into the same slots as used by the manufacturer. In the case of motors having a commutator and brushes, special attention must be given to the location of the stator winding, particularly if the brushes are not on an adjustable rocker ring. Barriers should be placed in the slots to insulate the starting winding from the main winding; these barriers should also insulate the end connections of one winding from the end connections of the other. The soldered or welded connections of the coils and external leads should be free from sharp points which might cut through the insulation; the connections themselves should be wrapped with insulating tape. Leads and windings should be tied down with Dacron cord.

Care should be taken to wind the coils uniformly and neatly in such a manner as to require a minimum amount of pounding and shaping of the final coil. The bore of the completed winding should be larger than the bore of the stator core so that the rotor can be inserted readily. The winding must be shaped enough that the ventilating fan or switch will not chafe the coils.

Many fractional horsepower motors today are wound with aluminum magnet wire. Because of the difficulty of reliably joining aluminum magnet wire to itself and to stranded coil leads it is recommended that the motor be rewound with copper magnet wire. To obtain the same performance, copper wire two gauges smaller should be substituted for the original aluminum. This decreases the slot fill and enhances the ease of inserting the coils, but it may be necessary to insert extra wedges in any slots that are too loosely fitted.

The magnet wire insulation should be of heavy build; Class A and B motors should preferably be wound with a polyester base coat and polyamide top coat magnet wire and Class F and H motors, with a polyester base coat and polyamide-imide top coat magnet wire. Plain enamel and cotton-covered magnet wire insulations are rarely used today in manufacturing.

3-14 Varnish Impregnation. Electric motors are usually varnish-impregnated for one or more of the following reasons:

1. Bonding of the windings into a rigid mass to prevent chafing and, in the case of rotating members, to prevent loss of balance

2. Removal of moisture from the insulation and addition of a permanent barrier against reentry of moisture, abrasive dirt, and dust

3. Reduction of the temperature rise

4. Lengthening the life of the insulation by reducing the rate of oxidation, thermal decomposition, and dehydration

5. Increasing the dielectric and mechanical strength of the insulation, as well as the resistance to certain chemicals

Almost invariably, moisture resistance and dielectric strength of a film of baked insulating varnish are proportional to the film thickness.

Selection of the proper insulating varnish for any particular motor in general is based upon a compromise between bond strength and heat life of the film. Air-drying varnishes, particularly those with a shellac spirit base, can be used in emergencies when time does not permit the use of a baked varnish, but baked varnishes are invariably better. Of the baking varnishes available today, the thermosetting ones are generally superior to the oleoresinous kinds. Silicone, polyester, epoxy, and polyimide varnishes are predominantly used today for Class B, F, and H insulation systems.

A new technique, referred to as "trickle coating," is being used to treat units with solventless varnishes. The unit is first heated by applying intermittent electric currents to the windings, and then solventless varnish is trickled onto the end turns and penetrates through the slots. This procedure eliminates the requirement for external heat. Curing of the resin is accomplished by the residual heat. It is recommended that the unit be rotated during the application of the resin, in order to obtain the best results.

Motors are also being treated for severe environmental conditions by encapsulating the windings in a 100 percent solid resin. Methods used to encapsulate include the following: casting the resin around the end windings, using a mold, using vacuum-pressure impregnation, using a thixotropic resin (a resin that, during cure, does not run out), and "buttering" the coil ends with a 100 percent solid resin.

3-15 Repetitive Surge Tests. Many motor manufacturers now use a repetitive-surge test as a part of their quality-control program in the manufacture of small motors. One instrument used for performing such tests consists of a repeating-type surge-voltage generator, cathode-ray oscilloscope, and synchronous switching equipment. The generator produces a rapid succession of voltage surges (60 per second when used with 60-Hz power) of very steep wave front and of only a few microseconds' duration for each surge. These surges are alternately applied to two coils or to two windings, and the pictures of each surge appear together on the screen of the oscilloscope, simultaneously, it appears to the eye. The tester thus compares two windings, that is, an unknown one against a control-test sample. Because of the steepness of the wave front, a high turn-to-turn voltage stress can be imposed upon the windings.

In quantity production of a large number of motors of a given design, the repetitive-surge test affords a quick and convenient way to detect the existence of any of the following potential faults in a winding:

1. Grounded winding
2. Short circuits between turns
3. Short circuits between windings
4. Incorrect connections
5. Incorrect number of turns
6. Misplaced conductors or insulation

Major faults are easily detected, but only a skilled operator can distinguish between minor faults.[2]

The repetitive-surge test is normally performed on a stator before varnish impregnation to allow for the repair of faults.

3-16 Dielectric Withstand Test. Any motor that has been rewound or serviced should be subjected to a dielectric withstand test applied between each circuit and the core or frame for a period of 1 min. Suggested values for such a test are the following, as taken from National Electrical Manufacturers Association (NEMA) publication MG1-1978, pt. 12.03.

Motors rated less than ½ hp. and 250 V or less. 1000 V

All other motors. 1000 V plus twice the rated voltage.

3-17 Leakage Current. Leakage current is the current that flows through, or across the surface of, insulation from the motor windings to the frame or ground. It consists of a resistive component and a capacitive component. The resistive component is a measure of the insulation resistance as determined by a megger; the capacitive component is the result of the capacitor effect between the windings and the stator iron, or ground. If the frame of any motor is not solidly connected to the ground, a human who is grounded can receive an electric shock from touching the frame.

Normally, the leakage current of a motor is about 90 percent capacitive. To reduce the capacitive component, which is determined by the length of stacking and the thickness of the slot insulation cell, it is necessary to increase the slot cell thickness.

References

1. Veinott, C. G.: *Fractional Horsepower Electric Motors,* 2d ed., McGraw-Hill Book Company, New York, 1948.
2. *NEMA Motor and Generator Standards,* NEMA Publ. no. MG1-1978, pt. MG1-12.04, National Electrical Manufacturers Association, 2101 L Street, N.W., Washington, DC 20037, 1978.

Bibliography

American National Standards Institute: *Leakage Current for Appliances,* ANSI C101.1, ANSI, 1430 Broadway, New York, NY 10018, 1973.

Braymer, Daniel H.., and A. C. Roe: *Rewinding Small Motors,* 2d ed., McGraw-Hill Book Company, New York, 1932.

Lanoy, Henry: *Les petites machines électriques,* vols. I, II, and III, Giradot et Cie, Paris, 1938–1941.

Underwriters Laboratories Inc.: *Room Air Conditioners,* UL 484, Underwriters Laboratories Inc., 333 Pfingsten Rd., Northbrook, IL 60062, 1982.

CHAPTER 4

SPLIT-PHASE INDUCTION MOTORS

The split-phase motor is one of the oldest types of single-phase motors ever built for commercial use.[1,*] Even today, it is one of the most important and most widely used of all the types of single-phase motors. It is used for such applications as washing machines, oil burners, blowers, centrifugal pumps, woodworking tools, business machines, bottle washers, churns, automatic musical instruments, buffing machines, grinders, machine tools, and a host of other applications. It is most widely used in the ratings from 1/20 to 1/3 hp.

4-1 Split-Phase Motor Defined. A split-phase motor is a single-phase induction motor that has a main and an auxiliary (starting) winding; the two windings are mutually displaced by 90 electrical degrees. The auxiliary winding has a higher ratio of resistance R to reactance X than the main winding to achieve a phase-splitting effect, and a starting switch cuts it out of the circuit as the motor approaches operating speed. A more complete and precise explanation of how it works is given in the following articles.

[*] For numbered references, see References at end of this chapter.

4-2 Essential Parts of a Split-Phase Motor. The essential parts of a split-phase motor are represented diagrammatically in Fig. 4-1. There are two separate and distinct windings on the stator: a *main*, or running, winding and an *auxiliary*, or starting, winding. (Sometimes the auxiliary winding is referred to as the phase winding.) Each winding is a com-

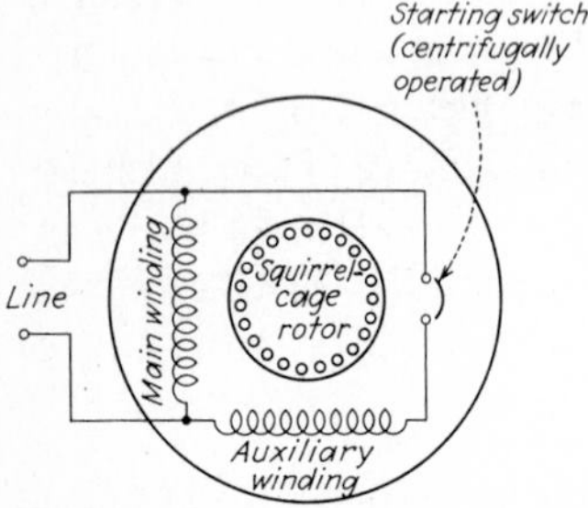

Fig. 4-1 The split-phase motor.

plete circuit in itself, consisting of as many sections or pole groups as there are poles (except in consequent-pole windings). Ordinarily the two windings are spaced 90 electrical degrees apart; i.e., the center of each pole group of the auxiliary winding is spaced halfway between the centers of two pole groups of the main winding. The two windings are drawn 90° apart in Fig. 4-1 to represent the space displacement of the windings. The rotor is of squirrel-cage construction.

For starting purposes, both main and auxiliary windings are connected in parallel across the line. (The sections or pole groups of either winding may be connected in series or in any number of parallel circuits, but the windings as a whole are connected in parallel with each other.) In series with the auxiliary winding is a starting switch which opens at approximately 75 to 80 percent of synchronous speed. This switch is usually and preferably operated by centrifugal means; for detailed information on such switches, refer to Arts. 17-5 to 17-9. Constructional details of a typical split-phase motor are illustrated in the cutaway view in Fig. 5-2, except that no capacitor is used on a split-phase motor.

4-3 Split-Phase Starting Principle. Like the two-phase motor, the split-phase induction motor has two primary (or stator) windings or phases displaced by 90 electrical degrees. An idea of how locked-rotor torque is developed may be gained by comparing these two types of motors. The two windings of a two-phase motor are connected to two different phases of the supply circuit which differ 90° in time phase. Therefore, the respective currents produced by these two voltages are necessarily 90° apart in *time* phase, and they flow through two windings displaced 90° apart in *space* phase. In Chap. 2 it was shown how, under such conditions, a rotating field is produced, causing the rotor to develop torque. The motor starts and runs—if the load is not too great—because the rotor tries to follow the revolving magnetic field.

If the two windings of a two-phase induction motor were both connected to the same single-phase voltage, the currents in the two stator windings would then be in phase, and no torque would be produced.* If, however, an external resistor is connected in series with one winding, *it will decrease the current in that phase, but it will bring this decreased current more nearly into phase with the voltage.* Thus, a phase displacement between the two currents will be obtained. This phase displacement is seldom large, being of the order of 20 to 30° in time. A phasor diagram of the locked-rotor currents of a typical split-phase motor is given in Fig. 4-2.

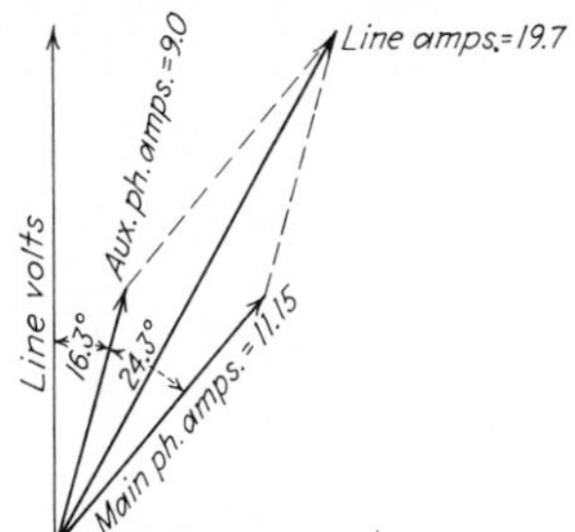

Fig. 4-2 A phasor diagram of the locked-rotor currents of a standard split-phase motor rated ⅙ hp, 115 V, 60 Hz, and 1725 rpm.

Split-phase motors do not use two-phase windings because better results are obtained with a different size of wire and a different number of turns in the two windings, whereas, in a true two-phase winding, both windings are identical. Rather than use an external resistor, it is more economical to obtain the additional resistance required in the auxiliary phase by using either a small size of copper wire or wire of a material having a specific resistance higher than that of copper. Either arrangement, particularly the former, which is the more common, results in a starting winding of light weight, requiring *less* than half the total slot space. Thus, *more* than half the total slot space can be used for the main winding, thereby permitting use of a larger size of wire for it than use of a two-phase winding would allow; the larger wire, of course, means lower resistance, better efficiency, and somewhat more breakdown torque.

Hence, in split-phase motors, the main winding is usually of heavier wire, is distributed in more slots, and is bulkier than the auxiliary wind-

*This point may perhaps be better understood when it is realized that, if the currents in the two windings were in time phase, the two windings would, in effect, be equivalent to a single winding and only a pulsating field could be set up, as explained in Art. 2-9.

ing. In addition, the main winding usually consists of more turns than the auxiliary winding and is wound almost invariably in the bottoms of the slots; i.e., it is wound first, and then the auxiliary winding is placed on top of it.

4-4 Purpose of the Starting Switch. At standstill, both windings must be in the circuit to develop torque, as explained previously. But after the motor has come up to approximately 75 or 80 percent of synchronous speed, the main winding alone can develop nearly as much torque as the combined windings; this point is shown in Figs. 4-3 and 4-4. At a higher speed, between 80 and 90 percent of synchronous, the combined-winding torque curve crosses the main-winding torque curve, so that, at speeds above this point, the motor develops less torque, at any given slip, with the auxiliary winding in the circuit than with it out of the circuit. Consequently, it would be advantageous, purely from a torque standpoint, to cut the auxiliary winding out of circuit exactly at the "crossover" point of the speed-torque curves. However, this point

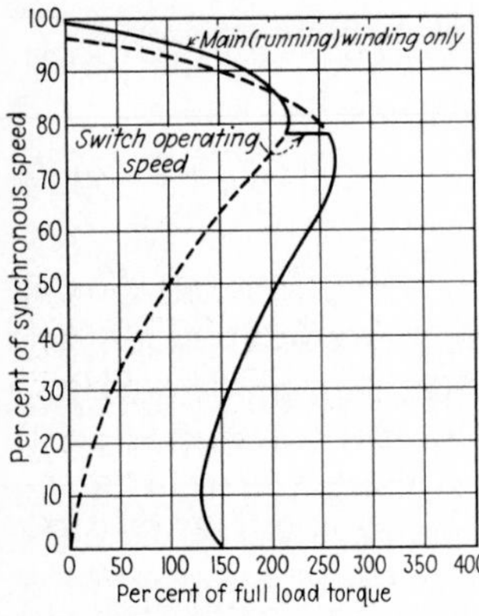

Fig. 4-3 Speed-torque curve of a standard split-phase motor.

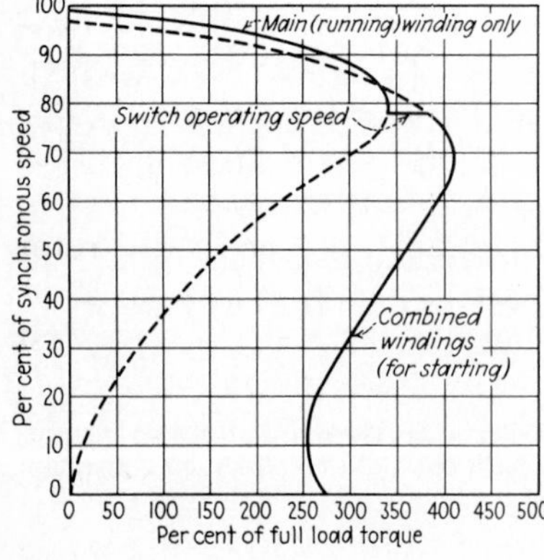

Fig. 4-4 Speed-torque curve of a special-service split-phase motor.

does not always occur at the same speed, even in individual motors of the same design, and varies in motors of different designs; moreover, switches vary in operating speeds, so that the usual practice is to make the average switch operate at a speed slightly below the average cross-over point.

There is a second reason why the starting switch is important; namely it prevents the motor from drawing excessive watts from the line and burning up the starting winding as it would do if it were left in the circuit continuously. Split-phase motors of low horsepower ratings, however, can be built for continuous operation with the auxiliary winding in circuit; they formerly were popular for small fans but are seldom used now.

The usual starting switch is a centrifugally operated mechanical device. (For more details on such switches, see Arts. 17-5 to 17-8.) Magnetically operated switches are used with hermetically sealed refrigerators where a switch inside the refrigerant would not be permissible, and for other special applications. (See Art. 17-9.)

4-5 Standard and Special-Service Motors. Applications for split-phase motors fall into two broad general classes:

1. Those that require frequent starting and a relatively large total running time per year, e.g., oil burners and furnace blowers

2. Those that require infrequent starting and a relatively small total running time, e.g., home-laundry equipment, home workshops, and cellar drainers

Standard split-phase motors are built in a wide variety of horsepower and speed ratings for the first class of applications and usually carry service factors as given in Table 1-3. Some manufacturers refer to these as general-purpose motors because of their wide usage, but this is not strictly correct since such motors do not meet the torque requirements specified by National Electrical Manufacturers Association (NEMA) for general-purpose motors. A typical speed-torque curve is shown in Fig. 4-3. *Special-service motors* (sometimes called high-torque motors) are used for the second class of applications. Figure 4-4 gives a typical speed-torque curve. Compared with standard motors, special-service motors have higher torques, lower efficiencies, higher locked-rotor currents, and a service factor of 1.0. Because they are built in high-volume production and in few models, they are generally lower in first cost. However, they should not generally be used on lighting circuits because of their high locked-rotor current. Moreover, their commercial availability in only a few horsepower and speed ratings further limits their range of usefulness.

4-6 Torque Characteristics of Split-Phase Motors. There are no NEMA torque requirements for standard split-phase motors, but the breakdown and locked-rotor torques that can generally be expected from commercially available motors are given in Table 19-1 (in Chap. 19) for fractional horsepower ratings. The breakdown torque of special-service split-phase motors is approximately equal to the breakdown torque of the capacitor-start motors given in Table 19-1 and their locked-rotor torques are approximately 175 percent of the values for standard split-phase motors.

Maximum permissible values of locked-rotor currents for standard split-phase motors are also given in Table 19-1.

CONNECTION DIAGRAMS FOR SPLIT-PHASE MOTOR WINDINGS

4-7 Motor Connections and Terminal Markings

(1) Motors with four tagged leads. Split-phase motors with four external line leads have had them identified in a wide variety of ways. They may be identified by tags, by colors, or by both. Different companies have changed their own system of marking leads from time to time. Standard terminal markings were developed and published by the American Standards Association (reorganized as The United States of America Standards Institute in September 1966 and later as the American National Standards Institute) in ASA C6 in 1944. These were used in an earlier work.[2] A revised edition of ASA C6 was published in 1956. ASA C6 has since been rescinded, and terminal markings and connections were added to NEMA MG1.[3] In this book, the NEMA diagrams are used; they differ little from those in ASA C6. In the NEMA-referenced diagrams the direction of rotation refers to the end of the motor opposite the drive end. The diagram for a split-phase four-lead single-voltage motor is given in Fig. 4-5. In this connection diagram, the leads bear the same tags as in ASA C6-1944, but the leads are connected differently for the same direction of rotation. When there is doubt as to

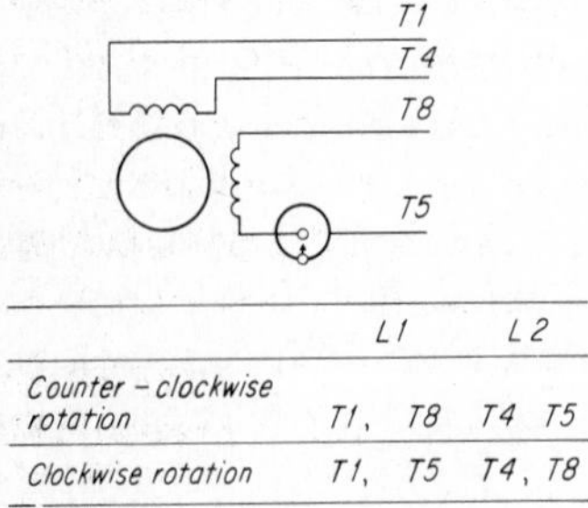

Fig. 4-5 Schematic wiring and line connection diagrams for a reversible (either-rotation) split-phase motor with four line leads. *(From NEMA Motor and Generator Standards, NEMA Publ. no. MG1-1978, National Electrical Manufacturers Association, Washington, 1978.)*

the correctness of the lead tagging, the leads can be identified and tagged by following the procedures of Art. 4-8.

(2) Motors with cast conduit box and built-in terminal board. Most split-phase motors are now supplied with the conduit box cast as an integral part of the front end shield; usually such motors are also provided with a built-in terminal board (see also Art. 5-7). Typical wiring and line connection diagrams for motors of this construction are given in Figs. 4-6 and 4-7. Basic methods of connection are essentially as shown, though

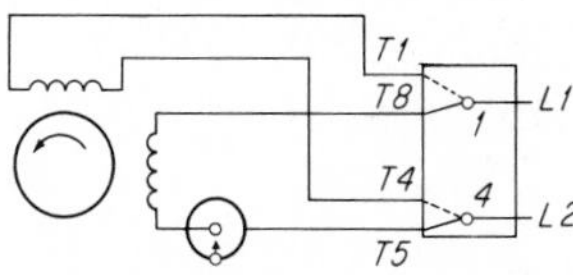

To obtain clockwise rotation, interchange leads T5 and T8

Fig. 4-6 Schematic wiring and line connection diagrams for a reversible (either-rotation) split-phase motor with terminal board. Main winding is permanently connected to the back of the terminal board. *(From NEMA Motor and Generator Standards, NEMA Publ. no. MG1-1978, National Electrical Manufacturers Association, Washington, 1978.)*

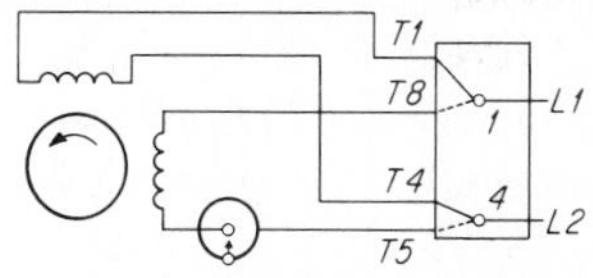

To obtain clockwise rotation, interchange leads T1 and T4

Fig. 4-7 Schematic wiring and line connection diagrams for a reversible (either-rotation) split-phase motor with terminal board. Auxiliary winding is permanently connected to the back of the terminal board. *(From NEMA Motor and Generator Standards, NEMA Publ. no. MG1-1978, National Electrical Manufacturers Association, Washington, 1978.)*

many variations may be encountered. For example, it is now a fairly common practice to mount the stationary member of the starting switch directly on the terminal board itself. As shown in Figs. 4-6 and 4-7, the motor can be connected for either direction of rotation simply by interchanging two leads connected to the outside of the terminal board, and it is not necessary to dismantle the motor to do this. In some motors the reversing leads have eyelets; in other motors, the reversing leads have quick-connects for ease of reversing.

(3) Motors with untagged colored leads. NEMA has developed color designations, corresponding to numbered tags, for single-phase motors as follows:

T1 = blue

T2 = white

T3 = orange

T4 = yellow

T5 = black

T8 = red

P1 = no color assigned

P2 = brown

Before dismantling a motor, the person servicing it should determine how the windings are connected to each other and to the line. These connections can best be recorded in the form of a wiring diagram such as Fig. 4-5, for example. It is also desirable to record the line connections if they are given on an instruction tag or card; if not, the person making the repair may have to determine these connections. In such a case, it is necessary to know how to tell the windings apart and how to identify the leads.

4-8 How To Tell the Windings Apart and How To Identify the Leads. If the motor is assembled and the lead markings are indistinguishable or not understood, and if there are four leads, the main winding usually can be differentiated from the auxiliary winding in one of two ways:

1. By measuring the ohmic resistance of each winding with a Wheatstone or Kelvin bridge or by the voltmeter-ammeter method. (See Art. 18-9.) For motors rated 1/12 hp or more, the resistance of the auxiliary winding is almost invariably—but not necessarily—greater than the resistance of the main winding; for 1/20-hp motors, the main winding is likely to have the higher resistance.

2. By measuring the locked-rotor power factors of the two windings separately at rated voltage and frequency with the rotor in place. The power factor of the auxiliary winding is invariably and necessarily higher than that of the main winding. The power factor is best measured with a voltmeter, ammeter, and wattmeter, though there are direct-reading power-factor meters.

If the end shields are removed, permitting an inspection of the windings, identification of the leads is simpler:

1. The auxiliary winding is usually wound on top of the main winding, i.e., nearer the stator bore.

2. The auxiliary winding has less bulk, usually fewer turns, and smaller-gauge wire.

COMMON WINDING CONNECTIONS

The more common winding connections will be discussed first, before some special and unusual ones are considered.

4-9 Both Windings Series-Connected. Split-phase motors are usually wound with two-layer concentric windings (except for two-speed motors). Stator connection diagrams for two- and four-pole windings are given in Figs. 4-8 and 4-9. In these two typical diagrams, both main and auxiliary windings are connected in series. Both of them show the lead T5, which goes to the starting switch, tied into the winding as a "dummy" lead. A motor connected by either of these diagrams will have terminals marked as in Fig. 4-5, which will then apply.

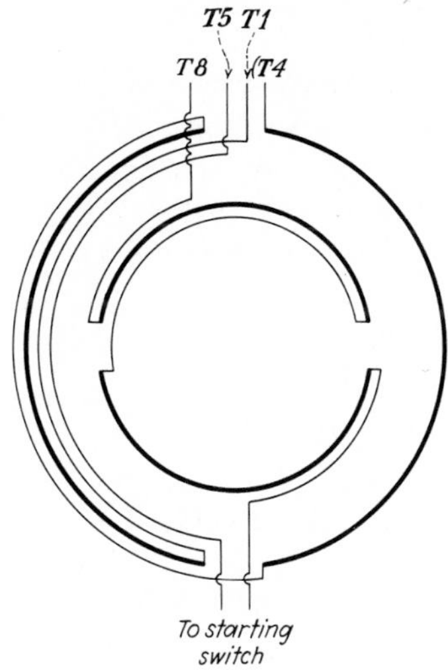

Fig. 4-8 Stator connection diagram for a two-pole winding with four line leads. Both windings are connected in series, and the motor is one phase and is externally reversible.

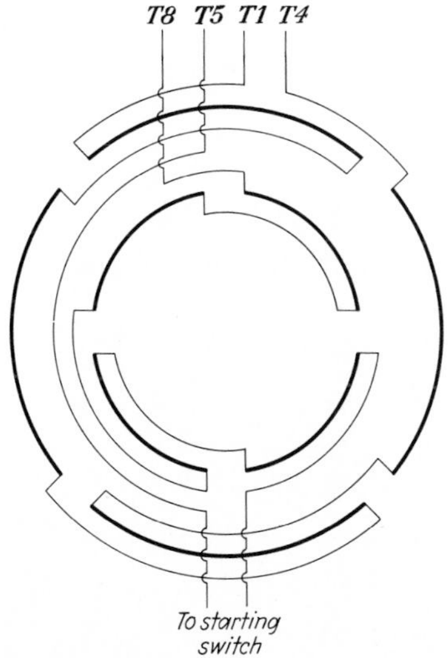

Fig. 4-9 Stator connection diagram for a four-pole winding with four line leads. Both windings are connected in series, and the motor is one phase and is externally reversible.

When a motor is encountered that does not use the present standard markings, it may save later confusion to re-mark them to agree with the present standard markings, using new tags if necessary. Care must be taken not to confuse T1 with T4 or T5 with T8, or rotation would be incorrect.

Direction of rotation in this book is specified the same way that NEMA specifies it, namely, from the front end, which is the end opposite the shaft extension. Unfortunately, this practice is not followed by all motor manufacturers.

4-10 Parallel-Connected Main Winding and Series-Connected Auxiliary Winding. The main winding of a split-phase motor is often connected in two or more parallel circuits for the sake of quietness. This practice is more common in motors having six poles or more. Since the auxiliary winding is energized only during starting, the more economical series-connected winding may be used.

Diagrams for motors with a parallel-connected main and a series-connected auxiliary winding are given in Figs. 4-10 and 4-11 for motors of four and six poles, respectively. These both use Figs. 4-5 to 4-7 for wiring and line connection diagrams. Note that these two stator connection diagrams both use equalizing connections (identified by *a* and *b*) so that

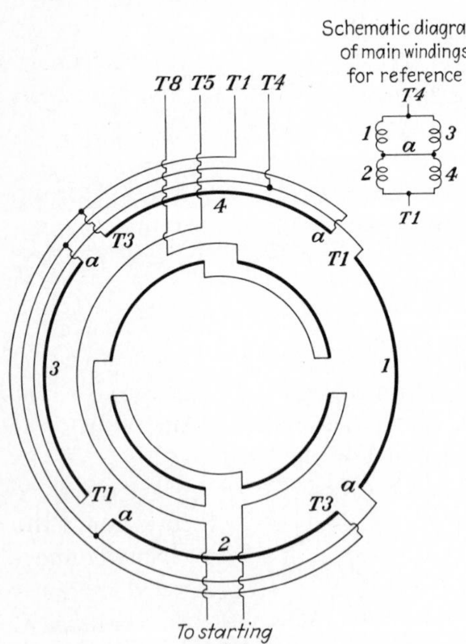

Fig. 4-10 Stator connection diagram for a four-pole winding with four line leads. The main winding is connected in parallel, and the auxiliary winding is connected in series. The single-phase motor is externally reversible.

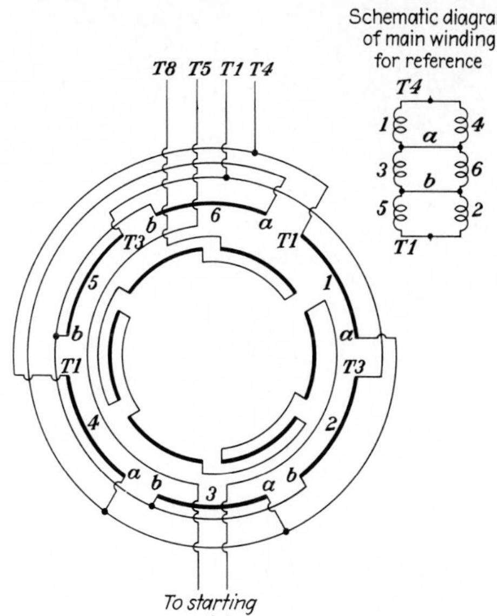

Fig. 4-11 Stator connection diagram for a six-pole winding with four line leads. The main winding is connected in parallel, and the auxiliary winding is connected in series. The single-phase motor is externally reversible.

each pole is connected in parallel with the pole diametrically opposite. The arrangement is best shown in the schematic diagram accompanying each figure; the numbers opposite the coils in the schematic correspond to the pole numbers shown on the main diagram.

Diametrically opposite poles are connected in parallel to equalize the flux in the respective poles and thereby minimize any unbalanced magnetic pull with consequent noise.

4-11 Both Windings Parallel and Cross-Connected. When starting noise is important, both main and auxiliary windings may be parallel-connected. Auxiliary-winding coils can be paralleled with each other, just as main-winding coils are. See Fig. 7-5.

4-12 "Short-Throw" and "Long-Throw" Connections. Consider Fig. 4-9: Each pole is connected directly to an adjacent one. This form of connection is usually the easiest and most logical way to connect the coils. It is called a "short-throw" connection because the throw from one coil to the next is short. In Fig. 4-12, on the other hand, connections are between alternate pole groups, necessitating a longer throw, hence the name "long-throw" connection. (Short-throw and long-throw connections for polyphase motors are discussed in Art. 14-6.) Electrically, it is not generally important whether the long or the short throw is used, except as discussed in Art. 3-12.

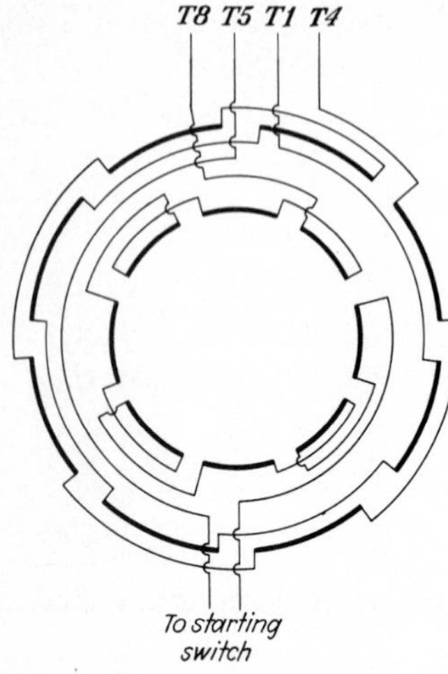

Fig. 4-12 Stator connection diagram for an eight-pole winding with four line leads. The main winding is a long-throw connection, and the auxiliary winding is a short-throw connection. Both windings are connected in series. The single-phase motor is externally reversible.

4-13 Switch in Center of Auxiliary Winding. In Figs. 4-8 to 4-12, the starting switch is connected to one end of the auxiliary winding and the other starting-switch lead goes directly to the line. Sometimes, as shown in Fig. 4-13, the starting switch is connected in the center of the winding.

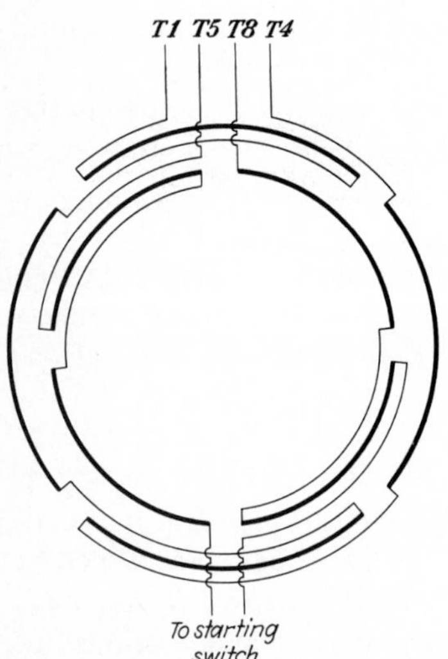

Fig. 4-13 Stator connection diagram for a four-pole winding with both windings connected in series and the starting switch in the center of the auxiliary winding.

4-14 Motors with Direction-Connection Plug. Westinghouse formerly supplied washing-machine motors which were provided with a square direction-connection plug. At the time of assembly of the motor in the washing machine, the plug was pushed into the front end shield in either one of two positions to give either counterclockwise or clockwise rotation, as needed by the machine in which it was being installed. More details of this arrangement were covered in an earlier work.[2]

4-15 Connecting Motor Windings for a Definite Direction of Rotation. Split-phase motors with four line leads, as shown in Fig. 4-5, can be connected to the power source for either direction of rotation. If the external connections shown in this figure are made inside the motor and only two line leads are brought out, the motor becomes nonreversible.

A stator connection diagram for a four-pole series-connected winding with two line leads, arranged for counterclockwise rotation, is given in Fig. 4-14. This diagram was derived from Fig. 4-9 by bringing one lead out from T4 to T5, and the other from T1 and T8.

"Direction of rotation," as used previously, refers to the end where the connections are made, which is usually the front end, or the end opposite the shaft extension.

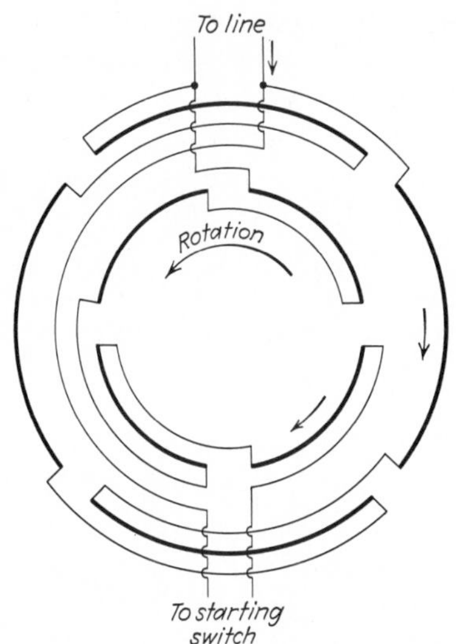

Fig. 4-14 Stator connection diagram for a four-pole winding with two line leads, arranged for counterclockwise rotation. Both windings are connected in series.

4-16 Predetermining Direction of Rotation. Of course, the simplest way to predetermine the direction of rotation of a split-phase motor is to connect it to a suitable power source momentarily and observe the direction in which it starts. Or, if a wiring and line connection diagram is available, the direction of rotation can be predetermined by comparing the lead connections with the diagram. Sometimes, however, it is desired to find the direction of rotation of a wound stator without assembling it into a motor. At other times the problem may be to find out or to verify the direction of rotation that would be produced by following a certain stator connection diagram. Both of these latter problems are discussed in the following articles.

(1) By testing the connected windings. Connect one end of the main winding and one end of the auxiliary winding to a common lead. Pass a small value of direct current through the auxiliary winding and, using a small compass in the stator bore, locate and mark a north pole on the stator core by the numeral 1. Now, using the same common lead as before, pass a small direct current through the main winding and locate the north pole nearest to the auxiliary north pole just marked; this distance should be 90 electrical degrees, and the pole should be marked 2. (Care must be taken not to reverse the magnetism of the compass needle in this process.) Since the direction of rotation of a split-phase motor is from an auxiliary pole to a main pole of the same magnetic polarity, the assembled motor will rotate in the direction from 1 to 2.

A variation of the preceding method is to connect the main and auxiliary windings to one pair of line leads, providing means for opening and closing the starting switch leads. With the starting-switch leads joined, pass a small value of direct current through the two windings. Place the compass in the stator bore with the needle free to assume a position. Now, open the starting switch leads and observe which way the needle moves, for it will move in the direction in which the assembled motor will rotate. It is easy to see why this is so if we stop to reflect that, with both windings energized, the needle will assume a position *between* the main and auxiliary windings; now, when the auxiliary winding is opened by opening the switch leads, the compass needle moves toward the main winding away from the auxiliary winding, which is, of course, the direction of rotation.

A simple variation of the method described in the preceding paragraph may be even more convenient for the small service shop. The windings are both connected to a high-frequency source, with the same provision for opening and closing the starting switch leads, and a paper clip suspended from a thread is used in place of a compass. The direction that the freely suspended paper clip moves in when the starting-switch leads are opened is the direction of rotation to be expected, for reasons given in the preceding paragraph.

(2) By inspection of the stator connection diagram. The way to predetermine the direction of rotation by inspection of the stator connection diagram is illustrated in Fig. 4-14. Assume current to be entering the line lead at the right, as indicated by the arrow. The current divides, and a portion of it flows through the first main-winding coil in a clockwise direction on the diagram, as shown by a curved arrow; now, following the other portion of the current through the starting switch to the first coil of the auxiliary winding, a second curved arrow is drawn as shown. Now, the two poles indicated are both of the same magnetic polarity, since the imaginary current traverses both poles in the same direction on the diagram. Therefore, the direction of rotation will be counterclockwise, i.e., from the auxiliary coil to the nearest main coil of the same magnetic polarity.

4-17 Nonreversible Motor with One Line Lead on the Starting Switch. Figure 4-15 is an interesting variation of Fig. 4-14. It shows a connection employed by at least one manufacturer to eliminate one soldered or welded joint and thereby reduce cost of manufacture. There is only one soldered connection, namely, where one side of the line connects to the common point of the main and auxiliary windings. The free or open ends of the main and auxiliary windings are connected to the two separate terminals of the starting switch. The remaining line lead is connected to the same switch terminal as is the main winding.

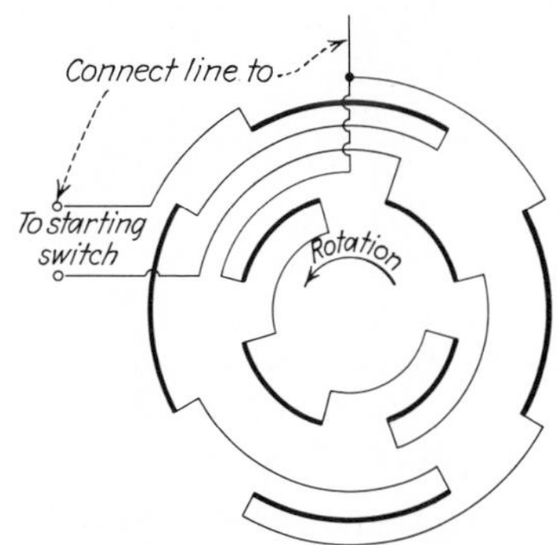

Fig. 4-15 Stator connection diagram for a motor with a line lead on the starting switch. The main and auxiliary windings have four-poles and are connected in series.

This method of connection is given here to illustrate one of the many types of connections that may be encountered; it is also given to warn the person making repairs that a switch lead is not necessarily a lead from the auxiliary winding.

4-18 Motor with Cord and Plug and Built-In Line Switch. Split-phase motors, especially for home-workshop use, are often provided with a cord and plug and a line switch. The line switch may be mounted on the end shield itself or on the conduit-box cover plate. The cord can be plugged into any convenience outlet, and the motor turned on and off by the motor-mounted toggle switch.

THERMALLY PROTECTED MOTORS

A general description of thermal protective devices and how they function is given in Arts. 17-16 to 17-21. Their use in split-phase motors is the subject of the following paragraphs.

4-19 Reversible: Five or Six Line Leads. Figure 4-16 shows wiring and line connection diagrams of a split-phase motor equipped with a thermal protective device which is mounted on or in the motor, thus becoming an integral part of the latter. The connections are the same as for the typical split-phase motor of Fig. 4-5 except that the thermal protective device is connected in series with the line. The stator winding connections can, of course, be the same as those of any split-phase motor not equipped with thermal protection. Five external leads are sufficient when protector post 3 and T1 are connected together inside the motor; without such an internal connection, six leads would be needed.

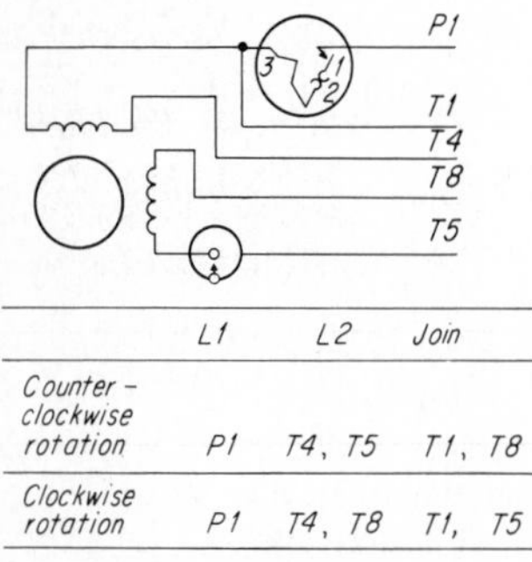

	L1	L2	Join
Counter-clockwise rotation	P1	T4, T5	T1, T8
Clockwise rotation	P1	T4, T8	T1, T5

Fig. 4-16 Schematic wiring and line connection diagram for a thermally protected reversible (either-rotation) split-phase motor with five line leads. *(From NEMA Motor and Generator Standards, NEMA Publ. no. MG1-1978, National Electrical Manufacturers Association, Washington, 1978).*

When a thermally protected motor is connected to a grounded circuit, the lead from the protector, P1 in this case, should be connected to the ungrounded side of the line so that, if the thermal protector opens, the motor windings will not be "hot." In Fig. 4-16, line lead L2 is the one to connect to the grounded side of the line. Most manufacturers state on their instruction tag which lead is to be connected to the ungrounded side of the line.

4-20 Nonreversible: Two Line Leads. A nonreversible motor, even if thermally protected, requires only two line leads, as does any other split-phase motor. Sometimes one lead is colored white to denote that it should be connected to the grounded side of the line. Some manufacturers, however, use a distinctive marking to indicate the protector lead; Robbins and Myers, for example, use a black lead with red tracer for just this purpose. Service personnel should always be careful to connect the proper lead to the grounded side of the line!

4-21 Motors with Terminal Board. Thermally protected split-phase motors, provided with a terminal board, may be wired up as shown in Fig. 4-17 or 4-18.

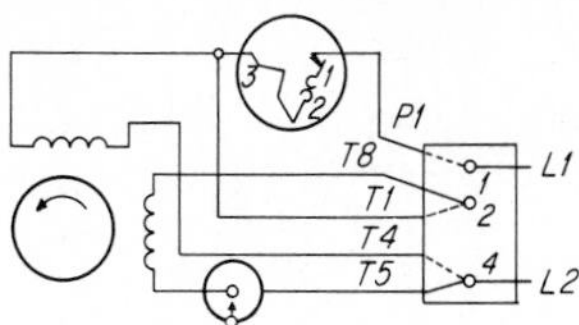

To obtain clockwise rotation interchange leads T5 and T8

Fig. 4-17 Schematic wiring and line connection diagram for a thermally protected reversible (either-rotation) split-phase motor with terminal board. Main winding is permanently connected to the back of the terminal board. *(From NEMA Motor and Generator Standards, NEMA Publ. no. MG1-1978, National Electrical Manufacturers Association, Washington, 1978.)*

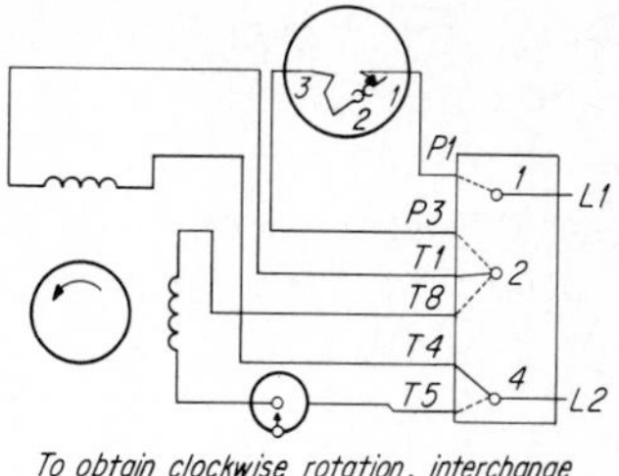

To obtain clockwise rotation, interchange leads T1 and T4

Fig. 4-18 Schematic wiring and line connection diagram that is similar to Fig. 4-17 except that the auxiliary winding is connected to the back of the board. *(From NEMA Motor and Generator Standards, NEMA Publ. no. MG1-1978, National Electrical Manufacturers Association, Washington, 1978.)*

CAUTION: *Post 2 must on no account be connected to either side of the line.* The same stator connection diagrams as for Fig. 4-6 apply.

4-22 Heating Coil of Thermal Protector in Auxiliary Phase Only. Built-in thermal protectors usually are provided with a heating coil, or its equivalent, so that the device is influenced both by temperature and by current passing through the heater coil. In the motors just discussed, line current passes through the heater coil. If a third terminal is added to the protector, the latter can be connected so that only the current of the auxiliary phase passes through the heater, although the contacts open the line circuit. Such an arrangement is shown in Fig. 4-19.

TWO-SPEED MOTORS

4-23 Two-Speed Pole-Changing Motors. Split-phase motors can be wound with pole-changing windings so as to give two different normal speeds of operation, either one of which can be selected by making changes in the external connections. Regardless of the speed selected, the motor will operate at a substantially constant speed within the range of its rating. The number of combinations is legion, and only a

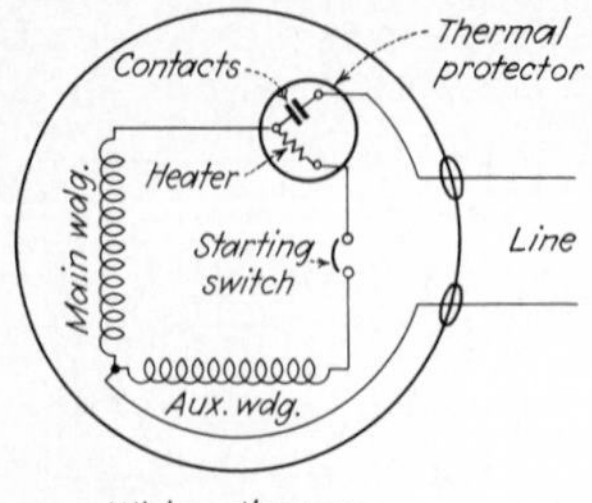

Fig. 4-19 Wiring and line connection diagram of a thermally protected split-phase motor with the heater in the auxiliary phase only.

few can be discussed here. For purposes of this discussion, pole-changing motors are broken down into two general classes: four-winding arrangements and three-winding arrangements. Because consequent-pole connections are often used, we will start by considering first what they are.

4-24 Consequent-Pole Windings. Briefly, a consequent-pole winding has only half as many pole groups as there are poles, and all these groups are connected for the same magnetic polarity. Why the resultant magnetic field has twice as many poles as there are coils was explained by Kincaid[4] with the use of a figure very similar to Fig. 4-20.

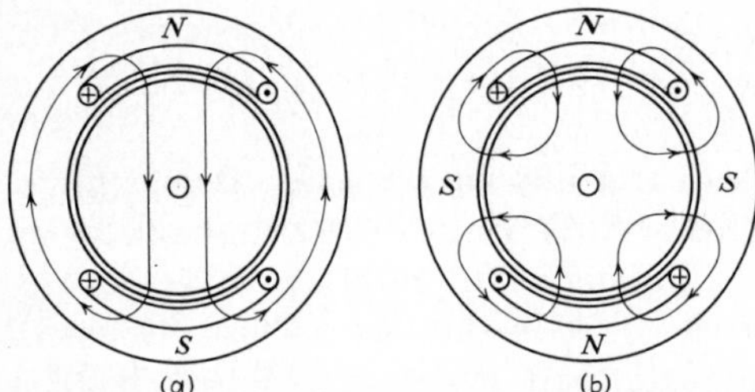

Fig. 4-20 A simple consequent-pole connection. (*a*) Normal connection, two poles; (*b*) consequent-pole connection, four poles.

In this figure, there are two coil groups: when these are connected for opposite magnetic polarities, as in Figure 4-20*a*, the magnetic field has two poles; when they are connected for the *same* magnetic polarity the two coils "buck" one another, but since the flux has to return across the air gap, two additional poles are formed. Thus, a four-pole magnetic field is produced as shown. In short, if all the coils are connected for the same magnetic polarity, there will be formed an equal number of consequent poles of the opposite magnetic polarity; the centers of the consequent poles will fall halfway between the coils. Figure 4-22 shows an interesting variation: the auxiliary winding for the six-pole connection has only two poles wound; note, however, that these two coils have the same span, positions, and magnetic polarities that two coils of a conventional six-pole winding would have. In other words, the auxiliary winding is like the six-pole winding with four coils omitted. [To those who wish to probe the subject further, let it simply be stated that, technically, this winding is a two-pole winding with coils of one-third pitch, producing a strong third harmonic or six-pole magnetomotive force (mmf) wave. The mathematical basis for this statement is contained in Eq. (14-12), as derived in Art. 14-3, of another work by Veinott.[5]]

4-25 Two-Speed Pole-Changing Four-Winding Motors. With this arrangement, there are two main windings and two auxiliary windings, one of each for each speed. Such an arrangement is shown schematically in Fig. 4-21, which shows the use of a double-pole double-throw switch to select the desired speed; a three-pole double-throw switch would be required to deenergize the motor completely. It will be noted that the connections are so arranged as to require but one starting switch. *This starting switch must operate at 75 to 80 percent of the synchronous speed of the low-speed winding.* To some extent, this requirement limits the switching torque on the high-speed connection.

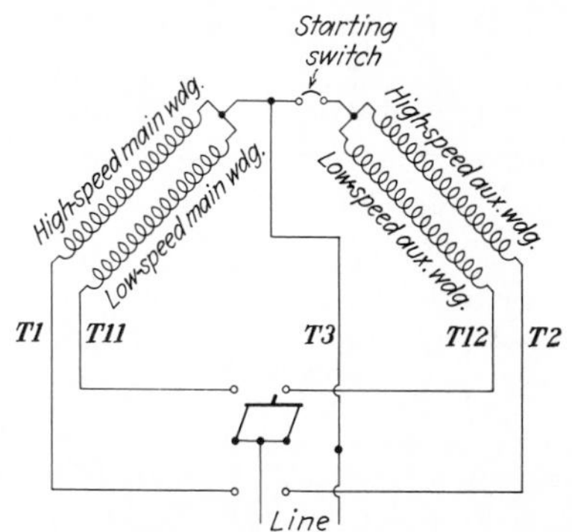

Fig. 4-21 Wiring and line connection diagram of a two-speed pole-changing split-phase motor with four windings.

The motor of Fig. 4-21 is not reversible, but it could be made so by bringing out the other ends of both main windings, at the expense of bringing out two more leads. It is also interesting to note that this motor could be arranged to operate in one direction on one speed, and in the other direction on the other speed; the same cannot be said for the three-winding motors discussed in Art. 4-26.

Because four windings are used, considerable winding space is required in the motor. For this reason, the physical dimensions of a two-speed pole-changing motor are necessarily greater than those of a

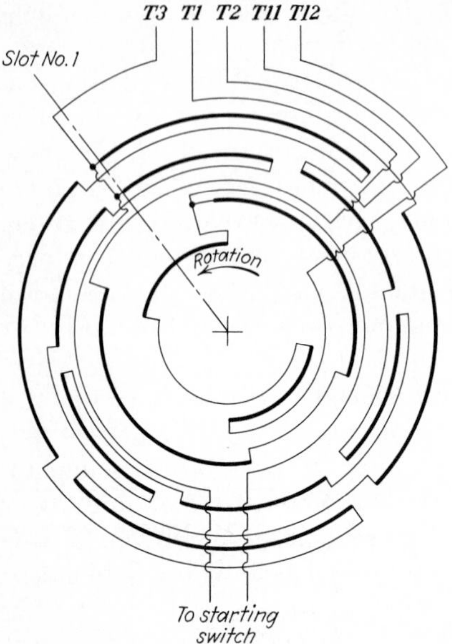

Fig. 4-22 Stator connection diagram for the motor of Fig. 4-21. The four- and six-pole windings are all series-connected. Both main windings are conventional; the four-pole auxiliary winding is consequent-pole, and six-pole auxiliary winding (extreme inner winding) has only two coils as shown.

single-speed motor of the same horsepower and speed rating. To facilitate winding, the low-speed windings often have a consequent-pole connection. Figure 4-22 is a typical stator connection diagram.

4-26 Two-Speed Pole-Changing Three-Winding Motors. Three-winding arrangements were developed for pole-changing single-phase motors to simplify the winding process by omitting one auxiliary winding completely. Schematic wiring and line connection diagrams for one such motor are shown in Fig. 4-23. This motor automatically starts on the high-speed connection, regardless of whether it is connected externally for high or for low speed. If connected for low speed, it starts on the high-speed connection and transfers automatically to the low-speed connection at a predetermined speed. Only a single-pole double-throw switch is actually necessary for speed control, but a double-pole switch is needed for complete deenergization of the motor. The direction of rotation can be interchanged by changing two leads on the outside of the terminal board. An interesting feature of this particular motor is the two-heater thermal protector; one heater is for the low-speed winding, the other for the high-speed winding.

HOME-LAUNDRY MOTORS

Home-laundry motors are specially designed and made for powering home washing machines, dryers, or combination washer-dryers. They

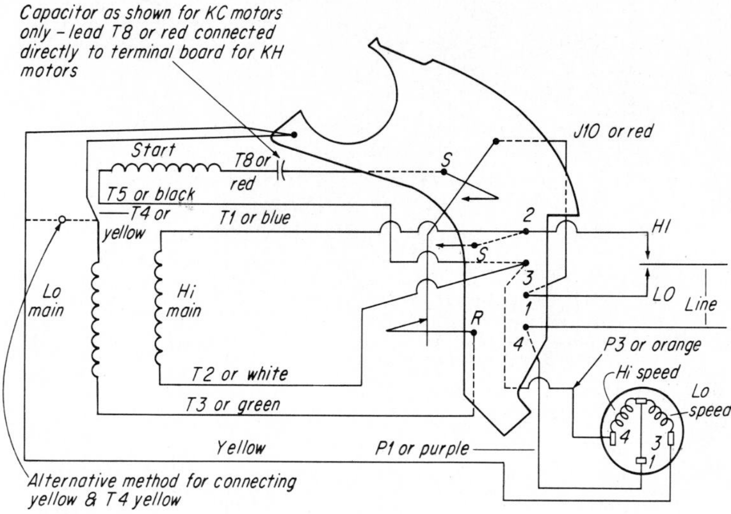

Fig. 4-23 Wiring and line connection diagram for a two-speed pole-changing split-phase or capacitor-start motor. Motor always starts on the high-speed connection, but if connected externally for low speed, it automatically transfers to the low-speed connection at a predetermined speed. Note the use of a thermal protector, which is provided with two heaters, one for each speed connection. *(General Electric Company.)*

are usually either split-phase or capacitor-start, single-speed or two-speed pole-changing. Sometimes they are provided with auxiliary terminals on the terminal board. A dryer motor that is bonded together with an epoxy is illustrated in Fig. 4-24.

4-27 Single-Speed Home-Laundry Motors. These are all four-pole motors, rated ⅙, ¼, ⅓, or ½ hp. The temperature rise in the appliance at an ambient temperature of 25°C is 75°C for motors with Class A insulation and 95°C for motors with Class B insulation. Locked-rotor current at 115 V may exceed 50 A, momentarily. Locked-rotor and accelerating torques for each appliance motor are unique to the appliance design and vary widely. Breakdown torques are generally as given for capacitor-start motors in Table 19-1 (see Chap. 19).

Although NEMA has developed standard connection diagrams for home-laundry-appliance motors, each appliance manufacturer uses a unique color code, therefore requiring unique motor connections. The NEMA schematic connections are shown in a previous work by Veinott.[2]

Fig. 4-24 A dryer motor that is held together with an epoxy. Note that the tongues on the die-cast and shields slide into an opening punched in the laminations. The rotor is held concentric in the air gap during curing of the epoxy, and after the epoxy cures the motor cannot be dismantled. *(Emerson Motor Division)*

4-28 Two-Speed Home-Laundry Motors. These motors are generally similar to the motors described in the preceding article, except that they are wound for 4/6 poles arranged generally as described in Art. 4-26. These motors invariably use the two four-pole windings for starting, with the auxiliary winding being cut out of the circuit by the centrifugal switch. After the centrifugal switch has operated, the motor may run on either the high-speed or the low-speed winding, depending upon the external connection.

One practical example of such a motor is illustrated in Fig. 4-25. In

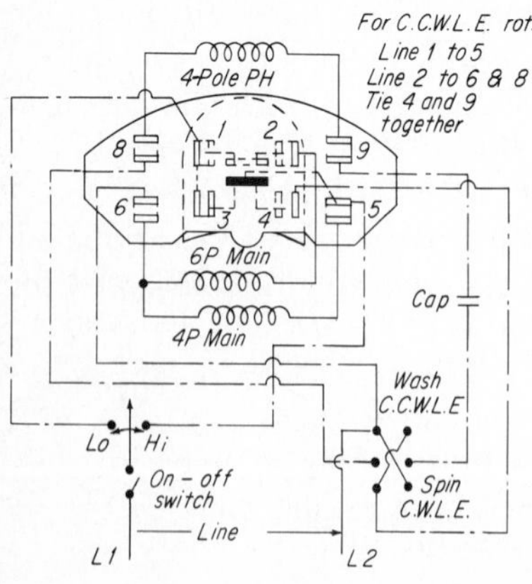

Fig. 4-25 Wiring and line connection diagram for a two-speed capacitor-start motor that has full automatic starting, with direction of rotation selectable for either speed. (See Art. 4-28.) The starting position at rest connects terminal post 5 to posts 3 and 4; in the running position, it connects terminal post 5 to posts 1 and 2. Posts 1 and 3 are permanently connected together. The double-pole double-throw switch in the lower right-hand corner reverses the connections between the auxiliary phase and the line to give the desired rotation. The lines broken by dashes (— - —) indicate customer leads. *(Delco Products.)*

this diagram, the external double-pole, double-throw switch shown provides the capability of starting the motor from rest in either direction of rotation on either speed connection.

As shown in Fig. 4-26, an innovation is the use of one pole of the four-pole main winding as conductors in the six-pole main winding. This scheme, called *duo-pole*, reduces the total winding and, therefore, the total active material in the motor.

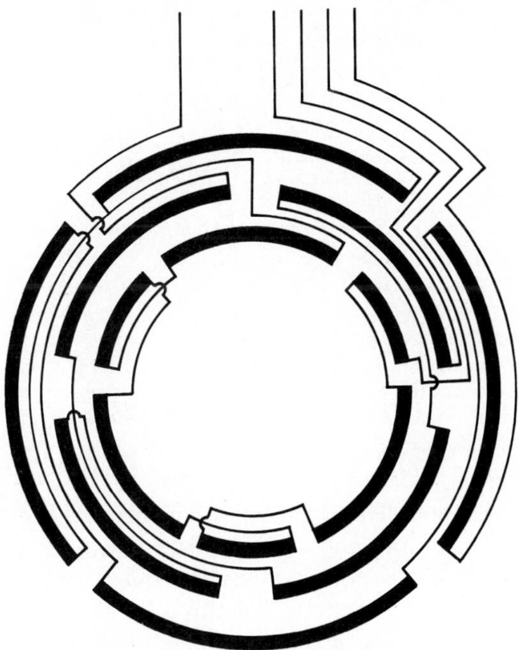

Fig. 4-26 Stator connection diagram for a duo-pole motor. The four- and six-pole windings are all series-connected and are all conventional. Reading from outside inward: four-pole main winding, four-pole auxiliary winding, six-pole main winding. *(Emerson Motor Division.)*

4-29 Laundry Motors with Auxiliary Terminals. Home-laundry motors are often provided with one or more extra terminals on the terminal board. These provide extra circuits which may or may not be common with the starting switch and may be open, or may be closed, when the motor is running. Such extra circuits make it easy for the appliance manufacturer to interlock the dryer heater with the blower motor, for example.

In Fig. 4-27, the use of such a circuit is illustrated. If the heater is connected to the line through posts 6 and 10, it can be energized only when the motor is running, for posts 6 and 10 are connected together

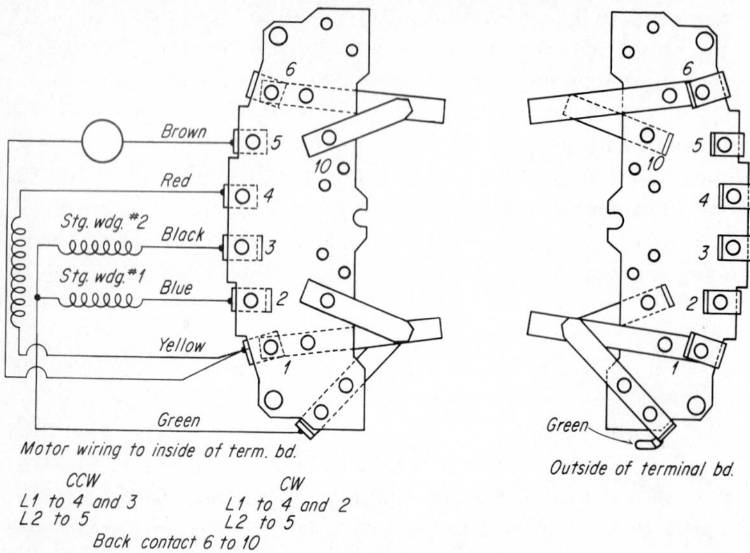

Fig. 4-27 A dryer motor with back contacts on the starting switch through which the heater can be connected so that it can be energized only when the motor is running. Another feature is the use of two starting windings, so that the motor can be reversed by means of a single-pole double-throw switch. *(Westinghouse Electric Corporation.)*

only after the motor is up to speed. This same motor incorporates an interesting feature: one main winding and two auxiliary windings to simplify reversibility. One auxiliary winding is used for one direction of rotation, the other for the other direction. (All three windings are four-pole.) Direction of rotation can be preset at standstill simply by connecting 4 to 3 or to 2; in either case, line power is applied to 4 and 5.

MISCELLANEOUS ARRANGEMENTS

4-30 Resistance-Start Split-Phase Motor. A resistance-start split-phase motor is a form of split-phase motor in which an external resistor is connected in series with the auxiliary winding. It is little used because it is more economical to obtain the resistance needed by dropping the size of wire in the auxiliary winding.

4-31 Reactor-Start Motor. A reactor-start motor is similar to a split-phase motor except that it is designed to have an external reactor connected in series with the *main winding* during the starting period. The effect of the reactor is to reduce the locked-rotor current in the main winding, at the same time increasing the angle of lag of main-winding current behind the impressed voltage. This increase in phase angle between the two currents tends to increase the locked-rotor torque, so that substantially the same locked-rotor torque is obtained, but with appreciably less locked-rotor current than would be obtained without the reactor. The starting switch has to be single-pole and double-throw, like the transfer switch used in the two-value capacitor motor shown in Fig. 6-5. Equations for calculating the reactor required for this type of motor have been developed by Veinott.[6]

4-32 Dual-Voltage Motors. Prior to the late 1960s the dual-voltage split-phase motor was generally frowned upon because of its cumbersome connections and inability to produce sufficient starting and accelerating torque on the higher voltage connection. The connections described in Chap. 5 have been used successfully since that time in dual-voltage split-phase motors with larger fractional horsepower ratings. The elementary principles, methods of winding, and thermal protection of dual-voltage split-phase motors are similar to those of capacitor-start motors, as described in Arts. 5-12 to 5-17. Dual-voltage split-phase motors are normally limited to soft-start applications, such as belt-drive or direct-drive fans and blowers.

4-33 Motors with Magnetically Operated Switch. Split-phase motors, especially those built in frame sizes too small to accommodate a centrifugal switch, often employ a relay to open the starting winding circuit when the motor approaches operating speed. The magnetic relay normally used in split-phase motors is current-operated. A discussion of the operation of a current-operated relay is given in Art. 5-20, with a typical arrangement as shown in Fig. 5-17*a* except that no capacitor is present in the auxiliary phase. Magnetic starting relays are further discussed in Art. 17-9. Many such devices are gravity-operated and must be mounted in a vertical position. Furthermore, if the motor is rewound as detailed in Arts. 4-36 to 4-38, or if a capacitor is inserted in the auxiliary phase as described in Art. 4-39, the relay must be reselected.

4-34 Reverse Turns in the Auxiliary Phase. In some designs, if one were to select the auxiliary-winding diameter so as to obtain an optimum R/X ratio, the current density in the starting winding might be excessive and lead to too many burnouts. One way to overcome this problem is to use the next larger wire size and increase the total length of wire by increasing the turns in the outside coil to obtain the original resistance; one-half of the additional turns are wound in the conventional, or forward, direction and one-half in the opposite, or reverse, direction. Thus, the original design R/X ratio is obtained but with a larger wire size. Such a scheme has been used in motors with a current-operated relay to allow the relay to drop out (open) as the motor approaches operating speed, especially at high values of line voltage.

One way to obviate the necessity for such an arrangement is to use wire of higher resistivity and larger diameter in the auxiliary winding. Still another way is to wind with a larger wire size, increase the winding mold perimeter above the minimum, and tuck the resultant excess auxiliary coil into the end turns.

MISCELLANEOUS SERVICE PROBLEMS

4-35 How To Reverse the Various Types of Motors. The fundamental principle involved in reversing split-phase motors is that the direction of current in the auxiliary phase, with respect to the current in the main phase, must be reversed. How to accomplish this change, in the case of four-lead motors, is shown in Fig. 4-5. How to connect the motor for a definite direction of rotation and how to predetermine this direction of rotation are fully covered in Arts. 4-15 and 4-16. It should be pointed out that if the line leads of Fig. 4-5 are connected to a snap switch for reversing, the motor cannot be *plugged;* it is necessary to wait at least until the rotor has slowed down sufficiently for the starting switch to close the auxiliary-phase circuit, and sometimes until the motor comes to rest, before the motor can be reversed.

4-36 Rewinding or Reconnecting Motor for a Different Voltage and the Same Performance. If a split-phase motor is series-connected in both windings, it can be reconnected for half voltage by connecting both windings in parallel. Sometimes it may be difficult in practice to reconnect the motor, particularly if the coils were originally gun-wound or machine-wound, because the connections between coils are very short, and it may be difficult to solder a lead to them.

If it is desired to make a 230-V motor out of a 115-V motor and if both windings are series-connected, it will be necessary to rewind both windings, using wire three sizes smaller and twice as many turns in each. It is inadvisable, however, to attempt to rewind any motor for more than 250 V, because of starting-switch limitations.

If the change in voltage is not to double or to halve it, both windings will have to be rewound. In general, the change in turns in each winding should be proportional to the change in voltage for which the motor is being rewound. The wire used in the main winding should be of the largest size that will go into the slots without omitting turns. For the auxiliary winding, however, it is definitely not advisable to choose the largest size that will fit into the slots; a fairly safe rule is to keep the difference in wire sizes (gauge numbers) between the main and auxiliary windings the same in the rewound motor as it was in the original motor.

An approximate rule of thumb which applies to split-phase fractional horsepower motors is that the auxiliary winding should have 70 to 75 percent as many turns as the main winding and should be wound with wire three sizes (gauges) smaller.

4-37 Rewinding for Different Torques at the Same Voltage. Probably it is not often that a service shop will be called upon to rewind a motor to obtain a different torque, but some general hints on what to do in such a case may be helpful. At a given voltage and frequency, the breakdown torque will vary approximately inversely as the square of the number of turns in the main winding. Therefore, to increase the breakdown torque, decrease the turns in the main winding.

A word of caution must be given here. Before rewinding for a different torque, some careful check tests should be made on the motor that is to be rewound. Determine experimentally how high the voltage must be increased to obtain the desired breakdown torque. With the motor operating at rated load, measure the watts input at normal voltage and at the increased voltage. If the input is appreciably higher, a temperature run should be taken at the operating load at this increased voltage to see whether the motor is capable of dissipating the increased losses. Instructions for making a temperature test are given in Art. 18-16(9). Also, listen carefully to the motor while it is operating at the higher voltage, as well as while it is operating at normal voltage, to make sure that with the increased torque the magnetic noise will not be excessive. If the temperature run indicates that the torque of the motor can be safely increased by as much as the ratio of, say, 110:120 V [the torque increase will be $(120/110)^2$, or 19 percent], the turns in the main winding can be decreased in the ratio of 120:110. For such a small change in turns, it will be unnecessary to change the auxiliary winding. If the main-winding turns are decreased, it is desirable to use a larger size of wire when space permits. Many motor manufacturers use half wire sizes, so it may be possible to increase the winding by one-half wire size to the next larger gauge.

If the motor is to be rewound for less torque, it is also important to check the input at reduced voltage, for the full-load losses may be either increased or decreased with a reduction in applied voltage. A good way to find out is to take a full-load saturation test, as described in Art. 18-16(8).

4-38 Rewinding for a Different Frequency. If the change in frequency is very small, say from 50 to 60 Hz, it may not be necessary to rewind at all, unless the motor is severely overloaded. If the change is from 60 to 50 Hz, a winding change is apt to be unnecessary. If the motor is a high-torque motor, it may be necessary to increase the turns of a 60-Hz motor 10 percent to obtain satisfactory operation on 50-Hz without overheating. When changing either a general-purpose or a high-torque motor from 60 to 50 Hz the rotating member of the starting switch should be changed, for the switching torque would otherwise be adversely affected. Should the change be from 50 to 60 Hz and should the torque requirements not be too severe, it may be unnecessary to change either the winding or the starting switch.

A 115-V, 60-Hz motor will have equivalent breakdown torque and temperature rise at rated horsepower when it is operated at 105 V on 50 Hz.

4-39 Use of a Capacitor in the Auxiliary Phase To Raise Locked-Rotor Torque or To Reduce Locked-Rotor Current. It has often been asked what beneficial effects can be obtained by inserting an electrolytic capacitor in series with the auxiliary winding of a split-phase motor. In Fig. 4-28 is shown the effect of inserting different values of capacitance in series with the auxiliary winding of a standard ⅙-hp, 115-V, 60-Hz,

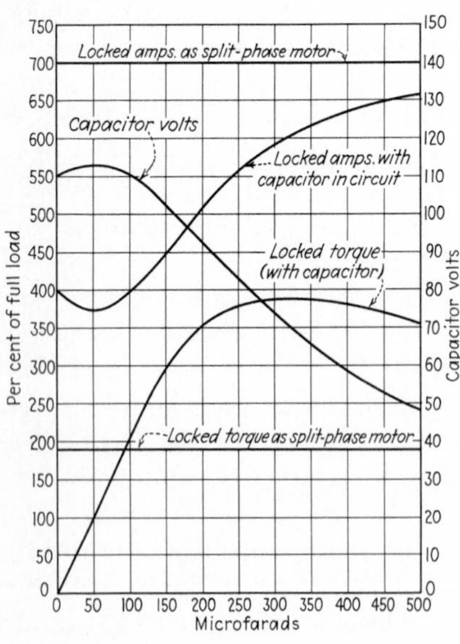

Fig. 4-28 Effect of inserting an electrolytic capacitor in series with the auxiliary winding of a ⅙-hp 115-V 60-Hz 1725-rpm standard split-phase motor.

four-pole split-phase motor. It will be noted that with 90 microfarads (mfd), the torque is the same as with a split-phase motor, but the locked-rotor current is reduced from 700 to 390 percent, or a total reduction of 44 percent. The use of a smaller number of microfarads reduces the locked current only slightly, but, within this range, the locked-rotor torque will be proportional to the number of microfarads used. It will be noted that, by use of 250 mfd, it is possible to double the locked-rotor torque with less locked-rotor current than is drawn as a split-phase motor. The conclusion may be drawn that use of a capacitor is both feasible and practical. If an electrolytic capacitor is available, it is suggested that an actual test be made to verify whether an important gain is effected or not. It would be well to check the locked capacitor volts to see that the voltage is not more than 20 percent higher than the rated voltage of the capacitor. (Commercial capacitors for 115-V motors are built to withstand 135 V for starting service.)

In Fig. 4-29 are shown corresponding curves for the use of a capacitor in series with the starting winding of a ¼-hp special-service split-phase washing-machine motor. It will be noted that it is necessary to use more actual microfarads to obtain a gain in locked-rotor torque than is necessary for a standard ⅙-hp motor.

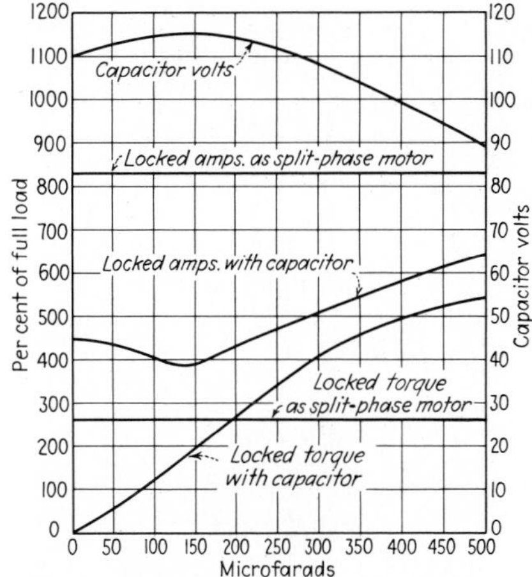

Fig. 4-29 Effect of inserting an electrolytic capacitor into series with the auxiliary winding of a ¼-hp special-service split-phase washing-machine motor.

It would be difficult to estimate the exact value of capacitance necessary to use with a complete line of motors, but the two curves just mentioned should give some indication of what may be expected.

4-40 Thermal Protection. If the motor has an integral thermal protector, no attempt should be made to rewind or reconnect the motor for a different voltage, torque, or frequency. Articles 17-16 to 17-20 cover details and requirements of thermal protectors. The selection of a protector for a given motor is an art unto itself and should not be attempted by a person without the expertise to do so.

4-41 Distribution Factor and Effective Conductors. Sometimes it is necessary to compare a single-phase winding with another one of a different distribution. In general, they would be equivalent if both windings had the same number of effective conductors. To calculate effective conductors, we must know the distribution factor. For a single-phase winding, the *distribution factor* may be defined as the ratio of the actual voltage generated in a distributed concentric winding to the voltage that would be generated in it if all the turns in the distributed winding were wound full-pitch. For a better understanding of the meaning of the distribution factor, refer to Fig. 3-2: less voltage is

TABLE 4-1 Distribution Factors for Single-Phase Concentric Windings

No.	Slots per pole	Distribution factor	Distribution 1	2	3	4	5	6	7	8	9	10	11	12	13
1	4	0.8536	1	1	x	1	1								
2	4	0.8047	1	2	x	2	1								
3	4	1.0000	1	x	x	x	1								
4	6	0.9659	1	x	x	x	x	1							
5	6	0.8365	1	1	x	x	1	1							
6	6	0.6440	1	1	1	1	1	1							
7	6	0.8080	1	2	1	x	1	2	1						
8	6	0.7887	1	1	1	x	1	1	1						
9	9	0.8186	1	2	2	1	x	x	1	2	2	1			
10	9	0.8823	1	2	2	x	x	x	x	2	2	1			
11	9	0.9114	1	2	1	x	x	x	x	1	2	1			
12	9	0.9254	1	1	x	x	x	x	x	1	1				
13	9	0.9373	3	2	x	x	x	x	x	2	3				
14	9	0.8689	2	2	1	x	x	x	1	2	2				
15	9	0.8312	1	1	1	x	x	x	1	1	1				
16	9	0.8557	3	3	2	x	x	x	2	3	3				
17	9	0.8581	8	8	5	x	x	x	5	8	8				
18	12	0.9250	2	1	1	x	x	x	x	x	x	1	1	2	
19	12	0.8721	2	2	1	1	x	x	x	x	1	1	2	2	
20	12	0.8294	1	1	1	1	x	x	x	x	1	1	1	1	
21	12	0.7400	1	1	1	1	1	x	x	1	1	1	1	1	
22	12	0.7763	3	3	2	2	2	x	x	2	2	2	3	3	
23	12	0.8387	1	2	2	1	1	x	x	x	1	1	2	2	1
24	12	0.9010	1	2	1	1	x	x	x	x	x	1	1	2	1
25	12	0.8223	1	2	2	2	1	x	x	x	1	2	2	2	1
26	12	0.9495	1	2	1	x	x	x	x	x	x	x	1	2	1

induced in the inner coil lying in slots 2 and 7 than in the outer coil lying in slots 1 and 8 because the latter coil embraces more flux. Hence the smaller coils in a concentric winding are less effective than the larger ones. If a given winding has, say, 100 turns and if the distribution factor is 0.850, the total number of effective turns is 85. In other words

$$\text{Effective conductors} = \text{actual conductors} \times \text{distribution factor} \quad (4\text{-}1)$$

where

$$\text{Actual conductors} = \frac{\text{total main-winding conductors}}{\text{number of main-winding circuits}} \quad (4\text{-}2)$$

For winding calculations, the number of effective conductors is always more significant than the number of actual conductors, except when figuring the resistance.

Distribution factors for a number of different single-phase windings are given in Table 4-1. A method for computing the distribution factor for any single-phase winding with a whole number of slots per pole is given in Fig. 4-30.

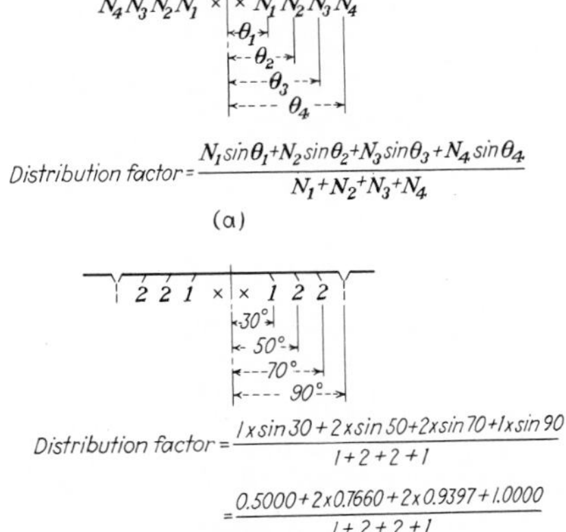

Fig. 4-30 How to compute the distribution factor of a single-phase concentric winding. (*a*) General formula; (*b*) specific example.

References

1. Veinott, C. G.: "An Old Tesla Motor, Grandfather of Modern Split-Phase Motors," *Electr. J.*, March 1931, p. 187.
2. ———: *Fractional Horsepower Electric Motors*, 2d ed., McGraw-Hill Book Company, New York, 1948.
3. *NEMA Motor and Generator Standards*, NEMA Publ. no. MG1-1978, National Electrical Manufacturers Association, 2101 L Street N.W., Washington, DC 20037, 1978.
4. Kincaid, C. W.: "Change-Speed Induction Motors," *Electr. J.*, August 1924, pp. 357–363.

5. Veinott, C. G.: *Theory and Design of Small Induction Motors,* McGraw-Hill Book Company, New York, 1959. (Out of print, but available from University Microfilms, P.O. Box 1647, Ann Arbor, Mich. 48106.)
6. ———: "Starting Windings for Single-Phase Induction Motors," *AIEE Trans.*, vol. 63, 1944, pp. 288–294.

Bibliography

Boothby, C. R.: "Discussion on Capacitor Motor Papers," *AIEE Trans.*, April 1929, p. 629.

Buchanan, L. W., and N. Maupin: "Auxiliary-winding Design for Split-Phase Motors," *AIEE Trans.*, pt. III, vol. 79, 1960, pp. 1183–1188.

Hanssen, I. E.: "Calculations of the Starting Torque of Single-Phase Induction Motors with Phase-Splitting Devices," *AIEE Trans.*, May 1908, p. 373.

Lloyd, T. C., and J. H. Karr: "Design of Starting Windings for Split-Phase Motors," *AIEE Trans.,* vol. 63, 1944, pp. 9–13.

CHAPTER

5

CAPACITOR-START MOTORS

The capacitor-start motor is by far the most popular type of single-phase motor for heavy-duty general-purpose applications requiring high starting torques. It is used for such applications as compressors, tire-changing tools, jet pumps, swimming pool pumps, farm and home-workship tools, and conveyors. Capacitor-start motors are generally available in ratings from as small as ⅛ hp and larger, although special applications may require less than ⅛ hp.

5-1 Capacitor Motor Types Defined. Since there are four distinctly different varieties of capacitor motor, each with its own particular set of characteristics, we shall start by first defining the capacitor motor. A *capacitor motor* has two windings, a main and an auxiliary winding; the auxiliary winding is displaced in space from the main winding, usually by an angle of 90 electrical degrees, and is connected in series with a capacitor. This much is common to the four types of capacitor motors described next.

(1) Capacitor-start motor. A capacitor-start motor is a capacitor motor that uses the auxiliary winding and capacitor only during starting, hence the easy-to-remember name "capacitor-start."

(2) Permanent-split capacitor motor. A permanent-split capacitor motor uses the auxiliary winding and capacitor continuously, without change in capacitance.

(3) Two-value capacitor motor. A two-value capacitor motor uses one value of capacitance for starting and a different value for running; that is, a two-value motor uses two values of capacitance.

(4) Split-phase capacitor motor. A split-phase capacitor motor uses the auxiliary winding during starting and running but switches a capacitor into the auxiliary circuit during running only; that is, it starts as a split-phase motor and runs as a permanent-split capacitor motor.

Since there are four recognized types of capacitor motors, the term capacitor motor should never be used by itself in specifying a motor, because the four types are so different from one another. A prospective motor user who writes to a manufacturer to make inquiries about a "capacitor" motor should be very explicit as to what kind of capacitor motor is wanted. It is best to state the application together with approximate torque requirements and reasons why a capacitor motor is desired.

5-2 Essential Parts of a Capacitor-Start Motor. We saw, in Art. 4-39, that a split-phase motor could be made into a capacitor-start motor simply by inserting a capacitor in series with the auxiliary winding. This arrangement is represented schematically in Fig. 5-1, which resembles

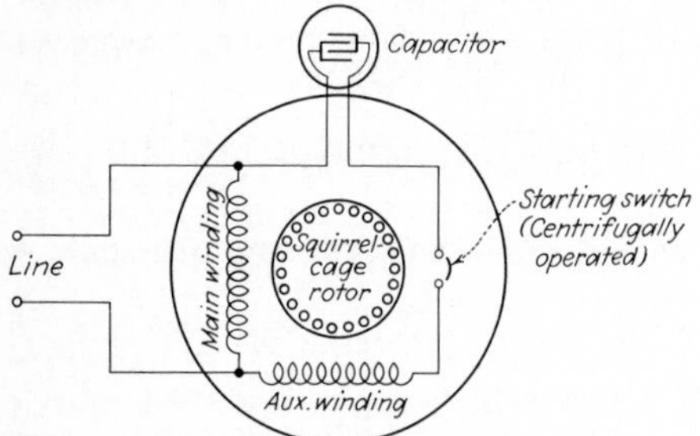

Fig. 5-1 The capacitor-start motor.

Fig. 4-1, the similar sketch for a split-phase motor, except for the addition of the capacitor in the auxiliary phase. This does not mean that commercial capacitor-start motors are merely split-phase motors with a capacitor added, for the windings have to be specially designed and proportioned. Moreover, the torques of capacitor-start motors are generally higher than the corresponding torques of standard split-phase motors.

Both types of motors have two electrically distinct windings, generally located 90 electrical degrees apart. As in the case of a split-phase motor, the main winding is the bulkier of the two. It is usually wound with a larger size of wire and with fewer turns than the auxiliary winding. The auxiliary winding of a capacitor-start motor, however, generally contains more copper (or aluminum) than the starting winding of a split-phase motor of the same rating. Articles 4-8, 4-15, and 4-16 on how to identify windings and how to predetermine rotation are equally applicable to capacitor-start motors. Capacitors are discussed in Chap. 8.

There are, therefore, two electrical circuits, or phases. The main phase consists of only the main winding connected across the line. The auxiliary phase, also called capacitor phase, comprises an auxiliary winding, a capacitor, and a centrifugal switch which opens at approximately 75 to 80 percent of synchronous speed; all three are connected in series, the whole being connected across the line in parallel with the main winding. Dual-voltage capacitor-start motors, on the higher-voltage connection, usually have the auxiliary-phase circuit connected across only half of the main phase, instead of across the line. These motors are discussed in Arts. 5-11 to 5-17. A cutaway view of a capacitor-start motor is shown in Fig. 5-2.

5-3 Capacitor-Starting Principle. As noted, the capacitor-start motor generally resembles the two-phase motor in that there are two stator windings, displaced 90 electrical degrees. In order to obtain the rotating field necessary to develop locked-rotor torque, it is necessary that the currents in these two windings be displaced in time phase. In the case of the polyphase motor (as pointed out in Art. 4-3), this phase displacement is obtained by connecting the two electrically similar windings to two voltages differing in phase. In both the split-phase and capacitor-start motors, the two circuits are connected to the *same voltage;* and the phase displacement of currents is obtained by a dissimilarity of the electrical constants of the two circuits.

In the split-phase motor, resistance is deliberately built into the auxiliary winding to bring the current more nearly in phase with the line voltage than is the main-winding current, as shown in Fig. 4-2. In a capacitor-start motor, the capacitor causes the auxiliary-phase current to *lead* the main-phase voltage, obtaining a large angle of displacement between the currents in the two windings. A phasor diagram of the locked-rotor currents of a typical capacitor-start motor is given in Fig. 5-3. The line current of this motor is only two-thirds the line current of the corresponding split-phase motor shown in Fig. 4-2; yet this

Fig. 5-2 Cutaway view of a 48-frame ball-bearing rigid-mounted capacitor-start motor. *a*, Electrolytic starting capacitor, in phenolic case; *b*, capacitor housing (pressed steel); *c*, quick-connect capacitor terminal; *d*, stator laminations (punchings); *e*, end windings; *f*, rotor laminations; *g*, die-cast end ring with cast-integral fan blades; *h*, pressed-steel cooling fan; *i*, rotating member of starting switch (see Fig. 17-7); *j*, contacts of stationary member of starting switch; *k*, terminal board for user line connections (see Fig. 5-5*b*); *l*, ball bearings; *m*, grease reservoirs; *n*, front end shield; *o*, back and shield; *p*, motor frame (steel). *(Reliance Electric Company.)*

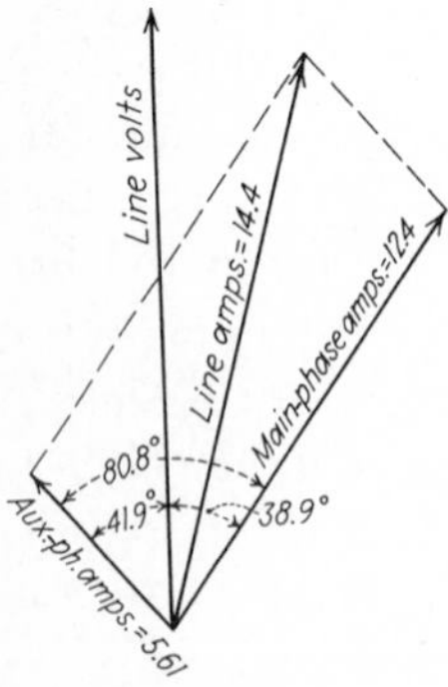

Fig. 5-3 A phasor diagram of the locked-rotor currents in a general-purpose capacitor-start four-pole motor rated at ⅙ hp, 60 Hz, 115 V, and 1725 rpm.

motor develops more than twice the locked-rotor torque of the split-phase motor. Thus a capacitor is seen to be a much more effective starting device than a resistor.

5-4 Capacitor Starting Compared with Split-Phase Starting. A capacitor-start motor develops considerably more locked-rotor torque per ampere of line current than the split-phase motor for a number of reasons. The locked-rotor torque of a single-phase induction motor with two windings displaced 90° is proportional, among other things, to the product of these three factors:*

1. The sine of the angle of phase displacement between the currents in the two windings

2. The product of the main-winding current multiplied by the auxiliary-winding current

3. The number of turns in the auxiliary winding

Each of these three factors is more favorable in the capacitor-start motor. (1) Using the figures from the typical examples chosen, the phase displacement between the currents is 24.3° in the split-phase motor and 80.8° in the capacitor-start motor. The sines of these two angles are 0.4115 and 0.987, respectively; i.e., because of this factor alone, the locked-rotor torque of the capacitor-start motor would be 0.987/0.4115 = 2.4 times that of the split-phase motor. (2) It will be noted that, in the split-phase motor, the line current is nearly equal to the numerical sum of the currents in the main and auxiliary windings, whereas, in the capacitor-start motor, the line current is considerably less than the numerical sum because of the greater phase displacement between the two currents. For this reason, more current can be allowed in either or both windings of the capacitor-start motor, obtaining more locked-rotor torque for the same line current. (3) In the auxiliary winding of a split-phase motor, the leakage reactance has to be kept small in order that the locked-rotor current be nearly in phase with the voltage. Because the leakage reactance varies as the square of the number of turns, only a few turns can be used. In the capacitor-start motor, the reactance of the auxiliary winding is more than neutralized by the capacitor, so that more turns can be used than in the auxiliary winding of the split-phase motor. If the auxiliary winding has more turns, the same current sets up more ampere-turns and hence more flux and more torque.

*For a technical explanation, see Ref. 1 of the References at the end of this chapter.

Another way to explain the greater effectiveness of starting is to point out that the flux conditions in a capacitor-start motor at standstill are similar to the flux conditions of a two-phase motor.

The main winding usually is of heavier wire, is distributed in more slots, and is bulkier than the auxiliary winding, although the latter frequently has more turns than the main winding. As in the case of split-phase motors, the main winding is usually wound into the slots first.

5-5 Purpose of the Starting Switch. It was explained in Art. 4-4 that a starting switch was necessary in split-phase motors to improve the torque characteristics at full-load speed, to keep down the watts input, and to prevent burnouts at normal operating speeds. All these considerations apply with equal force to capacitor-start motors, but there is an even more important reason, namely, to prevent burnout or breakdown of the capacitor. In Fig. 5-4 are shown the speed-torque charac-

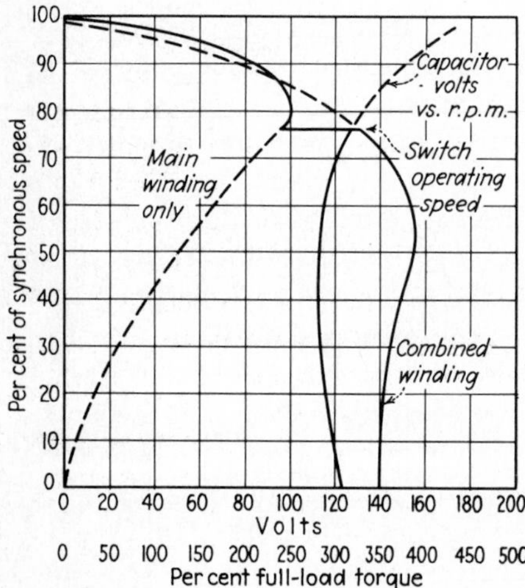

Fig. 5-4 Speed-torque curve of a capacitor-start motor.

teristics of a typical capacitor-start motor. In general, the torque characteristics are similar to those of the split-phase motor, except that the torques are much higher. Capacitor volts vs. rpm is also plotted on this curve. Observe that the capacitor volts increase rapidly above switch-operating speed. The switch must function at the proper speed; if it fails to operate and if the motor comes up to speed and is operated in this condition for an appreciable length of time, injury to the capacitor is certain.

The switch must, moreover, be positive in action; it must not flutter.

As high as double voltage can be momentarily impressed on the capacitor by a fluttering switch. The reason for this can be explained as follows: Suppose that the switch is fluttering and that it interrupts the circuit at such a time as to leave the capacitor fully charged; then suppose the switch happens to close when the voltage is of the opposite polarity. Double voltage will be impressed momentarily upon the capacitor. *Switches must not be allowed to flutter!*

5-6 Torque Characteristics of Capacitor-Start Motors. Capacitor-start motors have the torque characteristics of general-purpose motors, as shown in Table 19-1. Values given are minimum, and there is usually some margin over the figures given in the table.

GENERAL-PURPOSE SINGLE-VOLTAGE MOTORS

5-7 Terminal Boards and Integral Conduit Boxes. It has long been the general practice in the fractional horsepower industry to provide terminal boards with stud terminals to facilitate connecting the motors for a predetermined direction of rotation and also to provide terminals to which the motor user can readily connect the line leads. Examples of such practice are offered in Fig. 5-5. The examples shown there all use "quick-connect" terminals, which greatly shorten the time of making connection changes. These terminal boards are securely mounted inside the conduit box, which is cast as an integral part of the front end shield.

Cast-integral conduit boxes have the important advantage over attached steel boxes that they cannot possibly rattle. Moreover, the cast conduit box is usually provided with one hole large enough to accommodate a conduit connection, and with one or more small slots through which rubber-covered cable can be run.

5-8 Single-Voltage Motors: Without Thermal Protection. Single-voltage motors may be provided with four external line leads, or with a two-post terminal board; in the latter case, either the main winding or the auxiliary winding is connected to the back side of the board, and the other winding is connected to the front of the board.

National Electrical Manufacturers Association (NEMA) standard connections are shown in Fig. 5-6. Note that all three possibilities discussed above are covered. The arrangements shown in Fig. 5-6 are typical. For a certain special connection, using three leads, see Art. 5-18.

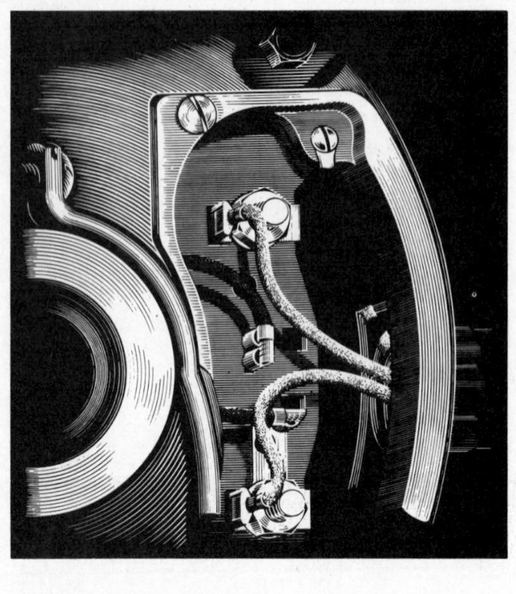

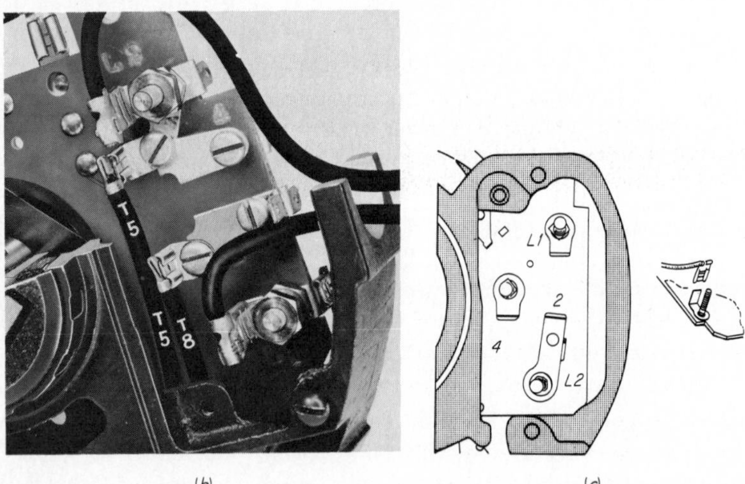

Fig. 5-5 Typical arrangements of terminal boards used in fractional horsepower motors. (*a*) General Electric Company; (*b*) Reliance Electric Company; (*c*) Westinghouse Electric Corporation.

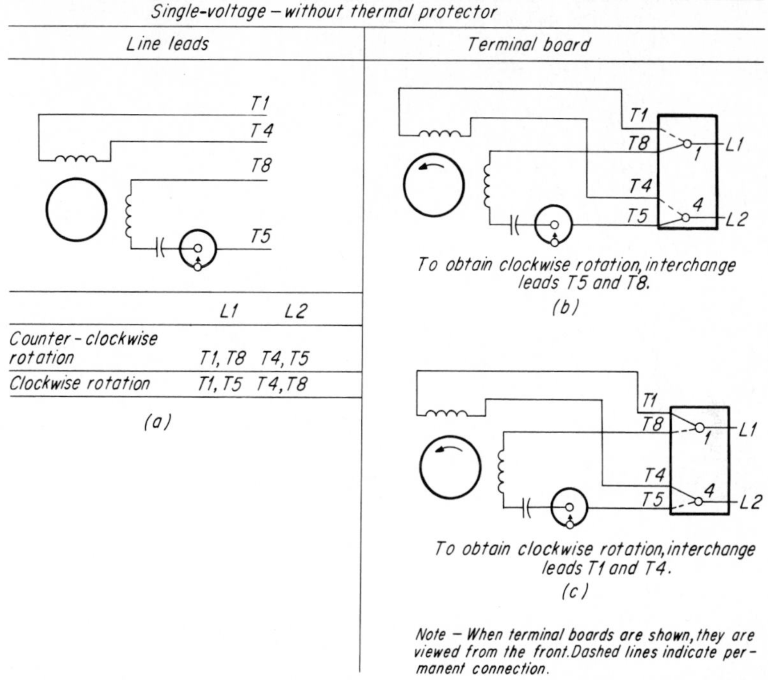

Fig. 5-6 Schematic wiring and line connection diagrams for single-voltage capacitor-start motors, using line leads or terminal boards. (*a*) Line leads; (*b*) terminal board, interchange auxiliary winding leads to reverse rotation; (*c*) terminal board, interchange main winding leads to reverse rotation. *(From NEMA Motor and Generator Standards, NEMA Publ. no. MG1-1978, National Electrical Manufacturers Association, Washington, 1978.)*

5-9 Single-Voltage Motors: Thermally Protected. When a thermal protector is added to a single-voltage capacitor-start motor, an additional lead, or an additional post on the terminal board, is usually required. The motor should always be connected to the line, on grounded circuits, so that the thermal protector is on the "hot" side.

NEMA standard connections are shown in Fig. 5-7. Compared with Fig. 5-6, an additional lead P1 from a protector terminal can be noted. When a terminal board is used—the usual case—an extra post, no. 2, is used.

Connections for a 230-V motor with a 115-V auxiliary phase are given in Fig. 5-8. In this diagram, the auxiliary phase is connected across half of the main winding, as in dual-voltage motors (discussed in the next section), in order to use 110-V capacitor elements.

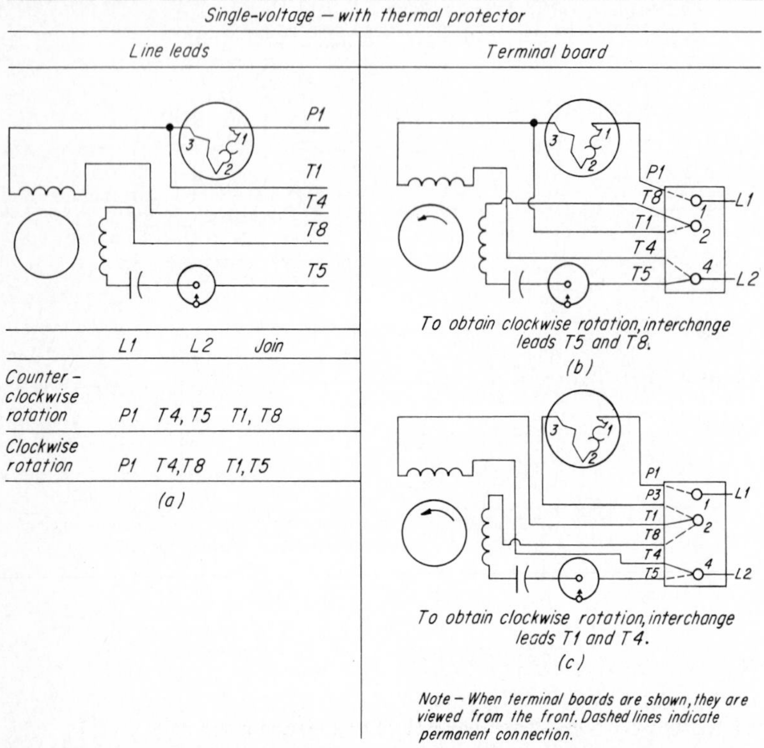

Fig. 5-7 Schematic wiring and line connection diagrams for single-voltage thermally protected reversible capacitor-start motors, using line leads or terminal board. (*a*) Line leads; (*b*) terminal board, interchange auxiliary winding leads to reverse rotation; (*c*) terminal board, interchange main winding leads to reverse rotation. *(From NEMA Motor and Generator Standards, NEMA Publ. no. MG1-1978, National Electrical Manufacturers Association, Washington, 1978.)*

Westinghouse uses a rubber terminal protector on one post (no. 2 in Fig. 5-8) to prevent connecting the power to posts 2 and 4, which would bypass the thermal protector, and also to prevent connecting the power to posts 1 and 2, which would blow a fuse or damage the thermal protector or both.

5-10 Stator Connection Diagrams. Stator connection diagrams for capacitor-start motors are generally the same as for split-phase motors, except that the capacitor element has to be connected in series with the auxiliary winding, usually in series with the T5 lead. Hence, the

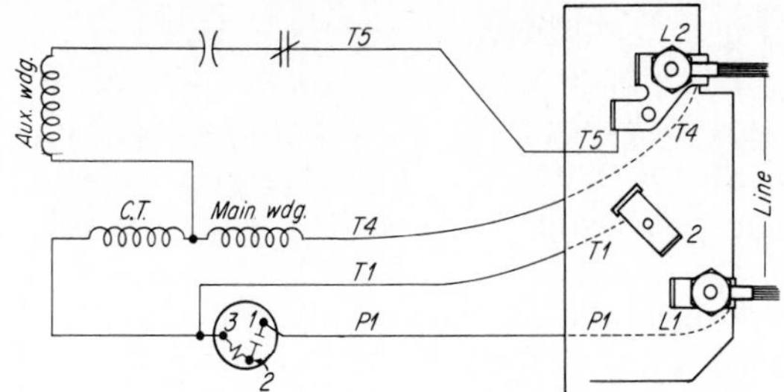

Note: When terminal boards are shown, they are viewed from the front. Dashed lines indicate permanent connections.

Connections shown are for CCW rotation, facing lead end. For CW rotation connect T5 to 2 position

Fig. 5-8 Connection diagram for single-voltage thermally protected reversible capacitor-start motors, using a half-voltage auxiliary phase for starting and using a terminal board. *(Reliance Electric Company.)*

stator connection diagrams of Chap. 4 are generally applicable, except for motors wired per Fig. 5-8.

DUAL-VOLTAGE MOTORS

5-11 Reasons for Dual-Voltage Windings. Capacitor-start motors rated ⅓ hp and above are usually wound for dual voltage so that they can be operated on either 115- or 230-V circuits. If a choice exists, it is generally preferable to connect a motor of this size to a 230-V line in preference to a 115-V line, in order to avoid the possibility of causing lights on the same feeder circuit to flicker. Some power companies will not permit connection of a ¾-hp motor to a 115-V circuit which is also used for lighting; others even extend this ban to ½-hp motors. A further reason for dual-voltage windings is that repulsion-start motors, formerly used for the same applications, were wound for dual voltage.

Even in ratings below ⅓ hp, two-, six-, and eight-pole motors are usually wound for dual voltage. These motors have relatively low activity, and use of the dual-voltage arrangement simplifies stocking problems.

5-12 Elementary Principles Involved. Windings for dual-voltage motors of the capacitor type use a rather special arrangement. The main winding is split into two sections, which can be connected in series or in parallel, as might be supposed. These two sections are both wound on the same magnetic axes; i.e., the windings are in space phase with each other. However, the auxiliary winding is in only one section, which is displaced 90 electrical degrees from the main winding. This arrangement is shown schematically in Fig. 5-9. In series with the auxiliary winding are the customary centrifugal switch and a single capacitor.

For a 115-V circuit, the two sections of the main winding and the auxiliary phase (consisting of winding, switch, and capacitor) are connected in parallel. Principles of operation are the same as for the single-voltage motor of Fig. 5-1.

For a 230-V circuit, the two sections of the main winding are connected in series, but the auxiliary phase is bridged across one section of the main winding instead of being connected across the line. Each main-winding section has 115 V across it, as does the auxiliary phase, which is across only one of the sections. This arrangement avoids splitting up the auxiliary winding and capacitor each into two sections, but it compels the main winding to perform a dual function when the motor is operating on the starting connection: to act (1) as an autotransformer and (2) as a main winding. Because of this autotransformer requirement, the series-parallel winding for a motor of this type is often somewhat special in order to obtain a close magnetic coupling between the two sections of the main winding.

It is possible, because of the autotransformer effect, that the locked-rotor torque may be as much as 10 percent lower on the 230-V connection than on the 115-V connection.

5-13 Winding Dual-Voltage Motors. Auxiliary windings, being actually single-voltage windings, are quite conventional. Practices vary in the way the main winding is divided into two sections.

(1) Two wires in parallel. The main winding is wound in the conventional fashion of a single-voltage winding, except that two wires (preferably of different colors) are wound in parallel and are later connected as the two separate sections. This procedure gives a winding with the closest possible coupling between the two main-winding sections, and, therefore, the torque characteristics on the 230-V starting connection approach closely the torque characteristics on the 115-V connection. However, this winding procedure means that on the average the voltage between adjacent strands is close to 115 V. When a motor is wound this way, the main winding looks externally exactly like a single-voltage main winding. See Fig. 5-14 for a connection diagram.

(2) Full-section two-layer main windings. Two complete main windings, each with coils on all the poles, are wound separately and inserted separately in the slots, forming two layers, which are usually insulated

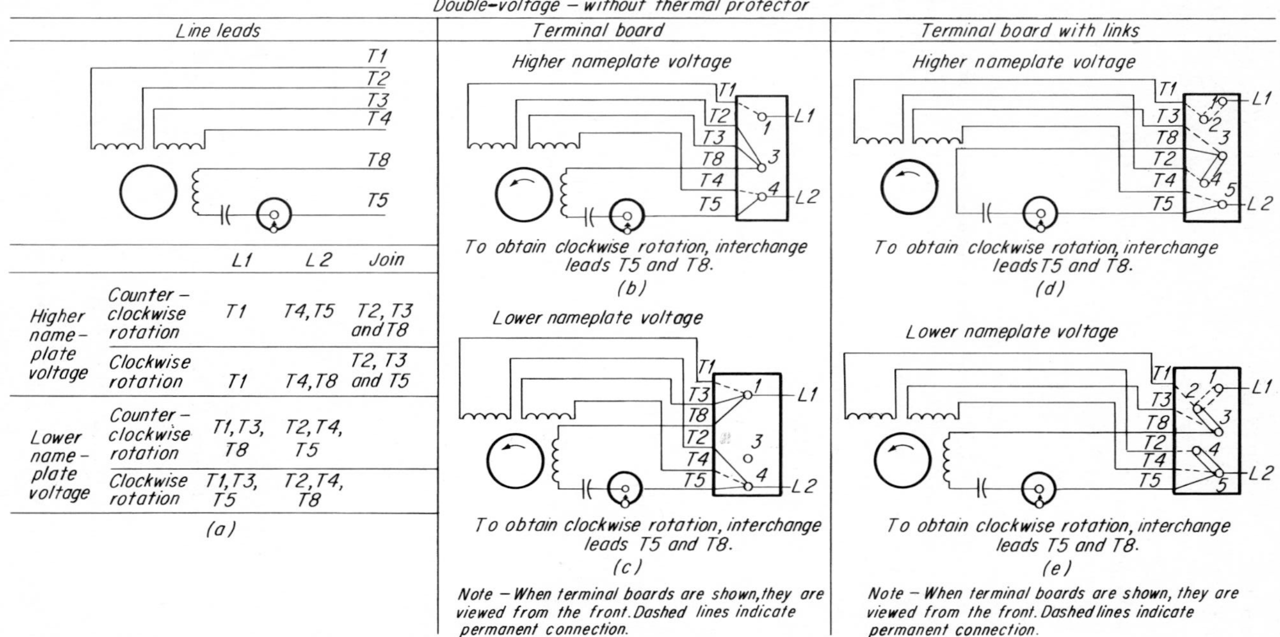

		L1	L2	Join
Higher name-plate voltage	Counter-clockwise rotation	T1	T4, T5	T2, T3 and T8
	Clockwise rotation	T1	T4, T8	T2, T3 and T5
Lower name-plate voltage	Counter-clockwise rotation	T1, T3, T8	T2, T4, T5	
	Clockwise rotation	T1, T3, T5	T2, T4, T8	

Fig. 5-9 Wiring and line connection diagrams for dual-voltage reversible capacitor-start motors, with leads and with three-post and five-post terminal boards. (*a*) Line leads; (*b*) three-post terminal board, higher-voltage connection; (*c*) three-post terminal board, lower-voltage connection; (*d*) five-post terminal board, higher-voltage connection; (*e*) five-post terminal board, lower-voltage connection. *(From NEMA Motor and Generator Standards, NEMA Publ. no. MG1-1978, National Electrical Manufacturers Association, Washington, 1978.)*

from each other. This gives very nearly the same coupling and the same torque characteristics as winding two wires in parallel, as discussed. However, the voltage between adjacent strands is very much lower and winding time is longer. This arrangement is readily identified visually because the two distinct layers of the main winding are plainly apparent. See Fig. 5-14 for a connection diagram.

(3) Use of half the poles in each section. Still another arrangement is to wind all the poles just the same as for a single-voltage motor, but to connect half of the poles in one section and the remaining half of the poles in the second section; usually alternate poles (i.e., poles of the same magnetic polarity) are connected together to form a section. Such an arrangement is entirely satisfactory on the running connection, but on the starting connection, the torque may be substantially less on the 230-V connection than on the 115-V connection because the coupling between the two sections of the main winding is not as good as it is for either arrangement 1 or arrangement 2 discussed previously. However, by careful attention to design, the difference in starting-connection torques on the two voltages can be minimized. Moreover, the regulation of a 230-V circuit is almost invariably better than that of a 115-V circuit, so that the percentage drop in voltage when the motor is started is less; hence, even if the motor does have a little less starting torque on 230 V, the fact may go unnoticed. See Fig. 5-15 for a connection diagram.

(4) Splitting the elements of a pole group. There is another method sometimes used for two-pole motors with four coil elements to each pole group. The two inner elements of one pole are connected in series with the two outer elements of the other pole group, and vice versa.

A canvass of four major producers of capacitor-start motors yielded interesting results: One said that arrangement 1 represented the company's general practice, another reported using arrangement 3, a third used both arrangements 2 and 3, the choice depending upon design considerations, and the fourth reported using arrangements 3 and 4.

5-14 Dual-Voltage Motors: Without Thermal Protection. Dual-voltage capacitor-start motors without thermal protection may have six line leads, a three-post terminal board, or a five-post terminal board with links to effect the voltage change. Direction of rotation is reversed by changing two leads (from the auxiliary phase) on the front of the terminal board. NEMA standard connections are shown in Fig. 5-9.

5-15 Thermal Protection of Dual-Voltage Motors. Thermal protection against running overloads poses no particular problem. In Art. 17-19, it is shown how a thermal protector makes use of motor current and temperature. The running winding of a dual-voltage motor is divided into two sections, each of which is presumed to carry the same current, for the same output, on either voltage connection. Thus, in order that the same protection be obtained on either voltage connection, the *auxiliary heater* of the protector is connected in series with only one of the two sections; the contacts are connected in series with the line in either case, so that when the device opens, the entire motor is cut off the line. Naturally then, on the lower-voltage connection, the contacts carry twice as much current as on the higher-voltage connection. This arrangement, then, requires that the protector be provided with a third terminal brought out between the contacts and auxiliary heater. Further, in a disk-type device such as that illustrated in Fig. 17-13, the thermal actuating element has to carry the same current as the contacts; naturally, the disk itself has some heater effect. The heater effect of the disk, compared with that of the heater, has to be relatively low in order to obtain comparable protection on both voltage connections. If the heater effect of the disk is substantial, the overload device will trip at lighter loads on the lower-voltage connection and will permit higher temperatures on the higher-voltage connection. In summary, to obtain comparable overload protection on both voltage connections (1) the heater effects of the contacts' circuit has to be small and (2) the currents in the two sections of the main winding have to be equal to each other.

Locked-rotor protection of the dual-voltage motor is somewhat more difficult, because the added complications of the auxiliary phase have to be taken into account. Under locked-rotor conditions, the currents handled by the protective device are much greater than under running conditions. An idea of the problem can best be understood by referring to Fig. 5-10. In this figure, I_m is assumed to be the total current in the main winding for the lower-voltage connection, so that each section carries 0.5 I_m. In Fig. 5-10*a*, the higher-voltage connection is shown; to the main-winding current, half the auxiliary-phase current is added in one section and subtracted in the other (just as in an autotransformer, magnetizing current is neglected). Contacts of a protective device would have to be connected at *n*, but the heater could be connected at *m*, *n*, or *o*. If the heater is inserted at *m*,

$$\text{Heater amperes} = 0.5(I_m - I_a) \tag{5-3}$$

at *n*,

$$\text{Heater amperes} = 0.5(I_m + I_a) \tag{5-4}$$

at *o*

$$\text{Heater amperes} = I_a \tag{5-5}$$

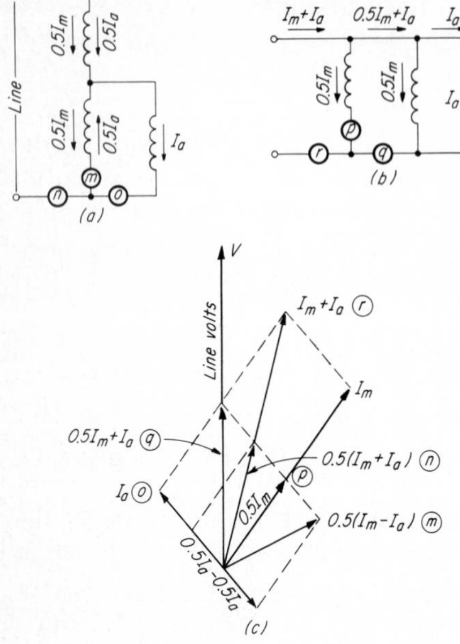

Fig. 5-10 Phasor diagram of the locked-rotor currents of Fig. 5-3 when wound and connected as a dual-voltage motor. (*a*) Higher-voltage connections; (*b*) lower-voltage connections; (*c*) phasor diagram of the currents.

On the lower-voltage connection in *b*, the contacts, of course, can be connected only at *r*, but the heater could be connected in at three possible places. If it is connected in at *p*,

$$\text{Heater amperes} = 0.5\ I_m \tag{5-6}$$

at *q*

$$\text{Heater amperes} = 0.5\ I_m + I_a \tag{5-7}$$

at *r*

$$\text{Heater amperes} = I_m + I_a \tag{5-8}$$

The additions indicated above are, of course, by phasors, and the relative magnitudes of these sums will vary with different designs. All six of these currents are shown to scale in Fig. 5-10*c*. In practice, the positions most commonly used are for the higher-voltage connections *n* or *m* and for the lower-voltage connections *p* or *q*. Correct application of a protector to obtain comparable protection for both voltage connections requires careful selection of the protector and choice of the most suitable method for connecting in the auxiliary heater, which involves taking into account the added heating effect in the disk on the lower-voltage connection due to the greater current through the disk.

5-16 Dual-Voltage Motors: Thermally Protected. It was shown in the preceding article that there are three different ways of connecting in

therefore, different currents are provided to actuate the thermal protector.

Note II – Motor starting switch shown in running position. All directions of rotation shown are facing the end opposite the drive.

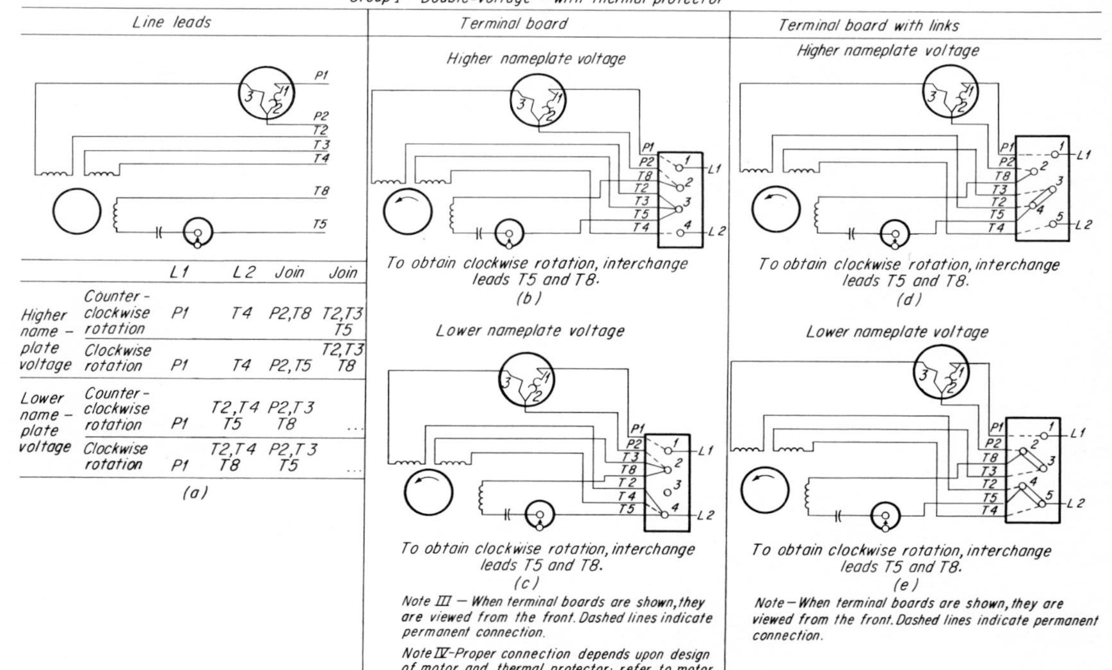

Fig. 5-11 Wiring and line connection diagrams for dual-voltage thermally protected capacitor-start motors. Group I: Heater coil is connected at *m* and *p* for the two voltage connections (See Fig. 5-10). (*a*) Line leads; (*b*) four-post terminal board, higher-voltage connection; (*c*) four-post terminal board, lower-voltage connection; (*d*) five-post terminal board, higher-voltage connection; (*e*) five-post terminal board, lower-voltage connection. *(From NEMA Motor and Generator Standards, NEMA Publ. no. MG1-1978, National Electrical Manufacturers Association, Washington, 1978.)*

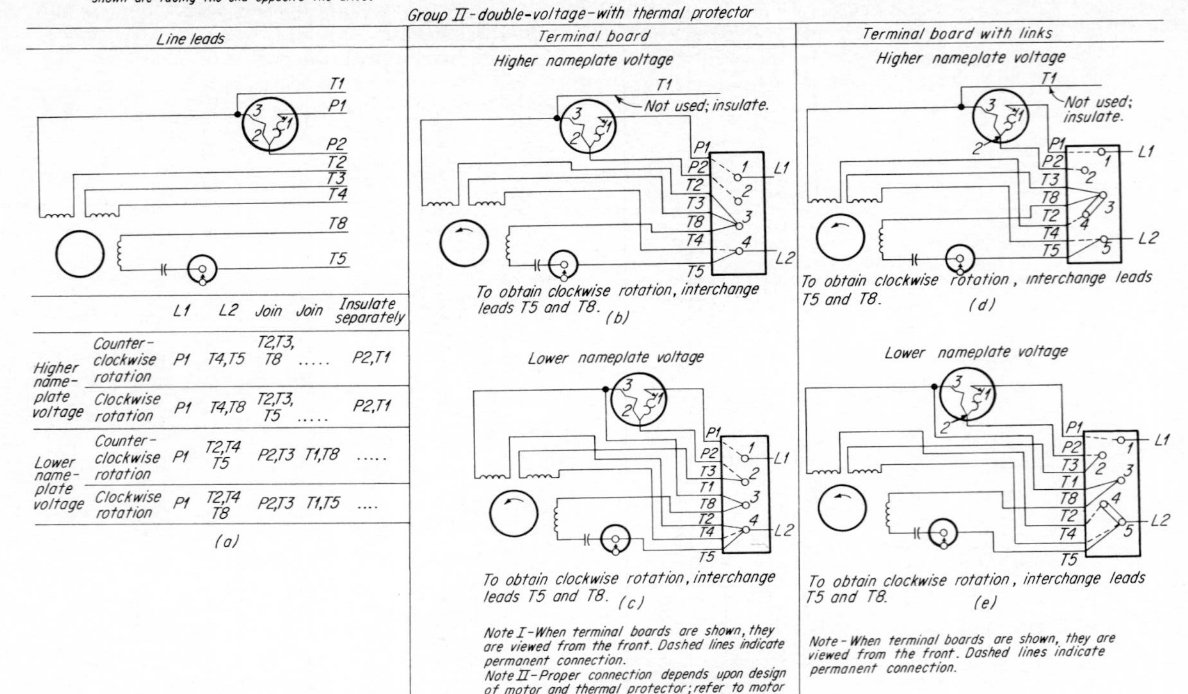

		L1	L2	Join	Join	Insulate separately
Higher nameplate voltage	Counter-clockwise rotation	P1	T4,T5	T2,T3, T8	.....	P2,T1
	Clockwise rotation	P1	T4,T8	T2,T3, T5	.....	P2,T1
Lower nameplate voltage	Counter-clockwise rotation	P1	T2,T4 T5	P2,T3	T1,T8	.....
	Clockwise rotation	P1	T2,T4 T8	P2,T3	T1,T5	....

Fig. 5-12 Wiring and line connection diagrams for dual-voltage thermally protected capacitor-start motors. Group II: Heater coil is connected at *n* and *q* for the two voltage connections (See Fig. 5-10). (*a*) Line leads; (*b*) four-post terminal board, higher-voltage connection; (*c*) four-post terminal board, lower-voltage connection; (*d*) five-post terminal board, higher-voltage connection; (*e*) five-post terminal board, lower-voltage connection. *(From NEMA Motor and Generator Standards,* NEMA *Publ. no. MG1-1978, National Electrical Manufacturers Association, Washington, 1978.)*

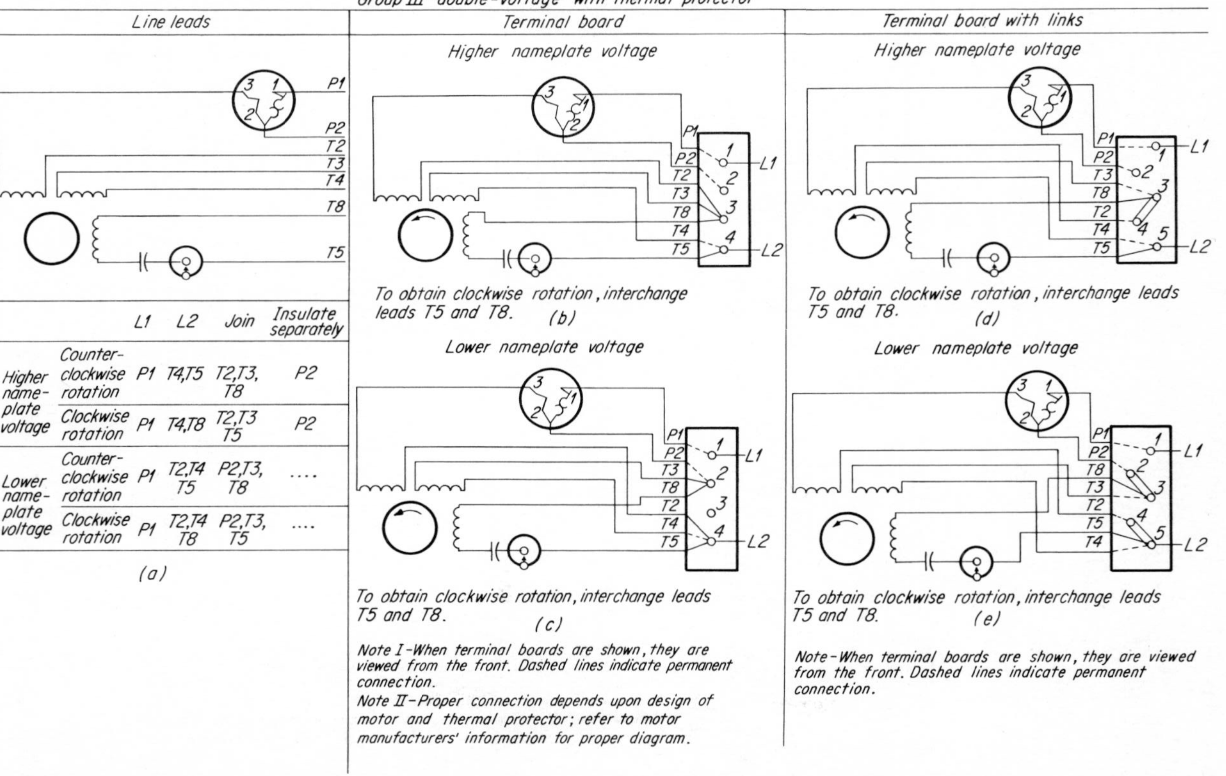

		L1	L2	Join	Insulate separately
Higher nameplate voltage	Counter-clockwise rotation	P1	T4,T5	T2,T3, T8	P2
	Clockwise rotation	P1	T4,T8	T2,T3 T5	P2
Lower nameplate voltage	Counter-clockwise rotation	P1	T2,T4 T5	P2,T3, T8	....
	Clockwise rotation	P1	T2,T4 T8	P2,T3, T5	....

Fig. 5-13 Wiring and line connection diagrams for dual-voltage thermally protected capacitor-start motors. Group III: Heater coil is connected at *n* and *p* positions. (See Fig. 5-10). (*a*) Line leads; (*b*) four-post terminal board, higher-voltage connection; (*c*) four-post terminal board, lower-voltage connection; (*d*) five-post terminal board, higher-voltage connection; (*e*) five-post terminal board, lower-voltage connection. *(From NEMA Motor and Generator Standards, NEMA Publ. no. MG1-1978, National Electrical Manufacturers Association, Washington, 1978.)*

the auxiliary heater on each voltage connection. None of them is universal. The NEMA connection diagrams for the three arrangements are given in Figs. 5-11, 5-12, and 5-13.

1. *Figure 5-11.* Heater connected in *m* and *p* positions of Fig. 5-10 (NEMA Group I).

2. *Figure 5-12.* Heater connected in *n* and *q* positions of Fig. 5-10 (NEMA Group II).

3. *Figure 5-13.* Heater connected in *n* and *p* positions of Fig. 5-10 (NEMA Group III).

5-17 Stator Connection Diagrams for Dual-Voltage Motors. In Art. 5-13 it was pointed out that there are a number of ways of splitting the main winding of a dual-voltage motor into two sections. Two of the methods involve having a coil group from every pole in both sections, whereas others do not. For the former, a stator connection diagram is given for a four-pole motor in Fig. 5-14. When each section contains half the poles, as discussed in Art. 5-13(3), the main winding would generally be connected in the same way as the stator illustrated in Fig. 5-15.

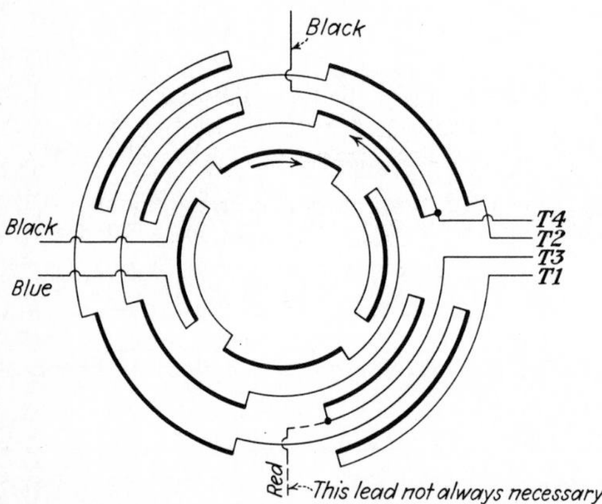

Fig. 5-14 Stator connection diagram for a four-pole dual-voltage capacitor-start motor with full-section main windings.

MISCELLANEOUS ARRANGEMENTS

5-18 "Swing" Connection (Half-Voltage Starting Winding). Single-voltage motors for 230-V operation are often wound the same way as dual-voltage motors, in order to use a half-voltage starting winding and

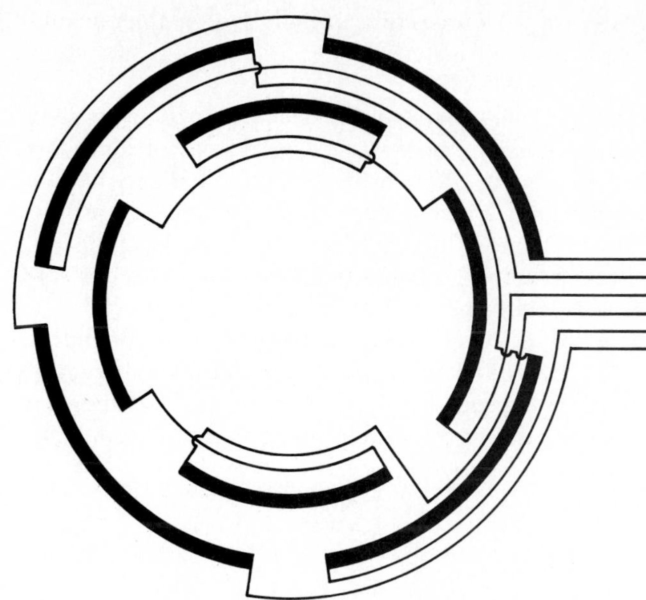

Fig. 5-15 Stator connection diagram for a four-pole dual-voltage capacitor-start motor with half the poles in each main winding section.

a 110-V capacitor. One such arrangement is illustrated in Fig. 5-8. Note that only one lead on the terminal board has to be moved to change direction of rotation. The arrangement is represented schematically in Fig. 5-16, where rotation is changed simply by swinging a link from one position to another, giving rise to the name "swing connection."

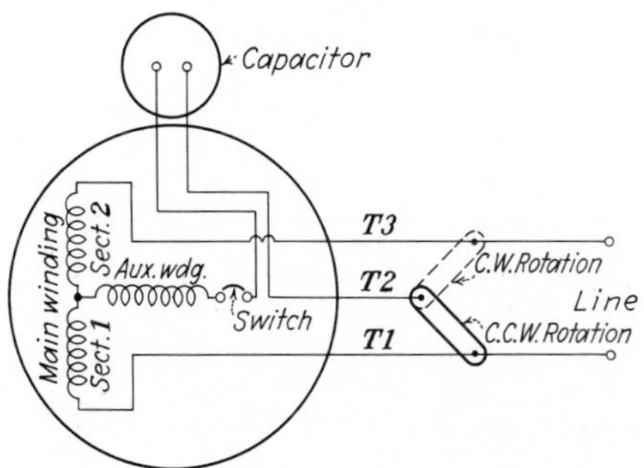

Fig. 5-16 The swing connection.

5-19 Two-Speed Pole-Changing Motors. Capacitor-start two-speed pole-changing motors are quite similar to their split-phase counterparts which were discussed in Arts. 4-23 to 4-26 and 4-28. The principal difference is the use of a capacitor in series with the auxiliary winding, which is therefore designed to make optimum use of the capacitor.

Locked-rotor torques are, of course, substantially higher than in split-phase motors.

5-20 Motors with Magnetically Operated Switch. Motors for hermetic refrigeration service commonly employ an external relay of some sort, located outside the sealed unit, to open the starting winding when the motor is up to near operating speed. Relays may also be used for other applications, such as for motors in small frame sizes or in appliances where motor length is limited. Magnetic relays for capacitor-start motors may be either current-operated or voltage-operated.

(1) Current-operated relays. These relays make use of the phenomenon that the current in the main winding decreases sharply as the motor approaches normal operating speed. A typical arrangement is shown in Fig. 5-17*a*. The magnet coil is connected in series with the main winding, and the switch contacts in series with the auxiliary winding. When there is no current in the relay coil, the contacts remain in the open position. When the motor is thrown on the line, the current drawn by the main winding under locked-rotor conditions is sufficient to close the contacts, thereby energizing the capacitor phase. As the motor comes up to speed, the main-winding current decreases; the relay is set so that when the current has fallen below a predetermined value, the relay drops out, opening the capacitor phase and allowing the motor to run as a single-phase induction motor. Such relays are discussed in Art. 17-9 and illustrated in Figs. 17-7 and 17-8.

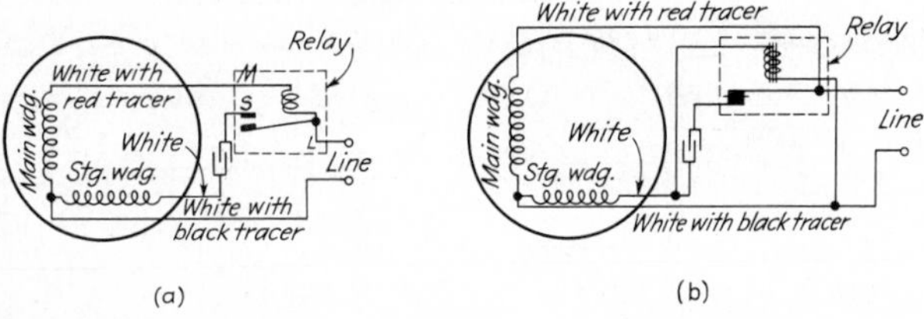

Fig. 5-17 Schematic diagrams of a capacitor-start motor with separate capacitor and magnetic relay, for hermetic refrigerator applications. (*a*) With current-operated relay; (*b*) with voltage-operated relay.

Many relays used for this purpose are not adjustable. If the relay does permit adjustment, it must be set so that it will always close under locked-rotor conditions with low voltage impressed. (By low voltage is meant the lowest circuit voltage likely to be encountered at the motor terminals.) It must also be set to drop out as the motor comes up to speed, even when driving its heaviest load, at any voltage, from the lowest to the highest likely to be encountered at the motor terminals. If there is an appreciable difference between these two adjustments, the proper setting becomes largely a matter of judgment; some intermediate setting can be used.

(2) Voltage-operated relays. These relays make use of the increase in auxiliary-winding voltage as the motor nears normal operating speed. A typical arrangement is shown in Fig. 5-17*b*. The operating coil of the relay is connected in parallel with the auxiliary winding (not including the capacitor), and the contacts in series with the capacitor phase. When not energized, the relay contacts are normally closed. When the motor is thrown on the line with the rotor at standstill, the voltage across the auxiliary winding is not sufficient to open the relay contacts; as the motor comes up to speed, the voltage across the auxiliary winding steadily increases and, finally, as the motor nears normal operating speed, the voltage becomes great enough to open the relay contacts, thereby deenergizing the auxiliary phase. It might be supposed that the voltage on the auxiliary winding would disappear and that the relay contacts would again close. However, when the motor is up to speed and energized, there is a voltage induced in the auxiliary winding,* and this voltage is sufficient to maintain the relay contacts open.

These relays may or may not be adjustable. If the relay does permit adjustment, it must be set so that (1) it will never pick up with the rotor at standstill, with maximum voltage impressed, (2) it will always pick up as the motor comes up to speed while driving its heaviest load, with minimum voltage impressed, and (3) once it has opened, it will not reclose with minimum voltage impressed. If all three of these conditions cannot be met with a single setting, a compromise setting has to be made.

5-21 Reversing Motors. Before discussing reversing motors, a clarification of terms is necessary.

A *reversible motor* is one which can be started from rest and operated in either direction of rotation. The direction of rotation is determined by the starting connection; once a capacitor-start motor is up to speed, the starting winding is out of the circuit, and hence reversing its connections will not affect the motor at all. But if the connection change is made while the motor is operating on the starting connection, it will usually reverse promptly. Synonymous terms are "externally reversible" (used in an earlier work[2]) and "either-rotation" (General Electric).

A *reversing motor* is one which can be reversed even while it is running at full operating speed. Such motors were referred to as "electri-

*In Art. 2-10, it was explained that the rotor of a single-phase squirrel-cage motor, running light, sets up a field which is displaced approximately 90° in both time and space from the field set up by the main winding. This cross field induces a voltage in the auxiliary winding of a capacitor-start motor by transformer action.

cally reversible" motors in Veinott's earlier work.[2] Some manufacturers distinguish between quick-reversing and instant-reversing motors; a *quick-reversing motor* requires a time delay of 1/25 s or more in switching; an *instant-reversing motor* requires no delay at all during switching.

A multitude of schemes have been used to achieve the reversing feature for capacitor-start motors. Two articles in the literature describe such schemes in detail.[3,4] In Art. 5-30 of Veinott's earlier work,[2] a reversing motor for hoist service was explained in great detail. Reversing motors offered today usually employ a direction-sensing centrifugal switch, a magnetic relay with special circuitry, or both. Such motors are offered by Franklin, General Electric, Reliance Electric, and others.

MISCELLANEOUS SERVICE PROBLEMS

5-22 Rewinding or Reconnecting Motor for a Different Voltage and Same Performance

(1) Changing from 115 to 230 V. (*a*) One way to do this is to double the number of turns in each winding by using wire three sizes smaller in each. It is then necessary to use a 220-V capacitor having one-fourth the microfarad rating of the capacitor used with the 115-V motor. (*b*) Another way is to rewind the motor for dual voltage or for the swing connection (see Arts. 5-11 to 5-18). In this case, the same capacitor and the same auxiliary winding are used, but each section of the main winding should use wire three sizes smaller and each section must have the same number of turns as the 115-V main winding. (*c*) If the main and auxiliary windings are both parallel-connected, it may be possible to reconnect them in series and use a 220-V capacitor of one-fourth the microfarad rating. If the main winding is wound in two parallel-connected sections, each section containing only half the total number of poles (i.e., using the diagram of Fig. 5-15), it is not recommended that such a winding be split into two sections to use for a swing connection or for a dual-voltage connection unless the torque requirements are relatively light. (*d*) If the main winding is wound of two strands of wire in parallel and if it is possible to separate these strands, a test between strands at 500 V should be tried. If the winding stands this test, it is safe and simple to reconnect the motor for 230 V using the proper swing connection diagram.

(2) Changing from 230 to 115 V. (*a*) If the motor is swing-connected, it is a simple matter to reconnect the two main-winding sections in parallel (if due care as to magnetic polarity is exercised), leaving the auxiliary winding and capacitor undisturbed. (*b*) If the stator connections are conventional (like those shown in Fig. 4-9), a 110-V capacitor of four times the microfarad rating of the 220-V capacitor must be used, and the windings can be either reconnected in parallel or rewound. If they are to be reconnected, each winding can be split into two sections, each section including every alternate pole. (Care must be taken to preserve proper magnetic polarities!) (*c*) If the windings are rewound, half as many turns and wire three sizes larger should be used in each winding.

(3) Changing from 230 to 460 V. The problem of changing a 230-V motor to a 460-V motor, if the motor has conventional connections, is similar to the problem of changing from 115 to 230 V, which is discussed in the first paragraph of this article, except that, because of switch and capacitor limitations, only the swing or dual-voltage connection is recommended.

(4) Other voltage changes. In general, changes for voltages other than those just described are not recommended, for a capacitor of the proper voltage and microfarad rating is seldom available. The rules for such a change would be as follows. If

V_r = rated voltage

V_d = desired circuit voltage

the number of turns of both main and auxiliary windings is

$$\frac{V_d}{V_r} \times \text{turns in old windings}$$

and the cross-sectional area of the copper wire should be

$$\frac{V_r}{V_d} \times \text{area of copper wire in present winding}$$

or as near as possible to this value. The capacitor should be rated for V_d volts, and the microfarad rating should be

$$\left(\frac{V_r}{V_d}\right)^2 \times \text{the microfarad rating of the present capacitor}$$

An approximate rule of thumb that applies to fractional horsepower capacitor-start motors is that the auxiliary winding should be wound with five wire sizes (gauges) smaller than the main winding and should have 150 to 175 percent as many turns. In some designs in order to obtain high locked-rotor torque, the ratio of auxiliary to main turns should be about 110 percent, using wire four sizes smaller for the auxiliary winding, and a capacitor of high capacitance rating.

5-23 Rewinding for Different Torques. In general, it is not wise for a repair shop to attempt to rewind for a different torque by changing the wire size or number of turns, as discussed for split-phase motors in Art. 4-37. The person making repairs may rewind a capacitor-start motor, actually obtaining more torque, but in so doing may subject the capacitor to higher voltages, a condition that may not show up in the test but will eventually show up in reduced life of the capacitor.

5-24 Rewinding for a Different Frequency. The redesign of a winding of a capacitor-start motor to operate on a different frequency is a problem for a design engineer, and no simple rules can be laid down that a person making repairs can follow with safety. Such a rewinding job is, therefore, not to be recommended unless the winding specification can be obtained from the manufacturer.

5-25 Changing a Capacitor-Start to a Split-Phase Motor. It was explained in Art. 4-39 that a split-phase motor could be converted to a capacitor-start motor by the addition of a capacitor in series with the auxiliary winding. By use of the capacitor it is possible to increase the locked-rotor torque and decrease the locked-rotor amperes. Likewise, it is usually possible to operate a capacitor-start motor as a split-phase motor by substituting an external resistance for the capacitor. Some capacitor-start motors will start under no-load conditions as split-phase motors, and others will not; even if a particular motor will start as a split-phase motor, a resistor in series with the auxiliary winding will usually improve the locked-rotor torque and reduce the locked-rotor current. For any given motor, there is a definite value of resistance that will give a maximum locked-rotor torque. This "best value" of resistance to give the most locked-rotor torque will probably be from one to two times the ohmic resistance of the auxiliary winding; different values of resistance in this range may be tried to determine the best value to use.

A rough idea of what might be expected is given in Table 5-1; these results are based on a ⅓-hp four-pole 60-Hz 115-V capacitor-start

TABLE 5-1 Capacitor-Start Motor Used as a Split-Phase Motor

	As a normal capacitor-start motor	As a split-phase motor, no external resistance	As a split-phase motor, 6.7 Ω external resistance
Locked amperes:			
Main winding	22.5	22.5	22.5
Auxiliary winding	8.05	9.71	6.38
Line	24.5	32.1	28.4
Locked-rotor torque, oz-ft	65.0	12.5	19.2
Full-load torque, oz-ft	16.0	16.0	16.0

motor. This example illustrates the point that an external resistance boosts the locked-rotor torque and cuts the line current, but with the best value of external resistance (which in this case happened to be 133 percent of the auxiliary winding resistance) the locked-rotor torque is very small compared with that of a capacitor-start motor. This torque is sufficient to start a fan or blower but would not start a compressor. The running performance, of course, is the same whether the motor starts as a resistance-start split-phase motor or as a capacitor-start motor.

5-26 Thermal Protection. As discussed in Art. 4-40 for split-phase motors, if the motor has an integral thermal protector, a person not trained in protector selection should not attempt to reconnect or rewind a capacitor-start motor for a different voltage, torque, or frequency.

5-27 Regeneration in Capacitor-Start Motors. Regeneration is an effect that can occur with a capacitor-start motor, particularly if the motor is driving a high-inertia load such as a fan or blower. The effect may be noticed in either of two ways:

1. While the motor is slowly coasting to a stop, the starting switch closes; immediately the motor vibrates, becomes noisy, and decelerates rapidly, as if it were being suddenly braked electrically, as, in fact, it is.
2. In an oil-burner application, a solenoid-operated oil valve is sometimes connected across the motor so that the valve is open only when the motor is energized. If regeneration occurs while the motor is coasting to a stop, a voltage appears at the motor terminals and may cause false opening or fluttering of the oil valve or any other relay connected across the motor.

(1) Causes of regeneration. It is well known that an induction motor can be made to act as an induction generator and supply power to the line to which it is connected, merely by driving it above synchronous speed. It is also well known that the power output of an induction generator must always have a leading power factor, never unity or lagging power factor; i.e., the current output of the generator must always have a leading component in order to provide the excitation to set up the magnetic field required in the generator. It has been shown that an induction motor can be made to act as an induction generator by connecting shunt capacitors of proper value across the output terminals; these capacitors draw the leading current necessary for the machine to build up as a self-excited generator, without the necessity for its being connected to an external supply line.[5-7] Suppose that the starting switch of the motor of Fig. 5-1 is closed, and further suppose that the motor is being driven mechanically at full-load speed with no external voltage applied to the terminals. There will always be a little residual flux in the rotor, which will cut the stator windings and induce a small voltage in them. As shown in Fig. 5-1, there is a voltage impressed across the capacitor which is equal to the sum of the two voltages induced in the main and auxiliary windings; this voltage causes a small leading current to flow through the capacitor. The leading current increases the field excitation in the motor, increasing the output voltage, increasing the leading current, and so on, until quite an appreciable output voltage is reached; the magnitude of this voltage is limited only by saturation of the magnetic circuit of the machine. This effect is so marked that it is probable that almost any general-purpose capacitor-start motor would burn out in a short time if it were driven at synchronous speed with the starting switch closed—so great is the regenerative effect under these conditions. This effect has been discussed more completely by Veinott, and some of the more important conclusions are merely summarized next without proof:[3]

1. For any given motor driven always at the same speed, there is a critical value of capacitance below which the machine will not build up at all. The more this capacitance is exceeded, the greater will be the voltage, current, and dynamic-braking torque.

2. For any given machine with a fixed value of capacitance, there is a critical speed below which the machine will not build up at all. The more the speed exceeds this critical value, the greater will be the current, voltage, and dynamic-braking torque.

3. For any given machine, the critical value of capacitance is inversely proportional to the square of the speed; the converse of this is likewise.

4. If the capacitor is charged at the instant that the starting switch closes, buildup is much more likely to occur, but buildup can occur with no initial charge in the capacitor.

5. If a resistor is connected in parallel with the capacitor, the tendency of the machine to build up is reduced, partly because the resistor bleeds off the capacitor charge, and partly because the resistor acts as a load on the generator—just as a load across a shunt generator decreases the critical field resistance above which the machine will not build up.

(2) Remedies. Probably the easiest remedy for servicing the motor is to install a small resistor—of the type found in a radio supply store—in the capacitor assembly, connecting it across the capacitor terminals.

If it is observed that the regenerative buildup occurs only if the motor is stopped within a few seconds after starting and does not occur if the motor has been running for some time, then a fairly high resistance, 1000 Ω, say, can be tried. If this does not cure the difficulty, a smaller value of resistance should be tried. This process can be repeated until a value of resistance that cures the trouble is found. This resistor should have the maximum resistance that can be used and still be effective. Care should also be exercised to use a resistor of a high enough watt rating that it does not overheat during a starting or stopping period. Use of a resistor tends to increase the locked-rotor current.

Lowering the reclosing speed of the starting switch enough will eliminate the difficulty. Ordinarily this can be achieved only by using a switch with a lower opening as well as a lower reclosing speed, and this combination would generally affect the switching torque adversely.

With a current-operated magnetic switch, regeneration difficulties cannot occur because the starting circuit never closes after the motor is deenergized. With a voltage relay, regeneration is almost certain to occur because the starting switch closes instantly after the power is shut off; however, these relays are used mostly for hermetic refrigerators, which stop very quickly so that there is little danger of serious trouble on this account.

References

1. Boothby, C. R.: "Discussion on Capacitor Motor Papers," *AIEE Trans.*, April 1929, p. 629.
2. Veinott, C. G.: *Fractional Horsepower Electric Motors*, 2d ed., 1948, *Fractional- and Subfractional-Horsepower Electric Motors*, 3d ed., 1970, McGraw-Hill Book Company, New York.
3. ————: "Discussion of Bassett and Potter Paper," *AIEE Trans.*, 1935, p. 1106.
4. Appleman, W. R.: "Twelve Ways of Reversing Capacitor-Start Motors," *Electr. Manuf.*, August 1950.
5. Bassett, E. D., and E. M. Potter: "Capacitive Excitation for Induction Generators," *AIEE Trans.*, 1935, pp. 540–545.
6. Angst, G.: "Self-Excitation of Capacitor Motors," *AIEE Trans.*, vol. 71, pt. III, 1952, pp. 557–562.
7. Oldenkamp, John L.: "Capacitive Regenerative Braking in Single-Phase Motors, *IEEE Trans. Power Appar. Syst.*, vol. PAS-86, no. 11, November 1967, pp. 1312–1316.

Bibliography

Bailey, Benjamin F.: "The Condenser Motor," *AIEE Trans.*, April 1929, p. 596.

Buchanan, L. W., and T. F. Winters: "Auxiliary Phase Design for Capacitor-Start Motors," *IEEE Trans. Power Appar. Syst.*, vol. 84, pp. 993–999.

Covo, A., and L. E. Lingo: "Graphical Determination of Starting Performance of Capacitor Motors," *AIEE Trans.*, vol. 75, pt. III, 1956, pp. 1309–1312.

London, Sol: "Reversing Single-Phase Motors," *Machine Design*, Dec. 3, 1964.

Morrill, Wayne J.: "The Revolving Field Theory of the Capacitor Motor," *AIEE Trans.*, April 1929, p. 614.

Rao, P. V.: "Switching Transients in Single-Phase Induction Motors, with Constant Speed," *AIEE Trans.*, vol. 78, pt. III, 1959, p. 713.

Spacht, H. C.: "Fundamental Theory of the Capacitor Motor," *AIEE Trans.*, April 1929, p. 607.

Trickey, P. H.: "Design of Capacitor Motors for Balanced Operation," *AIEE Trans.*, September 1932, p. 780.

Veinott, C. G.: "Starting Windings for Single-Phase Induction Motors," *AIEE Trans.*, vol. 63, 1944, pp. 288–294.

Veinott, C. G.: *Theory and Design of Small Induction Motors*, McGraw-Hill Book Company, New York, 1959. (Out of print, but available from University Microfilms, P.O. Box 1647, Ann Arbor, Mich. 48106)

CHAPTER

6

TWO-VALUE CAPACITOR MOTORS

The two-value capacitor motor, sometimes referred to as capacitor-start capacitor-run,* gained great commercial prominence in the early 1930s for applications such as refrigerators, compressors, and stokers, that require high locked-rotor and breakdown torques. It has since been largely replaced by the capacitor-start motor, although in recent years it has regained prominence because of its inherently higher efficiency; it is possible to increase the efficiency by as much as 10 percent with the addition of a properly designed running auxiliary winding-capacitor phase. It is necessarily more expensive; whether the additional cost is justified for any particular application depends a lot upon the total running time per year. The two-value capacitor design is often used to squeeze an additional rating or two into a given frame size.

6-1 Two-Value Capacitor Motor Defined. A two-value capacitor motor is a form of capacitor motor that starts with one value of capacitance in series with its auxiliary winding and runs with a different value. This

*This is an ambiguous term that should be avoided, for it applies to the permanent-split capacitor motor as well as to the two-value capacitor motor.

change in value of capacitance is automatic and can be effected by the use of two separate capacitors or by one capacitor and an autotransformer.

6-2 Essential Parts of the Two-Value Motor. Discussed in the following articles are the two major types of two-value capacitor motors.

(1) Motor using two capacitors. A two-value motor using two separate capacitors is illustrated diagrammatically in Fig. 6-1. The arrangement

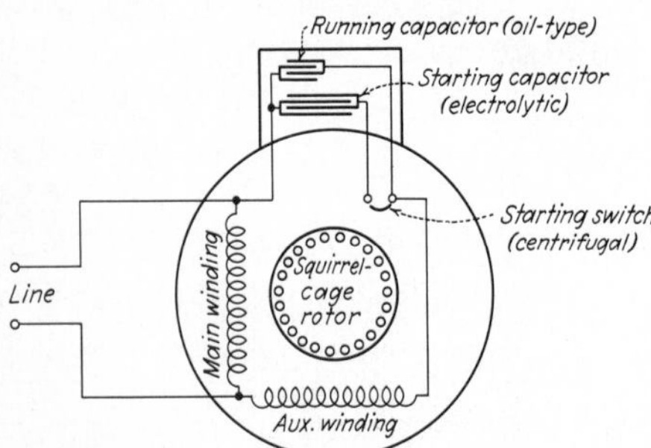

Fig. 6-1 Two-value capacitor motor using two capacitors.

of windings and switch in this motor is identical with the arrangement used in a capacitor-start motor. The difference between these two motors is that the two-value motor has a running capacitor permanently connected in series with the auxiliary winding; the starting capacitor is paralleled with the running capacitor only during the starting period. The running capacitor is usually of the paper-spaced oil-filled type and is rated at 370 (formerly 330) V for use with alternating current and for continuous operation. Capacitors range in rating from 5 to 25 mfd. The electrolytic capacitors for 115-V motors range in size from 85 mfd or less to 300 mfd or more. A motor of this type is illustrated in Fig. 6-2, where the two capacitor units are in two cases. Capacitors are discussed in Chap. 8.

(2) Motor using one capacitor and autotransformer. The two-value motor using one capacitor and an autotransformer is discussed in greater detail in earlier editions of this book.[1,*] The windings are the same as for the capacitor-start motor discussed in Art. 5-2. A "transfer switch" that is equivalent to a single-pole double-throw switch causes high voltage (in the range of 600 to 800 V) to be impressed across the capacitor during starting and then reduced to approximately 330 V for continuous running. This change in voltage is effected by changing taps on the primary of the autotransformer and corresponds to changing the effec-

*For numbered references, see References at end of this chapter.

Fig. 6-2 Two-value motor using two capacitors, an electrolytic one for starting and an oil-filled one for running. *(General Electric Company.)*

6-3 Effect of the Running Capacitor. It was pointed out in the preceding paragraph that the motor of Fig. 6-1 is merely the capacitor-start motor of Fig. 5-1 with a "running capacitor" permanently connected in the circuit. For any given motor, the effect of adding this capacitor is the following:

1. Increases the breakdown torque from 5 to 30 percent
2. Improves the full-load efficiency by up to 10 percent
3. Improves the full-load power factor by up to 5 percent
4. Reduces the noise under full-load operating conditions
5. Increases the locked-rotor torque 5 to 20 percent

Why does the running capacitor improve performance? The reader is referred to Chap. 2, Arts. 2-8 to 2-11, wherein it is pointed out that the rotating field of a single-phase induction motor may be considered as comprising two component pulsating fields, a main field and a cross field, spaced 90 electrical degrees apart and differing in time phase by 90°; the main-field component is set up directly by the main winding and the cross field by the rotor. When the auxiliary winding is connected across the line in series with a running capacitor of the proper value, the current drawn by the auxiliary winding leads the current drawn by the main winding by approximately 90°. The net effect is that the auxiliary winding sets up a part or all of the cross field, thereby reducing or eliminating the magnetizing currents in the rotor and the accompanying rotor copper losses.

The effect of the addition of a running capacitor is illustrated in Fig. 18-15, lines 38 to 47. The first column is an analysis of the full-load losses of a ¼-hp motor operating without a capacitor, and the second column is a similar analysis of the same motor operating with a running capacitor in the circuit. The reduction in losses and improvement in performance are marked.

Another effect of the running capacitor is to give 5 to 30 percent more breakdown torque. Still another effect is to reduce the double-frequency torque pulsations normally inherent in single-phase motors. (For a discussion of this phenomenon, see Art. 17-13.) This effect is noticeable at or near full load and may not be noticed at no load.

In short, the effect of the running capacitor is to make the motor perform more nearly like a two-phase motor, particularly at one value of load. It is not possible to duplicate two-phase motor performance at all load values with a single value of capacitance, however, because a different value of capacitance would be required for each different load. Trickey[2,3] shows how to proportion a winding and capacitor to obtain two-phase motor performance at any one desired load point. Morrill[4,5] shows how to calculate the performance of a capacitor motor at any load.

6-4 Purpose of the Starting Switch. The starting or transfer switch serves the same purpose as the starting switch of a capacitor-start motor, discussed in Art. 5-5. A representative speed-torque curve of a two-value capacitor motor is illustrated in Fig. 6-3. The curve of capac-

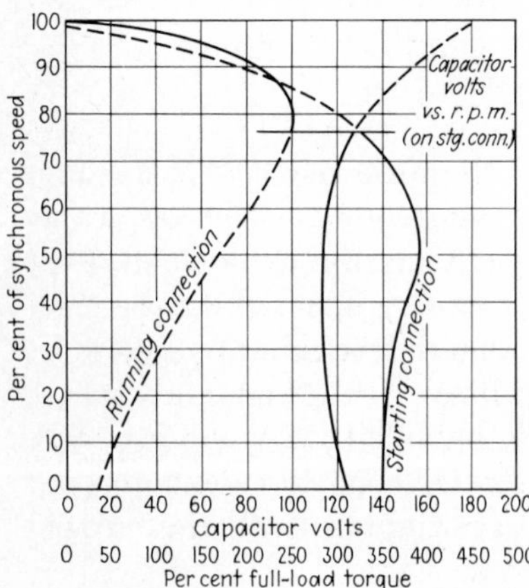

Fig. 6-3 Speed-torque curve of a two-value capacitor motor.

itor volts vs. rpm represents in the case of a motor such as that illustrated in Fig. 6-1 the actual voltage across the two capacitor elements when the motor is operating on the starting connection. In the case of the capacitor-transformer motor it represents the voltage across the primary of the transformer; the actual voltage across the capacitor element is approximately five to seven times the voltage across the primary, or up to 800 V.

6-5 Torques of Two-Value Capacitor Motors. By comparing Fig. 6-3 with Fig. 5-4, it will be noted that the capacitor-start motor has substantially the same torques as the two-value motor. The reason for this is simply that the main winding of the capacitor-start motor is designed for more torque, so that the torques of the two types of motors are practically identical. The two-value motor, however, will start a light load on the running connection, whereas the capacitor-start motor probably will not. The torques given in Table 19-1 apply also to two-value motors, as do the locked-rotor currents given there.

SINGLE-VOLTAGE MOTORS

6-6 Standard Connections for Reversible Motors. Figure 6-4 shows motor and line wiring diagrams for motors with and without thermal protection, with and without terminal boards, as developed by NEMA.

6-7 Motor with Dual-Section Auxiliary Winding. In Fig. 6-5 is illustrated schematically a motor which uses a two-section auxiliary winding. When starting, current flows through the first section of the auxiliary winding and the electrolytic capacitor; on the running connection, current flows through both sections of the auxiliary winding and the oil-type running capacitor. A natural question is to ask why such an arrangement is used in preference to the simpler arrangement of Fig. 6-1.

For economic reasons, it is usually advantageous to work electrolytic capacitors at a relatively low voltage, of the order of 115 to 130 V, whereas oil capacitors are best worked at higher voltages, of the order of 300 V or more. Using this arrangement of Fig. 6-5, it is possible to work the electrolytic capacitor at its own appropriate voltage; by proper choice of turns in section 2, the voltage on the running capacitor is increased, and the same auxiliary-winding current produces more effect because of the added turns in section 2. Thus, the arrangement of Fig. 6-5 enables the designer to obtain more of the beneficial effects cited in Art. 6-3 from the same number of running microfarads and to work the running capacitor closer to its rated voltage.

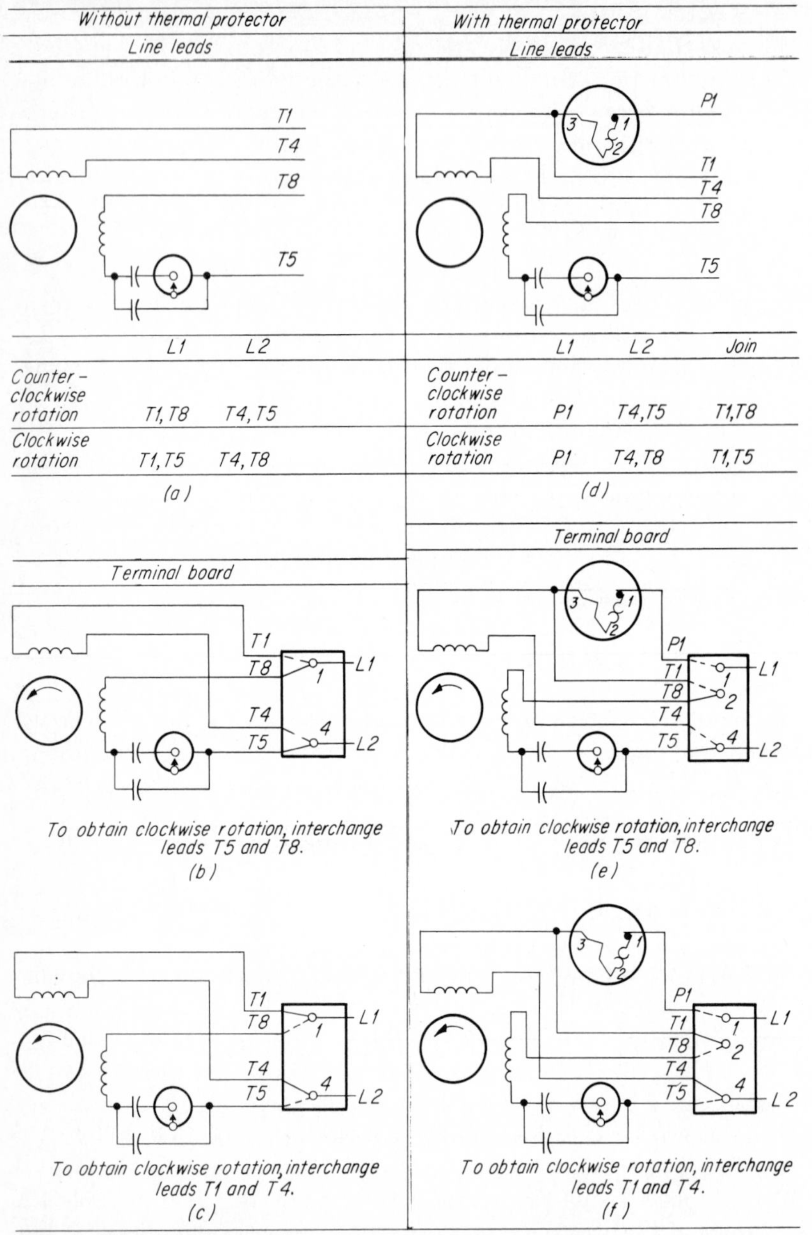

Fig. 6-4 Wiring and line connection diagrams for two-value capacitor motors. The motors are single-voltage and reversible, with or without a thermal protector. (*a*), line leads; (*b*), two-post terminal board; (*c*), three-post terminal board; (*d*), line leads; (*e*), two-post terminal board; (*f*), three-post terminal board. *(From NEMA Motor and Generator Standards, NEMA Publ. no. MG1-1978, National Electrical Manufacturers Association, Washington, 1978.)*

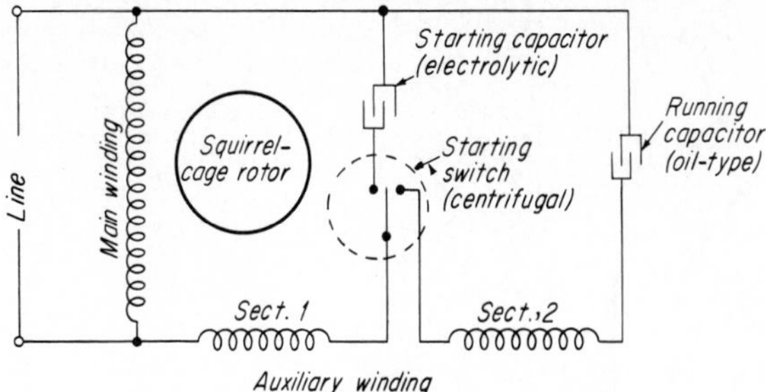

Fig. 6-5 Schematic arrangement of a two-value capacitor motor, using two capacitors and a two-section auxiliary winding.

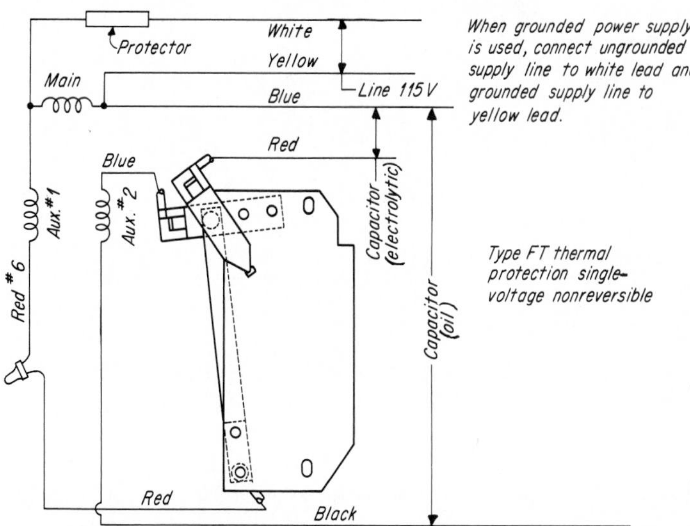

Fig. 6-6 Motor and line wiring diagram for a single-voltage two-value nonreversible capacitor motor, using a two-section auxiliary winding. Starting switch connects red to red for starting and red to blue for running. *(Westinghouse Electric Corporation.)*

One practical embodiment of this arrangment is shown in Fig. 6-6. A typical stator connection diagram—this one for six poles—is given in Fig. 6-7.

DUAL-VOLTAGE MOTORS

In general, dual-voltage two-value capacitor motors split the main winding into two sections, which are connected in series or in parallel, as do capacitor-start motors, and the same considerations, discussed in

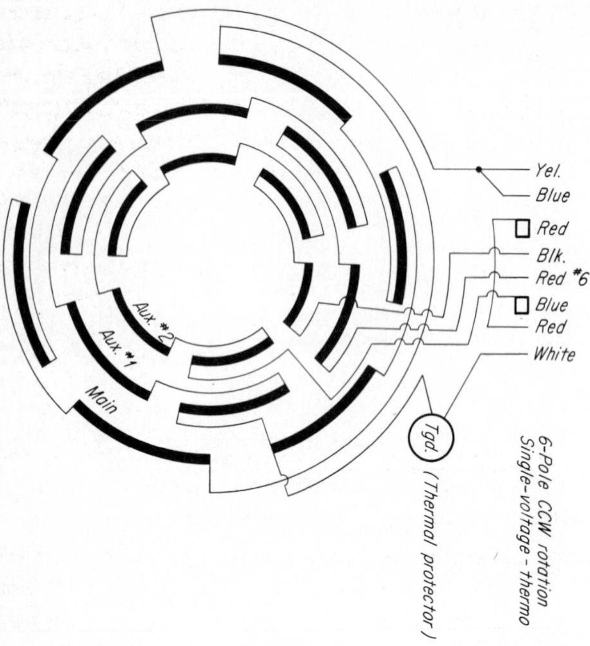

Fig. 6-7 Stator connection diagram for a six-pole two-value capacitor motor connected as in Fig. 6-6. *(Westinghouse Electric Corporation.)*

Chap. 5, generally apply. The auxiliary winding may have a single section, or two sections, as discussed in Art. 6-7 (not for series-parallel connection).

6-8 Dual-Voltage Motors with Single-Section Auxiliary Winding. Figure 6-8 is a motor and line wiring diagram for a dual-voltage (either-rotation) reversible thermally protected motor with a single-section auxiliary winding and a terminal board.

6-9 Dual-Voltage Motors with Two-Section Auxiliary Winding. A dual-voltage thermally protected motor with a two-section auxiliary winding is shown in Fig. 6-9. A four-pole stator connection diagram for such a motor is given in Fig. 6-10.

MISCELLANEOUS SERVICE PROBLEMS

6-10 Rewinding or Reconnecting Motor for a Different Voltage but for Same Performance.

(1) Changing from 115 to 230 V. About the only safe way to rewind for

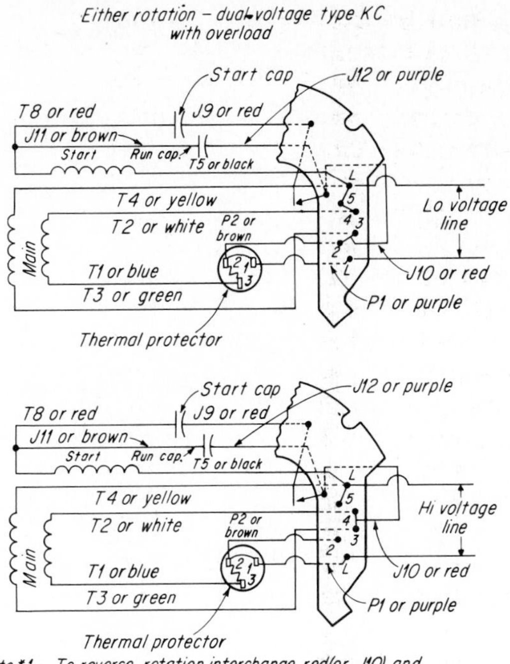

Fig. 6-8 Motor and line wiring diagram for a dual-voltage two-value reversible (either-rotation) thermally protected capacitor motor, using two capacitors. *(General Electric Company.)*

230 V is to rewind for the dual-voltage connection, such as that used in capacitor-start motors. In this case, the same double-unit capacitor and the same auxiliary winding are employed.

(2) Other voltage changes. Other voltage changes are not recommended unless rewinding information can be obtained from the motor manufacturer.

6-11 Rewinding for Different Torques, Speed, or Frequency. Because of the complications involved in the capacitor unit and the reasons outlined in Arts. 5-23 and 5-24, these types of changes are not recommended.

6-12 Changing a Two-Value Motor to a Capacitor-Start Motor. To change a two-value motor to a capacitor-start motor, all that is usually necessary is to remove the run capacitor. If the run capacitor is removed, certain adverse effects may occur: (1) the breakdown torque of the motor will be reduced from 5 to 30 percent, with a resultant

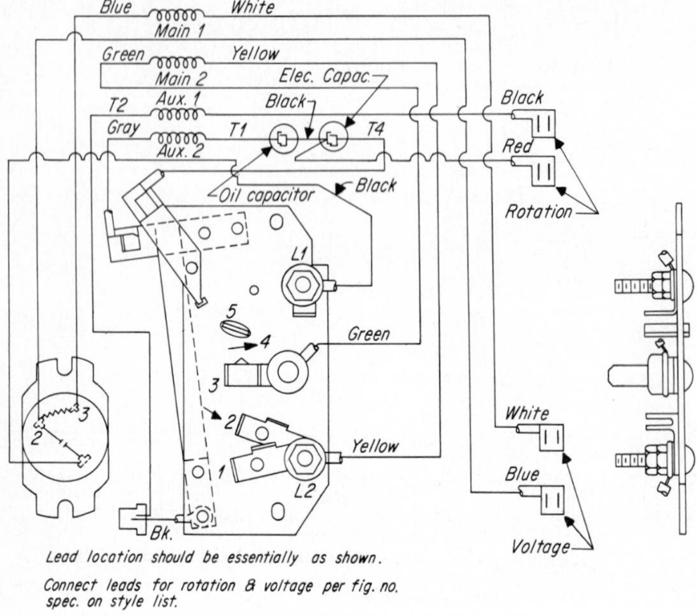

Fig	Rot.	Rot. conn.	Volt.	Volt. conn.
4	CW	Red → 2 Blk → 4	115	White → 1 Blue → 3 5
3	CCW	Blk → 2 Red → 4	115	White → 1 Blue → 3 5
2	CW	Red → 2 Blk → 4	230	1 White → 3 Blue → 5
1	CCW	Blk → 2 Red → 4	230	1 White → 3 Blue → 5

Fig. 6-9 Motor and line wiring diagram and style list for a dual-voltage two-value reversible thermally protected capacitor motor, using two capacitors and two sections of auxiliary winding. Starting switch connects black to T4 for starting and black to gray for running. *(Westinghouse Electric Corporation.)*

increase in slip at any given load, (2) the full-load losses, watts input, and resultant temperature rise will increase, though perhaps not to an injurious extent, (3) the noise under full-load operating conditions may increase, (4) the locked-rotor torque may decrease from 5 to 20 percent, (5) the capacitor voltage will increase slightly, sometimes enough to cause premature failure, and (6) a thermal protector, if provided, will not allow the motor to carry as much running overload.

If the motor has a dual-section auxiliary winding as shown in Fig. 6-5, removal of the running capacitor will neither reduce the locked-rotor torque nor increase the capacitor voltage.

Substitution of an electrolytic capacitor for a capacitor-transformer

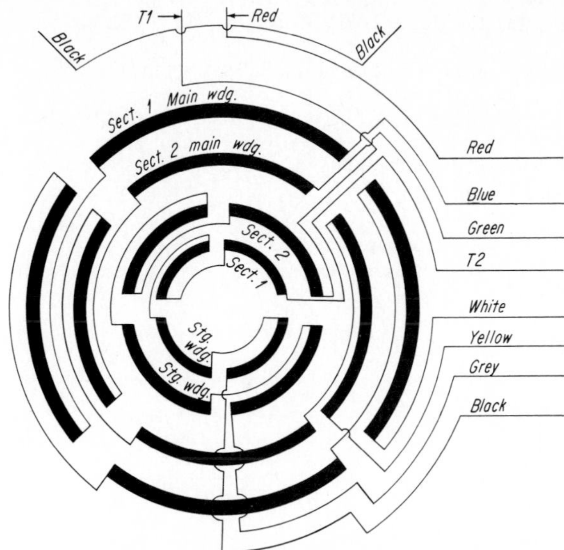

Fig. 6-10 Stator connection diagram for a four-pole dual-voltage two-value capacitor motor, using two capacitors and two sections of auxiliary winding. *(Westinghouse Electric Corporation.)*

unit is fraught with certain hazards in addition to the foregoing. Earlier editions of this book[1] should be consulted if this substitution becomes necessary.

If the motor uses two capacitors and the running capacitor develops a short circuit, the motor will not operate properly; however, if the defective capacitor is removed from the circuit, the motor should then operate satisfactorily as a capacitor-start motor (provided that there is no other defect), with the limitations discussed previously.

References

1. Veinott, C. G. *Fractional Horsepower Electric Motors*, 2d ed., 1948, *Fractional- and Subfractional-Horspower Electric Motors*, 3d ed., 1970, McGraw-Hill Book Company, New York.
2. Trickey, P. H.: "Design of Capacitor Motors for Balanced Operation," *AIEE Trans.*, September 1932, p. 780.
3. ————: "Equal Volt-Ampere Method of Designing Capacitor Motors," *AIEE Trans.*, vol. 60, 1941, pp. 990–992.
4. Morrill, Wayne J.: "The Revolving Field Theory of the Capacitor Motor," *AIEE Trans.*, April 1929, p. 614.
5. ————: "The Apparent-Impedance Method of Calculating Single-Phase Motor Performance," *AIEE Trans.*, vol. 60, 1941, pp. 1037–1041.

Bibliography

Bailey, Benjamin F.: "The Condenser Motor," *AIEE Trans.*, April 1929, p. 596.
McFarland, T. C.: "Current Loci For the Capacitor Motor," *AIEE Trans.*, vol. 61, 1942, pp. 152–156.

———: "Turn Ratio of the Capacitor Motor," *AIEE Trans.*, vol. 62, 1943, pp. 892–898.

Puchstein, A. F., and T. C. Lloyd: "Capacitor Motors with Windings Not in Quadrature," *AIEE Trans.*, vol. 54, 1935, pp. 1235–1239.

Specht, H. C.: "Fundamental Theory of the Capacitor Motor," *AIEE Trans.*, April 1929, p. 607.

Suhr, F. W.: "Symmetrical Components as Applied to the Single-Phase Induction Motor," *AIEE Trans.*, vol. 64, 1945, pp. 651–656.

CHAPTER

7

PERMANENT-SPLIT AND SPLIT-PHASE CAPACITOR MOTORS

Permanent-split capacitor motors (once called single-value capacitor motors) are generally used for special-purpose applications such as shaft-mounted fans and blowers, instruments (often using synchronous rotors), and servomotors. In the smaller sizes, they often compete with shaded-pole motors. They are more efficient, have better power factor, have more output per pound, but are more costly than shaded-pole motors. With shaded-pole motors they share the important advantage that they require no starting switch or relay. Generally speaking, they are not suitable for belted applications or for any other continuous-duty application requiring substantial locked-rotor torque. However, they do lend themselves well to high-torque reversing intermittent-duty service and have been used in European front-loading washing machines, where the agitation is accomplished by cyclic reversal. They are generally available in ratings from 1 mph to ⅓ hp and for shaft-mounted fans and blowers, up to ¾ hp.

Split-phase capacitor motors, so-called because their starting characteristics are similar to those of split-phase motors, as described in Chap. 4, and because their running characteristics are similar to those of permanent-split capacitor motors, have replaced split-phase motors in applications in which high efficiency for continuous operation is of primary importance.

Because the split-phase starting arrangement permits the use of lower rotor resistance, the split-phase capacitor motor usually operates at a higher efficiency and higher speed than a comparable permanent-split capacitor motor.

7-1 Permanent-Split Capacitor Motor Defined. This motor is defined in Art. 5-1 as a capacitor motor that uses its auxiliary winding and capacitor continuously, without change in capacitance. This definition shows why no starting switch or relay is required.

7-2 Essential Parts of a Permanent-Split Capacitor Motor. A permanent-split capacitor motor is represented schematically in Fig. 7-1. It is

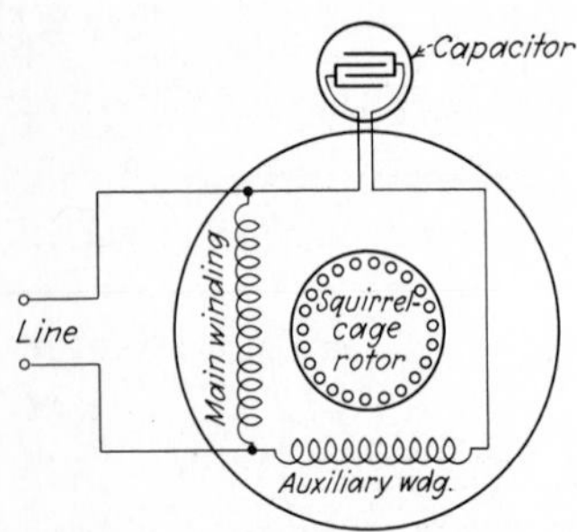

Fig. 7-1 The permanent-split capacitor motor.

to be noted that the arrangement of windings and connections is exactly the same as in Fig. 5-1, except that the starting switch is omitted. Further, it should be noted that it is exactly like Fig. 6-1, except that both starting switch and starting capacitor are omitted. Hence, the stator windings of a permanent-split capacitor motor generally resemble in form and arrangement those of a capacitor-start motor. That is, there are two windings, spaced 90 electrical degrees apart. A single-voltage motor will usually have a single-section main winding, and a dual-voltage motor a two-section main winding; but, in either case, the auxiliary winding will usually be a single-section winding. In many designs it is necessary to insulate between the main and auxiliary windings and even to sleeve the coil ends because of the high voltage developed between adjacent pole groups. The motor also uses a squirrel-cage rotor, almost invariably with a higher resistance than is found in a capacitor-start motor. As mentioned, no starting switch or relay is required. Some motors use tapped windings; these are discussed in Arts. 7-14 and 7-15. A continuous-rated capacitor is required with all arrangements, although some intermittent-duty motors use an electrolytic capacitor for starting and running. Continuous-rated capacitors are of the paper-spaced oil-filled, oil-filled polypropylene, or dry-film polypropylene type, as discussed in Chap. 8.

7-3 Principles of Operation. The effect of the running capacitor is to improve the performance by creating flux conditions at some one load approximating those of a polyphase motor. These polyphaselike flux conditions obtain at only one load, generally somewhere between half load and full load. The effect of this running capacitor is discussed at greater length in Art. 6-3, wherein it is pointed out that the capacitor increases the breakdown torque, efficiency, and power factor and reduces the noise. Thus, the net effect is to increase the available output per pound or per cubic inch of active material.

The speed-torque characteristic of a continuous-rated motor is quite like the running connection curve shown in Fig. 6-3, except that greater rotor resistance is customarily used in order to improve the locked-rotor torque; this higher rotor resistance increases the full-load slip over and above the slip shown in Fig. 6-3. Another typical speed-torque characteristic of a continuous-rated permanent-split capacitor motor is illustrated by the high curve of Fig. 7-7.

An intermittent-rated high-torque permanent-split capacitor motor would have a speed-torque characteristic more like that of the starting connection curve of Fig. 6-3, except that this motor, too, would have higher rotor resistance, which would increase the locked-rotor torque and cause the breakdown torque to occur at a higher slip than shown.

Use of a running capacitor makes it possible to reduce the speed of a motor driving a shaft-mounted fan by as much as 50 percent. It has been shown by Veinott[1,*] that the speed of a straight single-phase induction motor cannot be made to occur below about 70 percent of synchronous speed; hence, without a capacitor, a stable speed reduction of 50 percent cannot be achieved.

SINGLE-SPEED MOTORS

7-4 Performance Characteristics of Permanent-Split Capacitor Motors. The approximate full-load speeds of permanent-split capacitor motors used in heating, air-conditioning, and similar applications, as established by the National Electrical Manufacturers Association (NEMA), are given in Table 7-1.

*For numbered references, see References at end of this chapter.

TABLE 7-1 Full-Load Speed Rating of Fractional Horsepower Permanent-Split Capacitor Motors

	Poles			
Horsepower	2	4	6	8
1/20–1/3	3250	1625	1075	825
Less than 1/20	3000	1550	1050	800

Source: From *NEMA Motor and Generator Standards*, NEMA Publ. no. MG1-1978, National Electrical Manufacturers Association, Washington, 1978.

With the passage of time, the full-load speeds for motors up to 1/3 hp have gradually fallen to the same speeds as those shown for two-, four-, and six-pole motors of less than 1/20 hp. This change in operating speed came about as designers found a need to increase locked-rotor torques; the increase was accomplished by increasing rotor resistances, with a resultant increase in slip.

Permanent-split motors for shaft-mounted fans and blowers in heating, air-conditioning, and similar applications are predominantly six-pole, largely to reduce air noise. Breakdown torques and locked-rotor torques of permanent-split fan motors are given in Chap. 19 in Table 19-4.

Permanent-split motors with outputs of the order of 2 to 5 millihorsepower (mhp) are used in industrial instruments; such motors are generally rated on the basis of full-load torque and have a locked-rotor torque equal to rated torque, although the breakdown torque may be about 175 percent of full-load torque.

Some permanent-split motors are designed for high-torque intermittent duty. These are discussed in Art. 7-19.

Compared with capacitor-start or split-phase motors, permanent-split motors have lower full-load speeds, which are generally subject to wider variations among individual motors as well as to changes in load on any given motor. The problem is the same as it is for shaded-pole motors and is discussed in Art. 9-8.

7-5 Connections and Diagrams: Single-Voltage Motors. Wiring and line connection diagrams, as specified by the National Electrical Manufacturers Association (NEMA), are given in Fig. 7-2. Note that the external connections are identical with those for a split-phase motor, as shown in Fig. 4-5, or for a capacitor-start motor, as shown in Fig. 5-6.

NEMA connection diagrams for permanent-split motors for industrial instrument service are given in Figs. 7-3 and 7-4.

Stator connection diagrams for split-phase motors, as given in Chap. 4, are generally applicable to permanent-split motors. However, for

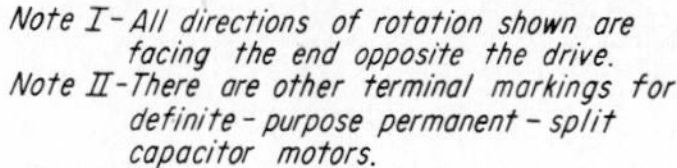

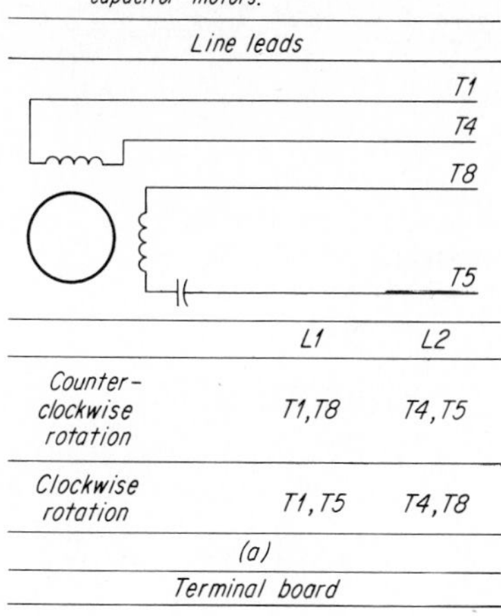

	L1	L2
Counter-clockwise rotation	T1,T8	T4,T5
Clockwise rotation	T1,T5	T4,T8

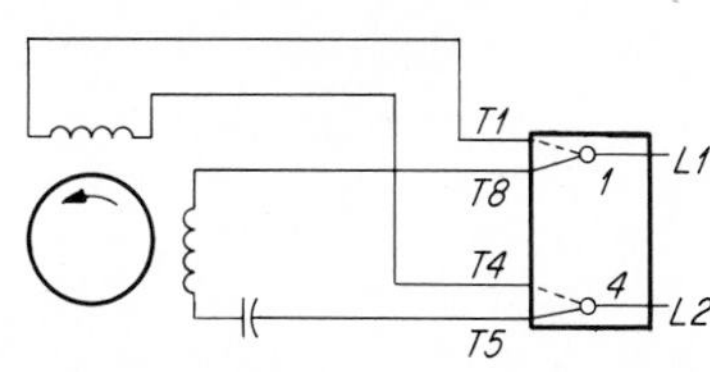

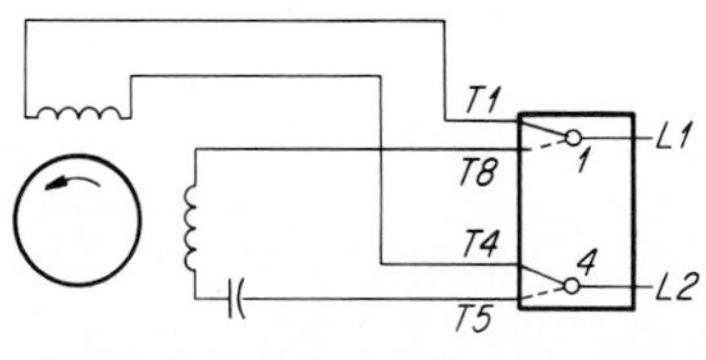

Fig. 7-2 Wiring and line connection diagrams for single-voltage permanent-split capacitor motors with tagged leads and with two-post terminal board. (*a*) Line leads; (*b*) terminal board, interchange auxiliary winding leads to reverse rotation; (*c*) terminal board, interchange main winding leads to reverse rotation. *(From NEMA Motor and Generator Standards, NEMA Publ. no. MG1-1978, National Electrical Manufacturers Association, Washington, 1978.)*

greater quietness, both main and auxiliary windings of a permanent-split motor are often parallel-connected. A diagram for a parallel-connected six-pole motor is given in Fig. 7-5; this diagram shows diametrically opposite poles connected in parallel in order to minimize any unbalanced pull on the rotor due to unbalanced magnetic fluxes on opposite sides of the shaft.

7-6 Connections and Diagrams: Dual-Voltage Motors. Dual-voltage single-speed permanent-split motors can be arranged in substantially

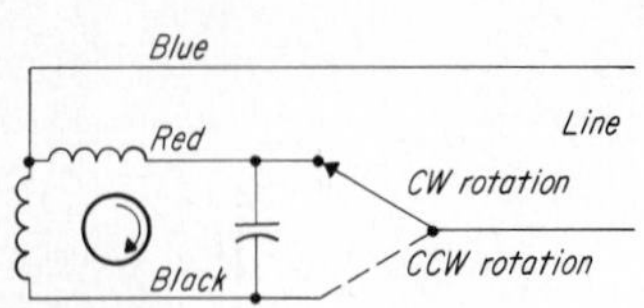

Fig. 7-3 Connection diagram for a three-lead single-voltage reversing permanent-split capacitor motor. *(From NEMA Motor and Generator Standards, NEMA Publ. no. MG1-1978, National Electrical Manufacturers Association, Washington, 1978.)*

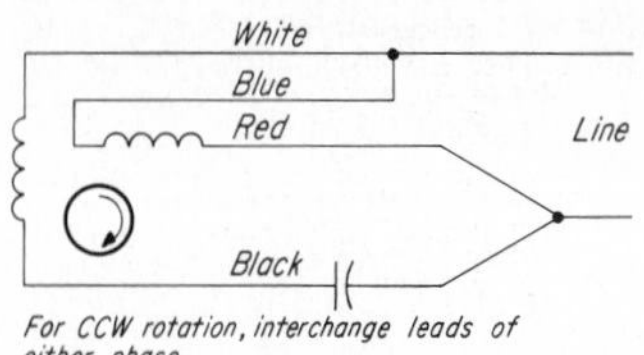

Fig. 7-4 Connection diagram for a four-lead single-voltage reversible permanent-split capacitor motor. *(From NEMA Motor and Generator Standards, NEMA Publ. no. MG1-1978, National Electrical Manufacturers Association, Washington, 1978.)*

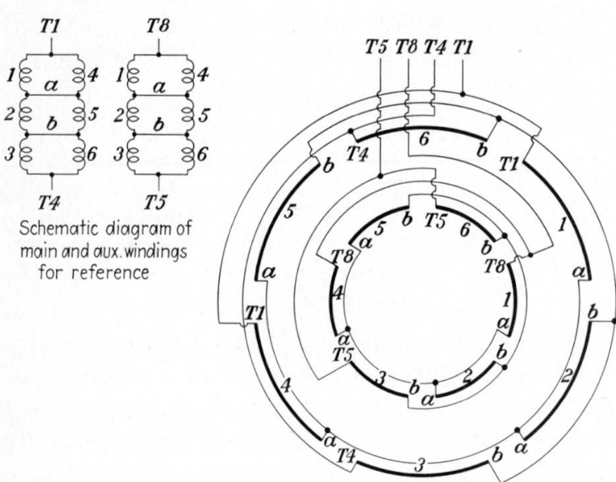

Fig. 7-5 Stator connection diagram for six-pole four-lead externally reversible permanent-split capacitor motor. Both windings are parallel and cross-connected.

the same way as capacitor-start motors (omitting the switch, of course), as discussed in Arts. 5-11 to 5-17, inclusive.

Wiring and line connection diagrams for an Emerson motor are given in Fig. 7-6. The motor shown in Fig. 7-6 may be made externally reversible by breaking the auxiliary circuit at the common connection to the protector and bringing an additional lead out from the auxiliary at this point so that the auxiliary winding can be reversed.

7-7 Rewinding a Permanent-Split Capacitor Motor for a Different Voltage. A permanent-split motor that is rated 115 V may be rewound for 230 V by doubling the number of turns in both the main and aux-

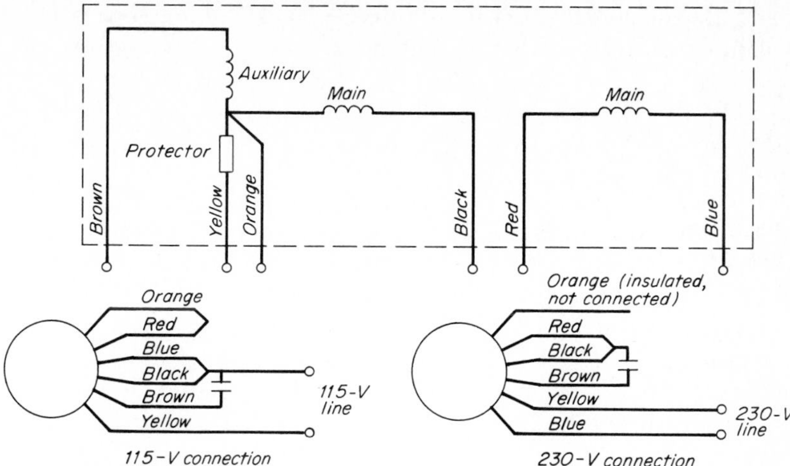

Fig. 7-6 Wiring and line connection diagrams for a dual-voltage permanent-split capacitor motor. Recommended capacitor of 5 mfd and 370 V for both connections. *(Emerson Motor Division.)*

iliary windings and using wire three sizes (gauges) smaller in each; it is then necessary to use a capacitor with one-fourth the microfarad rating and twice the voltage rating. Since most permanent-split motors rated 115 V employ a capacitor rated 370 V, this would lead to quite a problem in obtaining a capacitor for a permanent-split motor rated 230 V. A method used by design engineers is to double the number of turns in the main winding using wire three sizes smaller, wind the auxiliary with about 80 percent of the turns but with the original wire size, and use the same capacitor as with the 115-V motor. Although the line current is not exactly halved, the operating characteristics are very nearly the same in the same appliance.

As discussed in Art. 4-37, the breakdown torque of a split-phase motor may be increased by reducing the number of turns in the main winding. The same is true of the permanent-split motor; the number of turns in the main winding is reduced to strengthen the motor. However, the converse is true of the auxiliary winding. Turns must be *added* to the auxiliary winding in order to increase the breakdown torque. However, caution must be exercised, since increasing the auxiliary turns will increase the voltage across the capacitor.

SPEED-CONTROL METHODS FOR PERMANENT-SPLIT CAPACITOR MOTORS

Speed control is often desirable in a large number of fan and blower applications in heating and air-conditioning installations. One such typical use is for unit-heater service, in which it is desirable to operate the fan at a high speed in order to obtain a quick initial warming up; once the room is warm, a quieter and more uniform flow of heat can be achieved by reducing the speed. Moreover, it may often be found necessary to increase the normal operating speed when the outside temperature falls a great deal.

7-8 Basic Methods of Speed Control. Speed control of permanent-split capacitor motors is obtained by adjusting the field strength in the motor and thereby changing the slip. Pole changing is not used; when pole-changing motors are wanted, split-phase or capacitor-start motors should be used.

There are a number of different methods of changing the field strength and thereby the slip and operating speed, of a permanent-split motor:

1. Changing the voltage impressed on one or both windings by means of autotransformers. Several such arrangements were discussed and reviewed in previous editions of this book;[2] these arrangements permitted either a continuously adjustable speed or several discrete speed settings. They have now been superseded in practice by more economical tapped-winding arrangements.

2. Adjusting the field strength by using tapped-winding arrangements, which adjust the field strength by a change in the number of turns between line terminals.

3. Inserting an external impedance, preferably in series with only the main winding.

All three of these methods give somewhat similar operating characteristics except that the impedance-control method is more apt to give starting difficulties. (See Art. 7-13.)

7-9 Fundamental Relationship between Slip, Field Strength, and Torque. In Arts. 2-2 and 2-3 it was shown that the rotor of an induction motor is caused to rotate by the action of a magnetic field revolving about it. The revolving field induces in the rotor bars currents which react with the revolving field and produce torque that causes the rotor to follow the rotation of the revolving field at a speed always less than that of the revolving field; the difference in speeds is called the *slip*.

For the moment, let us assume the slip to be fixed, and let us look at the relationship between field strength and torque. Clearly, the voltage induced (BLV principle)* in rotor conductors will be directly propor-

*Voltage induced in a conductor = flux density × length of the conductor × velocity of the conductor, or symbolically, $E = BLV$.

tional to field strength (flux density in the air gap), and the resultant currents in those conductors will therefore also be proportional to field strength. Now, since the force on any conductor in a magnetic field is proportional to the product of field strength multiplied by the current in the conductor (BLI principle)* and since the current is proportional to the field strength itself, it follows that *at any slip the torque developed by the rotor is proportional to the square of the field strength.*

From the preceding statement, two corollaries are at once deducible:

1. At any slip, the torque developed is proportional to the square of the voltage applied to the main winding.

2. At any slip, if the voltage is held constant, but the effective number of turns is changed, the torque will be inversely proportional to the square of the number of conductors in the main winding.

These two relationships are not exact, for the torque developed also depends to a lesser extent upon what happens to the auxiliary phase.

7-10 Speed Control by Means of Field Strength Control. On Fig. 7-7 are plotted three speed-torque curves for an adjustable-speed perma-

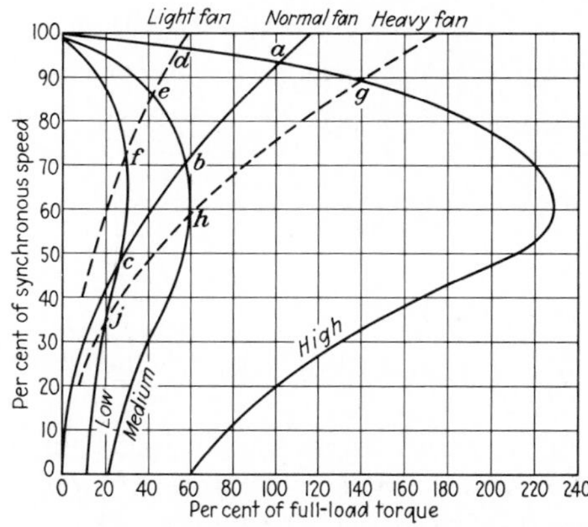

Fig. 7-7 How speed control of a fan is obtained with an adjustable varying-speed capacitor motor. The high, medium, and low curves are the three speed-torque curves on the respective speed connections.

nent-split capacitor motor. The curves are labeled high, medium, and low, representing three different field strengths achieved by one of the three methods outlined in Art. 7-8. A speed-torque curve for a normal fan is likewise plotted. The operating speeds of the motor driving this fan are at the points where the fan curve crosses the motor curve, or at points *a*, *b*, and *c*, respectively, for the three different field strengths. In the figure, a maximum speed reduction of about 50 percent is achieved.

*Force on a conductor = flux density × length of conductor × current in the conductor, or symbolically, $F = BLI$.

7-11 Limitations of Field-Strength Systems of Speed Control. Motors with this type of speed control have been very successful when used with shaft-mounted fans and blowers, but this method does have limitations.

(1) Speed depends upon the load. At no load, the motor will operate at practically the same speed on the high, medium, and low connections. If a light fan requiring one-half the power of the normal fan is used, the motor will operate at speeds as shown by points *d*, *e*, and *f* on Fig. 7-7. It should be noted that even on the low connection, a speed reduction of less than 25 percent is obtained instead of the 50 percent obtained when the normal fan is used. If a heavy fan, i.e., one requiring 50 percent more power than the rating of the motor, is used, the three operating speeds are *g*, *h*, and *j*.

(2) Locked-rotor torque necessarily low. The locked-rotor torque on the low-speed connection has to be lower than the fan torque at the desired low speed. At half speed, the fan torque is about 25 percent of full-load torque, and the locked-rotor torque of the motor has to be less than this value.

(3) Not suited to belted drives. Because of the inherently low locked-rotor torque and also because there is apt to be a considerable variation in belt friction, particularly if a V belt is used, and because the low speeds would, therefore, be uncertain, this type of speed control is not suited to belt-driven applications.

(4) Unstable low-speed connection. Because on the low-speed connection the fan-torque and motor-torque curves cross at a very small angle, speed characteristics tend to be unstable, and the speed is sensitive to changes in voltage and to restrictions in the inlet and outlet ducts of the unit heater or fan the motor is driving. Restrictions in the path or ducts of a propeller fan generally increase the load requirements and slow down the motor; restrictions in either the inlet or exhaust ducts of a centrifugal fan or blower unload the fan, causing the motor to operate at a higher speed. In general, this characteristic of instability is more pronounced in a motor having a low slip at full load than it is in a motor having a higher full-load slip.

Also, manufacturing and voltage variations can create the kinds of problems encountered by shaded-pole motors, as discussed in Chap. 9 in Art. 9-8 and illustrated in Fig. 9-10.

MULTISPEED MOTORS

7-12 Motors for Adjustable-Speed Operation. It was mentioned earlier that adjustable speed is a desirable feature for unit-heater motors in order to provide a smooth, steady flow of heat quietly under normal conditions, and quick initial warm-up when needed, or to adjust for abnormally low outdoor temperatures. Earlier editions of this book[2] went into great detail on this, as well as on other lines of such motors and their speed controllers which were in greater vogue then. Pole-changing split-phase motors, tapped-winding permanent-split capacitor motors, and tapped-winding shaded-pole motors have almost completely replaced them.

Adjustable speed control is a desirable feature for shaft-mounted blowers in central heating furnaces. In one such scheme a plenum chamber sensor transmits a variable signal, whose magnitude depends on the plenum temperature, to a voltage controller mounted in a permanent-split motor that controls the motor speed. When the plenum is heated, the fan motor speeds up; when the plenum cools, the fan motor slows to a minimum speed. This gives continuous air flow but reduces the air velocity when little or no heat is required, thereby adding to comfort.

A method of field-strength control, and hence speed control, using solid-state devices such as a triac is discussed at length in Art. 13-8.

7-13 External Impedance Control. If an external impedance, which can be either a resistor or a reactor, is connected in series with the main winding, the voltage impressed across the main winding itself is thereby reduced, reducing the flux, thereby increasing the slip when loaded, and reducing the operating speed. The effect is not exactly the same as reducing the voltage by means of an autotransformer, for in the latter case, the voltage impressed across the winding does not change with load, whereas the series impedance causes a drop in voltage proportional to the load current drawn. Hence, when an external impedance is used, the low-speed connection is likely to be a little less stable than it would be if a transformer were used. When the speed reduction is obtained by a connection change, the effect is comparable to the use of a transformer. However, the external impedance can be added later to almost any high-slip permanent-split capacitor motor to obtain speed reduction when the motor is driving a shaft-mounted fan. A reactor has the advantage that it has negligible losses in it, whereas a resistor would have substantial losses in it. A connection diagram for a motor with external impedance control is given in Fig. 7-8. It should be noted that the external impedance is connected in series with only the main winding, in order to obtain better stability on the low-speed setting, for reasons discussed previously.

7-14 Three Kinds of Tapped-Winding Arrangements. For shaft-mounted fans and blowers, tapped-winding permanent-split capacitor motors, giving up to as many as five different speeds, are now available.

Either rotation - single-voltage
Type KCP with thermal protector

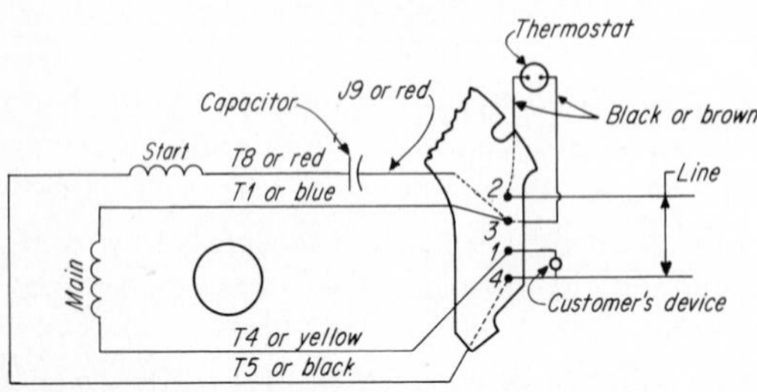

*Note *1: To reverse rotation, interchange blue (or T1) and yellow (or T4) leads.*

Fig. 7-8 Motor and line wiring diagram for a reactor-controlled adjustable-speed single-voltage permanent-split capacitor motor with terminal board. (Customer's device is the reactor controller, which can be adjustable.) *(General Electric Company.)*

Such motors effect speed control by field strength control, accomplished primarily by changing the impressed volts per turn on the main winding. Tapped-winding motors vary the volts per turn, and hence the flux, slip, and speed, by changing the number of series conductors in the main phase.

A two-speed tapped-winding motor has three windings, which we shall call here the *main winding, intermediate winding,* and *auxiliary winding.* The main and intermediate windings are wound in space phase with one another, i.e., one is wound on top of the other, in the same slots, with the same distribution but not necessarily with the same number of turns or wire size. In principle, they are similar to the dual-voltage windings used in dual-voltage capacitor-start motors, discussed in Arts. 5-11 to 5-17, except that the two sections of the tapped-winding motor are seldom, if ever, identical. The third winding is the auxiliary winding, which is displaced 90 electrical degrees from the main and intermediate windings. The rotor, of course, is of squirrel-cage construction, and no starting switch is used. The intermediate winding is also variously identified as an extra main, a booster, or a low-speed winding. For more than two speeds, the intermediate winding itself is tapped. There are three recognizable systems for connecting the three windings and capacitor, and as these give somewhat different conditions, they will be discussed separately.

(1) Auxiliary phase across low-speed taps (L connection). Figure 7-9 shows one arrangement, identified as the L connection. Consider first the high-speed connection, with 115 V impressed across the main winding.

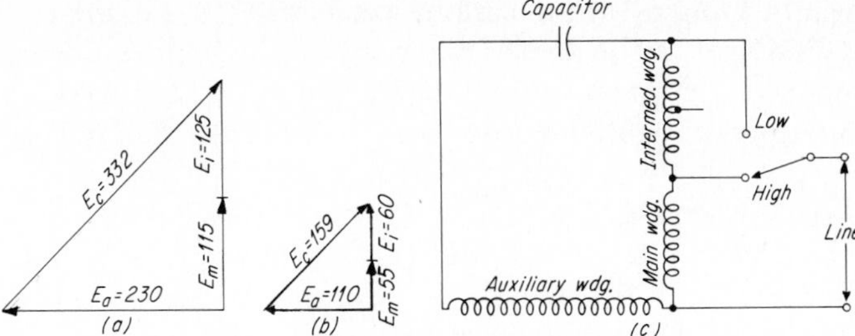

Fig. 7-9 Schematic arrangement of a two-speed tapped-winding capacitor motor with auxiliary phase connected across the low-speed taps (L connection). (*a*) Phasor diagram at high speed; (*b*) phasor diagram at low speed; (*c*) schematic arrangement of windings.

Let us assume, for purposes of discussion, that the number of turns in the intermediate winding is such as to give a voltage of 125, and that in the auxiliary winding to give 230 V, when 115 V is impressed across the main and the motor is operating at full speed. Now the voltages on the main and intermediate windings are in time phase because the windings are in space phase, and the voltage across the auxiliary winding is 90° out of phase. Figure 7-9*a* represents this condition; by drawing the phasor diagram to scale, or by calculation, it can be found that the voltage across the capacitor element itself is 332 V. On the low-speed connection, the 115 V is impressed across the main and intermediate windings in series with one another; hence the voltages on all three windings are reduced in the ratio of 115:240 (240 = 115 + 125). The resultant phasor diagram is given in Fig. 7-9*b*.

(2) Auxiliary phase across high-speed taps (T connection). Figure 7-10 shows the auxiliary phase connected across the high-speed taps. This arrangement is called the T connection for convenience, because the auxiliary phase was connected to the point between the main and inter-

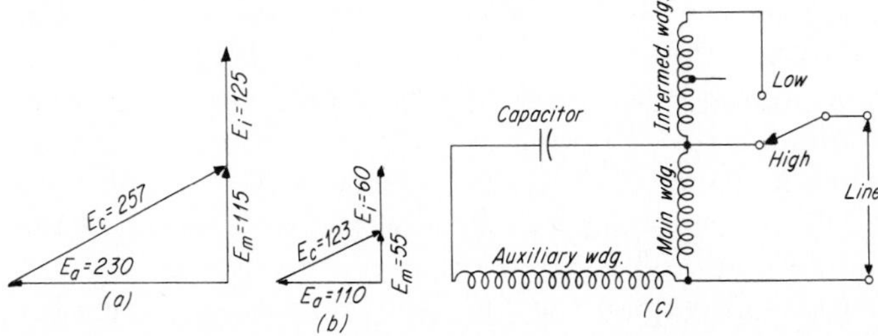

Fig. 7-10 Schematic arrangement of a two-speed tapped-winding capacitor motor with auxiliary phase connected across the high-speed taps (T connection). (*a*) Phasor diagram at high speed; (*b*) phasor diagram at low speed; (*c*) schematic arrangement of windings.

mediate windings, forming a T (in this figure on its side). Assuming the same windings as in Fig. 7-9, the phasor diagrams for the high- and low-speed connections were constructed in Fig. 7-10*a* and 7-10*b*. It should be noted that the capacitor voltages are substantially lower than in Fig. 7-9.

(3) Auxiliary phase across the line. In Fig. 7-11, the same windings used in the previous two figures are shown, but the capacitor phase is connected across the line. The corresponding phasor diagrams, Fig. 7-11*a* and *b*, show the approximate capacitor voltages encountered for the two speeds.

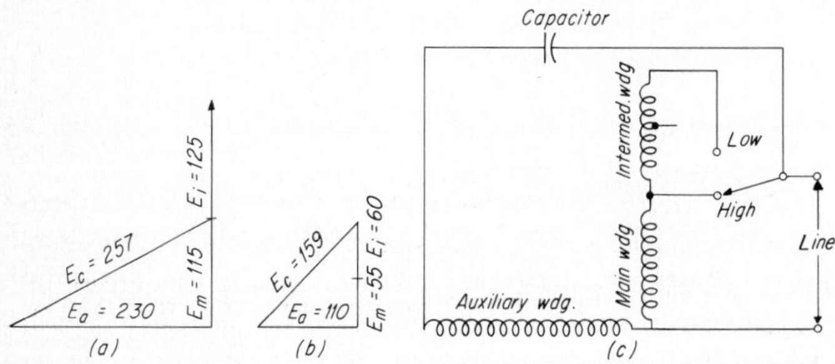

Fig. 7-11 Schematic arrangement of a two-speed tapped-winding capacitor motor with auxiliary phase connected across the line. (*a*) Phasor diagram at high speed; (*b*) phasor diagram at low speed; (*c*) schematic arrangement of windings.

Motors can be designed and built to operate successfully using any one of the three arrangements just discussed, and the motor specifier need not be concerned with which arrangement is used. However, in service, the motor must be connected for the system for which it is designed, meaning that the manufacturer's instructions must be followed carefully. As an example of what could happen, suppose a motor were designed to be connected per the T connection, but in service, it was inadvertently connected for the L connection, and let us further suppose that the turns ratios were the same as in the three figures. Now, reconnecting from T to L would raise the capacitor voltage from 257 to 332 V, an increase of 29 percent. This might or might not be above the safe voltage rating of the capacitor and might or might not damage the capacitor. A more serious effect would be to increase the capacitive volt-amperes by 66 percent or more ($1.29^2 = 1.66$); this increase in capacitive volt-amperes would likely cause any one or all of the following effects: increased operating speed, losses, temperature rise, current and power consumption, and inability to obtain normal speed reduction on the low-speed connection. Conversely, if the motor were designed to be operated on the L connection and if it were con-

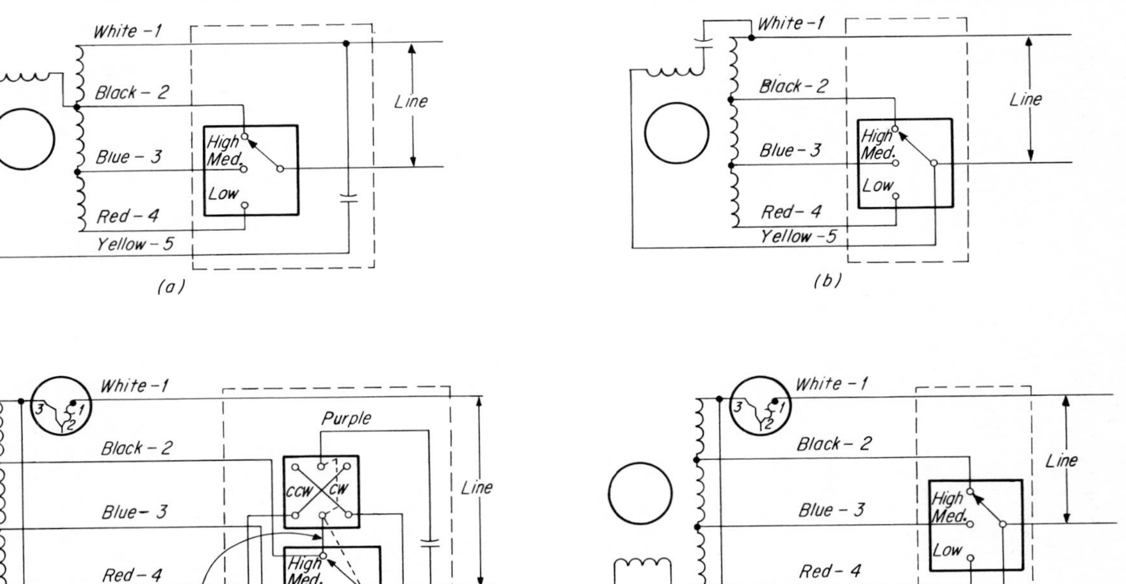

Note I — Parts shown within the dotted area are not a part of the motor. They are included in the diagram to clarify the motor terminal connections made by the user.

Note II — For two-speed motors, omit terminal 4 (red) and the corresponding winding.

Fig. 7-12 Wiring and line connection diagrams for three-speed tapped-winding permanent-split capacitor motors for shaft-mounted fans and blowers. (See Art. 7-15.) (*a*) Nonreversible, T connection; (*b*) nonreversible, auxiliary phase across line; (*c*) reversible and thermally protected with solid lines showing T connection and dotted lines showing auxiliary phase across line; (*d*) nonreversible and thermally protected with auxiliary phase across line. *(From NEMA Motor and Generator Standards, NEMA Publ. no. MG1-1978, National Electrical Manufacturers Association, Washington, 1978.)*

nected in service for the T connection, the motor would operate at too slow a speed on both connections, but especially on the low-speed one, and might not even start at all on the low-speed connection.

Hence, it is most important that the appropriate connection diagram be used when connecting the motor to the line.

Another important point is that, when the auxiliary phase is connected across the line, the external line switch needs an extra pole in order to disconnect the auxiliary phase from the line; with the other two arrangements, as shown in Figs. 7-9 and 7-10, the motor windings are both deenergized when the single-pole double-throw switch is in the neutral position.

More than two speeds are obtained by bringing extra taps out from the intermediate winding. In Figs. 7-9 to 7-11, a single tap from the intermediate winding is shown; this would give a third speed, if used. More taps may be added.

7-15 Connection Diagrams for Tapped-Winding Motors. Many arrangements and modifications are variously used.

NEMA diagrams for three-speed motors are given in Fig. 7-12.

1. In Fig. 7-12*a* the T connection is shown and the motor is nonreversible.

2. In Fig. 7-12*b* the auxiliary phase is across the line, and the motor is nonreversible.

3. In Fig. 7-12*c* the solid lines show the T connection, and the dotted lines show the auxiliary phase across the line and the motors in both are reversible and are thermally protected.

4. In Fig. 7-12*c* the auxiliary phase is across the line, and the motor is thermally protected and nonreversible.

A Westinghouse diagram is given in Fig. 7-13 for a four-speed nonreversible motor. A General Electric diagram is given in Fig. 7-14 for

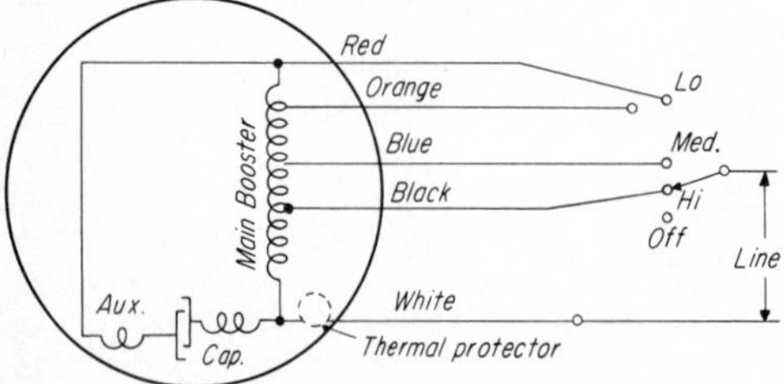

Fig. 7-13 Motor and line wiring diagram for a four-speed single-voltage nonreversible tapped-winding capacitor motor. *(Westinghouse Electric Corporation.)*

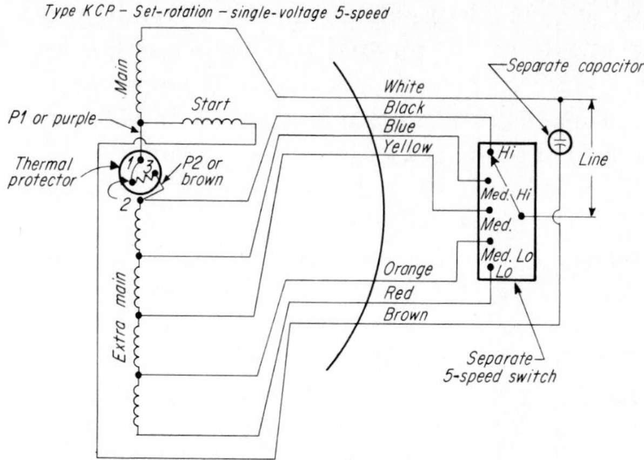

Fig. 7-14 Motor and line connection diagram for a five-speed single-voltage nonreversible (set-rotation) tapped-winding capacitor motor with thermal protection. *(General Electric Company.)*

a five-speed nonreversible (set rotation) motor. A Westinghouse diagram is given in Fig. 7-15 for a five-speed nonreversible motor.

It is obvious that the number of possible stator connection diagrams for tapped-winding capacitor motors is legion—certainly too many to include here. For two-speed motors, the stator connection diagrams could be essentially the same as for dual-voltage capacitor-start motors, as shown in Fig. 5-14, for example, except for tagging of the leads. When more than two speeds are needed, it generally is not feasible or

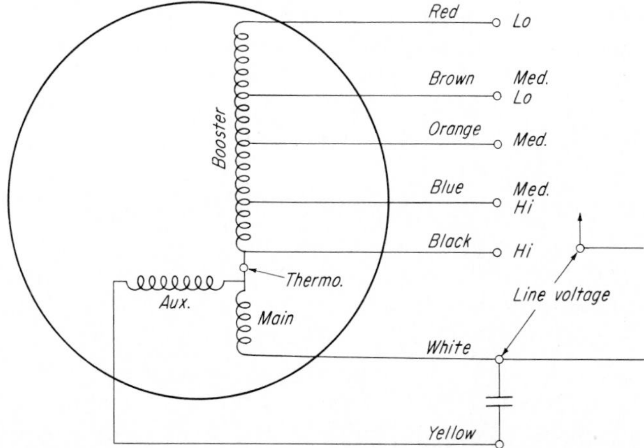

Fig. 7-15 Motor and line connection diagram for a five-speed single-voltage nonreversible tapped-winding capacitor motor with thermal protection. *(Westinghouse Electric Corporation.)*

practicable to use several layers. Figure 7-16 shows the stator connection diagram for a six-pole five-speed motor, by way of illustration; it is for the motor of Fig. 7-15. Note that this stator connection diagram shows two booster (intermediate) windings, both of which are tapped at an intermediate point, not at the center of either winding.

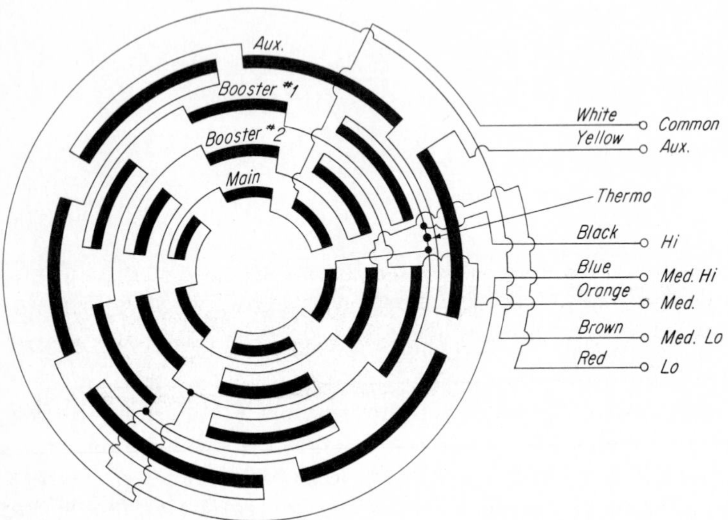

Fig. 7-16 Stator connection diagram for a six-pole five-speed motor wired and connected as in Fig. 7-15. Rotation is counterclockwise. *(Westinghouse Electric Corporation.)*

7-16 Special Mechanical Features of Fan Motors. Motors used for propeller fan applications almost invariably have the fan draw air over the motor, although in some condenser fan applications the air flow is reversed; that is, it is blown over the motor. In instances where the environment is such as to draw dirt, grease, or other foreign matter through the motor, such motors are often totally enclosed. Outdoor applications are often required by the Underwriters Laboratories Inc. (UL)[3] to pass a rain test, which may necessitate a totally enclosed or drip-proof motor; or, if the motor is mounted vertically with the shaft up, it may require a rain shield on the shaft to prevent water from running down the shaft and into the bearing.

Since the fan is mounted on the shaft, provision to accommodate the axial thrust of the fan must be made. In early fan motors the use of thrust spring washers, steel balls, and other devices predominated. In recent years manufacturers have gone to lubricated thrust plates, such as those detailed in Fig. 17-1, and use schemes similar to those described in Art. 17-1 to attain all-angle operation.

Capacitors may be mounted on the fan motor frame, although today most equipment manufacturers supply the capacitor mounted remotely from the motor. This allows for a bellyband to be wrapped around the motor for assembly in the unit. Through bolts may extend out the rear of the motor; these may be used for either mounting the motor itself or attaching a fan guard.

7-17 Direct-Drive Blower Motors. Shaft-mounted blowers are often used in heating and air-conditioning applications to move air through a conditioned space. A gas-fired or electric furnace, with an evaporator coil for air conditioning, may employ a four- or five-speed motor. During the air-conditioning cycle the motor is operated on one of the two highest-speed connections to give high air velocity. During the heating cycle the motor is operated on one of the low-speed connections, the speed selected depending on the furnace rating, back pressure in the duct system, and heated air temperature rise. Normally, the speeds are selected after installation of the appliance is complete. These motors are usually the open construction type. Since the blower is of the centrifugal type, there is little axial thrust and, therefore, no special thrust arrangement is necessary.

7-18 Ceiling Fan Motors. Although large-diameter blade ceiling fans have existed since the turn of the century, they gradually lost popularity; the energy shortage in the 1970s played a significant role in their reemergence. These fans are now widely used in homes as well as in places of business. In summer they conserve energy because, when properly used, they can substantially reduce the amount of air-conditioning needed. In wintertime some homeowners have found that circulating the warm air near the ceiling downward somewhat reduces the amount of heating needed. (For such an application the motor is usually reversed to prevent the fan from blowing drafts onto the occupants of the room.)

These fans are driven by multipolar ball-bearing motors designed to operate at about 300 rpm or less. Many of these motors are wound as 18-pole motors, using a 36-slot stator. The poles are single-coil poles and are often skein-wound, as described in Art. 3-11.

Ceiling fans often have a means of attaching a light fixture or control switch beneath them. This requires the motor to have a hollow shaft with a stationary tube or wireway extending through the shaft for leads to supply power to the light fixture. Figure 7-17 shows a stationary-shaft design that allows for additional leads for speed control as well as a light fixture. The fan hub extends into the motor and is supported by the ball bearings; the rotor core is pressed onto the fan hub, and the

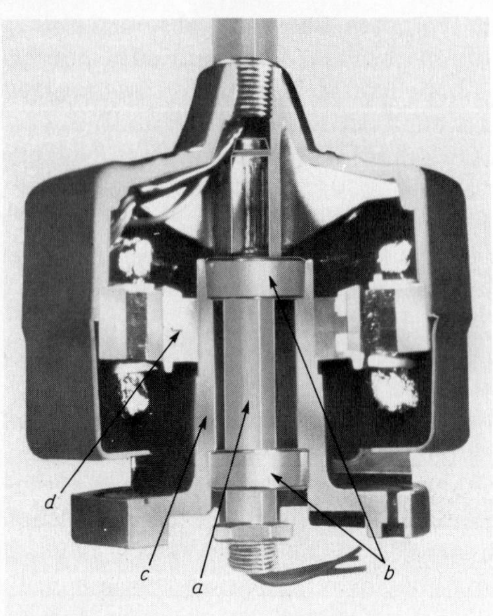

Fig. 7-17 Cutaway view of a ceiling-fan motor. (*a*) Stationary shaft; (*b*) ball bearings; (*c*) die-cast fan hub; (*d*) rotor core. *(Emerson Motor Division.)*

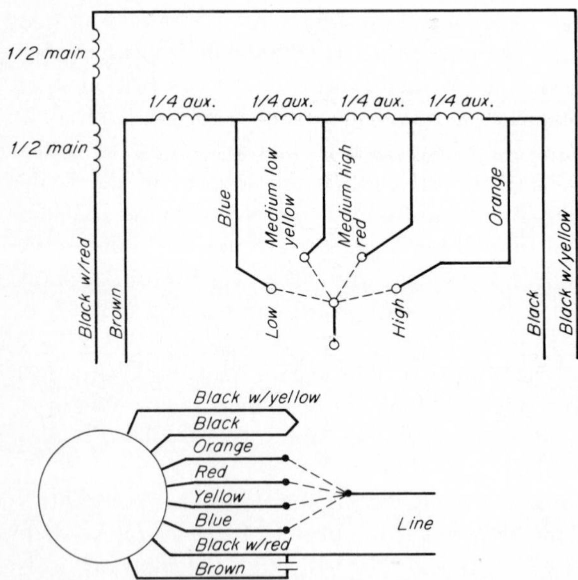

Fig. 7-18 Wiring and line connection diagrams for a reversible four-speed permanent-split capacitor tapped-auxiliary motor. *(Emerson Motor Division.)*

entire assembly rotates about the stationary shaft, which is hollow. A box for attaching a light fixture is connected to the bottom of the shaft. The motor leads are routed through the upper pipe to the power supply.

In Fig. 7-18 is illustrated a schematic diagram of a permanent-split tapped-auxiliary winding for a four-speed ceiling fan motor wound for 18 poles in a 36-slot stator. The stator connection diagram is shown in Fig. 7-19. The windings are calculated for a single-speed motor; the main winding is divided into two equal sections, and the auxiliary winding into four equal sections. Each winding section consists of nine single coils, as shown in Fig. 7-19, each wound over two teeth. The auxiliary winding is tapped between each section. The main winding has full voltage impressed across it only on the high-speed connection. As the connection is changed from high to a lower speed, the portion of the auxiliary winding not in the auxiliary phase is connected in series with the main winding, thereby reducing the main phase voltage per turn. The auxiliary phase consists of the capacitor plus one to four sections of the auxiliary winding, depending upon the speed setting.

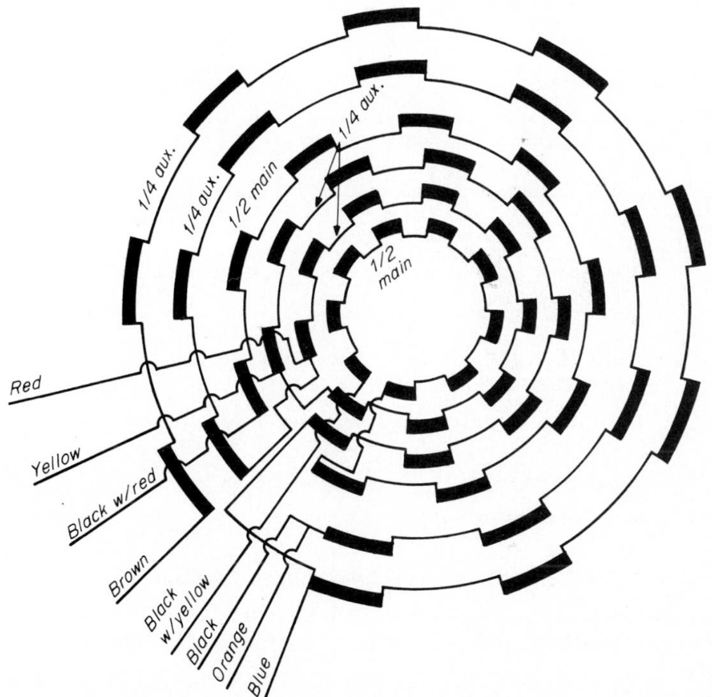

Fig. 7-19 Stator connection diagram for an 18-pole four-speed reversible permanent-split capacitor motor connected as in Fig. 7-18. *(Emerson Motor Division.)*

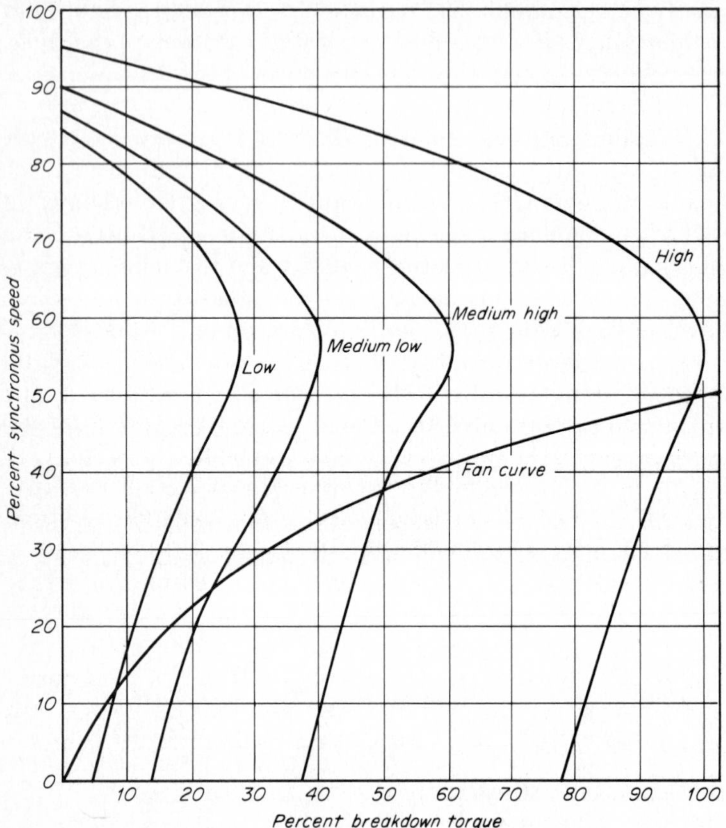

Fig. 7-20 Speed-torque curve of a 1/80-hp four-speed ceiling fan motor connected as in Fig. 7-19. *(Emerson Motor Division.)*

The speed-torque curve for a four-speed motor connected as in Fig. 7-18 is shown in Fig. 7-20. The fan curve is representative of a 52-in ceiling fan. The fan motor operates at 200, 150, 100, and 50 rpm, with input watts of 105, 77, 52, and 38 W, respectively. The slope of the speed-torque curves indicates a high rotor resistance, with resultant high slip and high locked-rotor torque.

7-19 Reversing, High-Torque, Intermittent-Rated Motors. The permanent-split capacitor motor lends itself ideally to high-torque reversing duty, intermittently rated. Essentially such motors have a two-phase winding or equivalent and are designed for high locked-rotor torque. The speed-torque characteristic is similar to the starting connection curve of Fig. 6-3, except that the breakdown torque occurs down closer to standstill. Such a motor is usually connected as shown in Fig. 7-3. It can be reversed instantly and requires only a single-pole double-throw switch; neither winding is energized in the off position of this switch. The control could hardly be simpler!

For applications requiring different torques in the two directions of rotation, these motors are designed with windings having more turns in one phase than in the other.

Typical applications for permanent-split high-torque intermittent-duty motors are garage-door openers, hospital beds, and dental chairs. For many applications, an integral reduction gear is furnished.

7-20 Shifted Auxiliary Winding. The two windings of a permanent-split motor are normally spaced 90 electrical degrees apart. If, however, the auxiliary winding is shifted from its normal position in the direction opposite to rotation of the rotor (i.e., more than 90°), the locked-rotor torque is increased and the current at the operating load is decreased. It is possible to decrease the number of turns in the auxiliary winding and obtain the same performance as with the conventionally spaced windings. However, caution must be exercised because of slot fill due to the additional shared slots of the main and auxiliary windings.

7-21 Rewinding or Reconnecting. In general, rewinding or reconnecting, except to duplicate the original windings, is discouraged unless the specifications for the new winding are obtained from the motor manufacturer.

7-22 Changing Capacitor. Continuous-duty motors are normally designed for optimum running performance, whereas intermittent-duty motors are usually designed for optimum locked-rotor torque. The design includes the auxiliary phase winding and capacitor combination. One objective is to work the capacitor at its rated voltage. If a capacitor is open-circuited, or otherwise faulty, it should be replaced by one of the same microfarad rating and of the same or higher voltage rating. If the load on a permanent-split motor is reduced, the voltage across the capacitor increases; if the load is removed, the voltage increases, in many cases quite dramatically, and could exceed 150 percent of the capacitor voltage rating.

SPLIT-PHASE CAPACITOR MOTORS

7-23 Split-Phase Capacitor Motor Defined. This motor is defined in Art. 5-1 as a capacitor motor that uses the auxiliary winding during starting and running but switches a capacitor into the auxiliary circuit during running only; that is, it starts as a split-phase motor and runs as a permanent-split capacitor motor.

7-24 Essential Parts of a Split-Phase Capacitor Motor. A split-phase capacitor motor is represented schematically in Fig. 7-21. It should be noted that the arrangement of windings and connections is the same as that represented in Fig. 4-1, except that the switch is a single-pole double-throw type and a capacitor has been added. Hence, the stator windings are the same in form and arrangement as those of a split-phase motor. That is, there are two windings, spaced 90 electrical degrees apart. When the motor is up to speed and the capacitor element is connected in the auxiliary phase, the arrangements of windings and connections are the same as those of the permanent-split capacitor motor in Fig. 7-1.

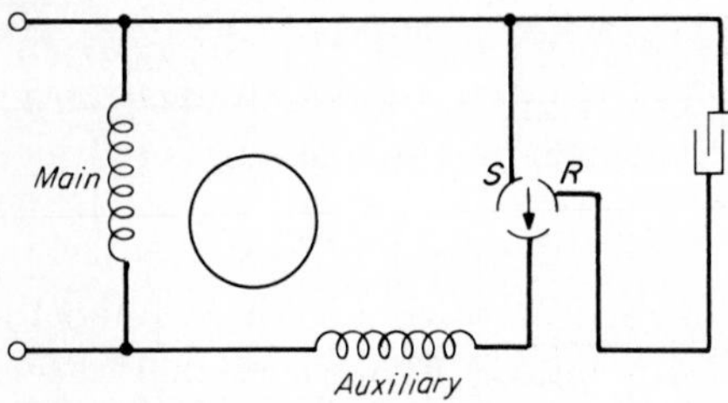

Fig. 7-21 Schematic diagram of a split-phase capacitor motor with a single-section auxiliary winding. *(Emerson Motor Division.)*

A disadvantage of the motor shown in Fig. 7-21 is that in order to obtain locked-rotor torque similar to a split-phase motor, the auxiliary winding is normally wound with fewer turns than the main winding. This results in relatively low voltage across the capacitor during running, with a resultant lower effect than with full voltage. The arrangement in Fig. 7-22, which embodies a dual-section auxiliary winding, is

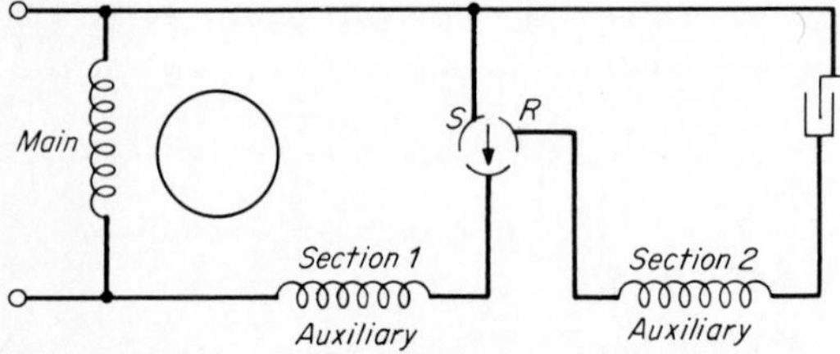

Fig. 7-22 Schematic diagram of a split-phase capacitor motor with a dual-section auxiliary winding. *(Emerson Motor Division.)*

the same as that in Fig. 6-5, except that there is no starting capacitor. By proper design of the turns in section 2, the voltage on the capacitor is increased to its normal operating value, and the same auxiliary-winding current has more effect because of the added turns in section 2. Thus, the arrangement of Fig. 7-22 enables the designer to obtain more of the beneficial effects of a permanent-split motor, as discussed at length in Art. 6-3. Most split-phase capacitor motors are supplied with a dual-section auxiliary winding for reasons given previously.

References

1. Veinott, C. G.: *Theory and Design of Small Induction Motors,* McGraw-Hill Book Company, New York, 1959, Art. 8-9, p. 163. (Out of print, but available from University Microfilms, P.O. Box 1647, Ann Arbor, Mich. 48106.)
2. ————: *Fractional Horsepower Electric Motors,* 2d ed., 1948, *Fractional- and Subfractional-Horsepower Electric Motors,* 3d ed., 1970, McGraw-Hill Book Company, New York.
3. *Room Air Conditioners,* UL 484, Underwriters Laboratories Inc., 333 Pfingsten Rd., Northbrook, IL 60062, 1982.

Bibliography

Trickey, P. H.: "Performance Calculations on Tapped-Winding Capacitor Motors," *AIEE Trans.,* vol. 62, 1943, pp. 1–3.

Veinott, C. G.: "Performance Calculations on L- and T-Connected Capacitor Motors," *IEEE Trans.,* vol. PAS-96, no. 4, July–August 1977.

CHAPTER 8

CAPACITORS

Capacitor motors employ two basic types of capacitors: an electrolytic type for intermittent duty for motor starting and a paper-spaced oil-filled type or a metallized polypropylene film type for continuous duty for capacitor-run motors.

AC ELECTROLYTIC CAPACITORS

AC electrolytic capacitors were used for motor-starting service as far back as 1892, but it is only since 1930 that capacitor-start motors have attained commercial importance. It is interesting to note that in 1896 the repulsion-start motor wiped the capacitor-start motor out of existence. Now, some decades later, the capacitor-start motor has made the repulsion-start motor obsolete in fractional horsepower ratings.

8-1 Construction. The modern dry-type electrolytic capacitor is formed by rolling two sheets of high-purity aluminum foil (electrodes) into a cylindrical shape. The electrodes are etched (roughened) chemically to increase their surface area and thereby the capacitance by a factor of 2 or 2½ times. An anodic film is produced on the surface of the etched electrode by an electrochemical process. The two foil electrodes are separated by a common electrolyte-impregnated electrode, usually two or more layers of thin paper. Ethylene glycol, or a derivative, is often used as the electrolyte. The anodic film is the dielectric of the capacitor and is also the insulation between the aluminum electrodes and the common (paper) electrode.

The capacitor is provided with suitable terminals and placed in a case, usually of phenolic material. A potting compound, such as tar, is poured into the space between the capacitor and its case to reduce vibration of the capacitor. The open end is closed by a phenolic plug that also supports the terminals. A vent is provided to prevent explosion. A section view of a typical element with a phenolic case is shown in Fig. 8-1.

Formerly, the enclosing can was made from aluminum. Although the active element inside the can was insulated, good engineering practice demanded that the aluminum case be insulated electrically from the motor frame, from ground, and from any live parts.

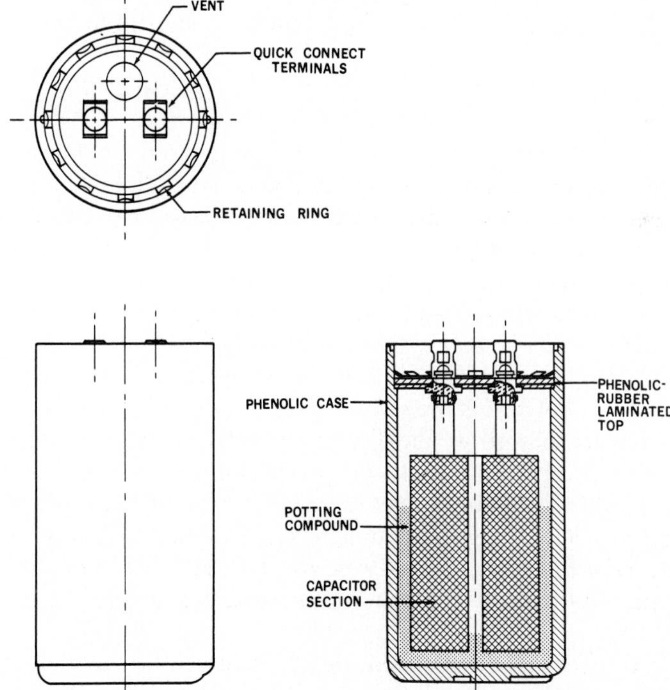

Fig. 8-1 An ac dry electrolytic capacitor for motor-starting service. It has a phenolic case and quick-connect terminals. *(Sprague Electric Company.)*

8-2 Characteristics of Motor-Starting Capacitors. These capacitors are designed for use on alternating current and for intermittent service only. They must not be confused with the dc electrolytic capacitors used in radios and other electronic devices. For the same voltage and microfarad rating, ac capacitors are larger and bulkier, and the terminals do not bear any polarity markings. Present commercially available ac electrolytic capacitors have characteristics as follows:

(1) Voltage rating. The voltage rating is usually stamped on the capacitor itself. This may or may not be the same as the voltage rating of the motor itself. Both 115-V and 115/230-V capacitor-start motors commonly use 110-V capacitors. Also, 230-V motors often use 110-V capacitors when the windings are suitably arranged (see Art. 5-18). Sometimes the capacitor carries a higher voltage rating than the motor; for example, 125-V capacitors may be used on 115-V motors. The proper voltage rating is determined more by the design and arrangement of the windings than by the voltage rating of the motor. Good electrolytic capacitors will withstand, within the limits of the prescribed duty cycle, 125 to 130 percent of the rated voltage. An increase in operating voltage increases the power factor and slightly increases the capacitance.

(2) Capacitance rating. Electrolytic capacitors are usually stamped with a range of capacitance values. Standard capacitance ratings, together with case size and minimum and maximum current values, are given in Table 8-1. These current values represent what the capacitor should draw at rated voltage at a frequency of 60 Hz and can be used for a routine check of capacitance.

(3) Temperature rating. Electrical characteristics of capacitors are normally tested at 25°C, but they are rated to operate in ambient temperatures up to 65°C (150°F) and will function successfully in ambients up to 80°C (176°F). However, their life is shorter at elevated temperatures. What really affects their life is internal temperature, which is a function of duty cycle as well as ambient temperature: 95°C is safe if not held continuously. Operation at very low temperatures does not harm the capacitor, but at temperatures lower than 0°C the capacitance falls off, and at −50°C it may drop as low as 50 percent of normal. However, at these reduced temperatures, the power factor becomes quite high so that the capacitors draw more watts from the line and warm up quickly.

(4) Duty cycle. Motor-starting capacitors are rated on the basis of 20 three-second periods per hour, or an equivalent duty cycle; 60 one-second periods per hour would be one equivalent duty cycle.

The 20 three-second duty-cycle rating is based on the use of a normal (type 2) electrolytic capacitor. Manufacturers also furnish a type 1 elec-

TABLE 8-1 Ratings, Case Sizes, and Test Limits for Normal Performance AC Electrolytic Capacitors*

Range, mfd	110 V (140 max.)			125 V (160 max.)			165 V (210 max.)			220 V (280 max.)		
	Case size	Amperes Min.	Amperes Max.	Case size	Amperes Min.	Amperes Max.	Case size	Amperes Min.	Amperes Max.	Case size	Amperes Min.	Amperes Max.
21–25	0, 1	0.87	1.04	0, 1	0.99	1.18	0, 1	1.31	1.56	0, 1	1.74	2.07
25–30	0, 1	1.04	1.25	0, 1	1.18	1.42	0, 1	1.56	1.87	0, 1	2.07	2.49
30–36	0, 1	1.25	1.50	0, 1	1.42	1.70	0, 1	1.87	2.24	0, 1	2.49	2.98
36–43	0, 1	1.50	1.79	0, 1	1.70	2.03	0, 1	2.24	2.68	1	2.98	3.57
43–52	0, 1	1.79	2.16	0, 1	2.03	2.45	0, 1	2.68	3.24	1, 2	3.57	4.32
47–56	0, 1	1.95	2.32	0, 1	2.22	2.64	0, 1	2.93	3.49	1, 2	3.90	4.65
53–64	0, 1	2.20	2.65	0, 1	2.50	3.02	0, 1	3.30	3.99	1, 2	4.40	5.31
64–77	0, 1	2.65	3.20	0, 1	3.02	3.63	1	3.99	4.79	2, 3, 4	5.31	6.39
72–87	0, 1	2.99	3.61	0, 1	3.29	4.10	1	4.48	5.42	3, 4	5.97	7.22
88–106	0, 1	3.65	4.40	1	4.15	5.0	1, 2	5.49	6.60	3, 4	7.30	8.80
108–130	1	4.48	5.40	1	5.10	6.13	2	6.74	8.10	4	8.95	10.8
124–149	1	5.15	6.18	1	5.85	7.03	2	7.73	9.28	5	10.3	12.35
130–154	1	5.40	6.39	1	6.13	7.26	2, 3	8.1	9.59	5	10.8	12.8
145–175	1	6.02	7.26	1	6.85	8.25	2, 3	9.05	10.9	5, 6, 7	12.0	14.5
161–193	1	6.67	8.01	1	7.60	9.10	2, 3	10.05	12.0	5, 6, 7	13.4	16.0
189–227	1	7.85	9.42	2	8.92	10.7	2, 3	11.8	14.1	7	15.7	18.8
216–260	2	8.96	10.8	2	10.2	12.25	3	13.45	16.2	7, 8	17.9	21.6
233–280	2	9.67	11.6	2	11.0	13.2	3, 4	14.55	17.4	7, 8	19.3	23.2
243–292	2, 3	10.1	12.1	2, 3	11.45	13.8	3, 4	15.15	18.2	7, 8	20.2	24.2
270–324	2, 3	11.2	13.45	2, 3	12.75	15.3	3, 4	16.85	20.2	8	22.4	26.9
324–389	2, 3	13.45	16.15	2, 3	15.3	18.35	3, 4	20.2	24.2	8	26.9	32.3
340–408	3, 4	14.1	16.9	3, 4	16.0	19.25	4, 5	21.2	25.4	8	28.2	33.85
378–455	3, 4	15.65	18.9	3, 4	17.85	21.45	5	23.7	28.3	8	31.4	37.8
400–480	3, 4	16.6	19.9	3, 4	18.9	22.6	5, 6, 7	25.0	29.9	8	33.2	39.8
430–516	3, 4	17.85	21.4	3, 4	20.3	24.3	5, 6, 7	26.8	32.1	8	35.7	42.8
460–553	3, 4	19.1	22.95	3, 4	21.7	26.1	5, 6, 7, 8	28.7	34.4	8	38.2	45.9
540–648	5	22.4	26.9	4, 5	25.5	30.55	7, 8	33.6	40.3			
590–708	5	24.5	29.4	5, 6	27.8	33.4	7, 8	36.7	44.1			

Note: Voltages in this table are all root-mean-square values at 60 Hz.
*Data from Mallory Capacitor Company.

trolytic capacitor rated for extra duty for applications in which the on time is appreciably longer or for intermittent-duty applications, such as dental chairs, beds, and door operators, to replace oil-filled types of capacitors.

(5) Power factor. At normal room temperatures, the power factor of electrolytic capacitors is of the order of 5 to 6 percent, but it may be as high as 8 percent for small units, and 9.5 percent for large ones. At low temperatures, it goes up markedly.

8-3 When Replacing Electrolytic Capacitors. *When a defective capacitor is replaced, it is imperative that the new capacitor be of the same voltage and microfarad rating.* A 124-mfd capacitor may sometimes be substituted for the 130-mfd size without apparently impairing the ability of the motor to start its load, particularly if the machine is broken in and requires less power to drive than it did when new. However, reducing the microfarads of the capacitor in any given motor normally increases the voltage across the capacitor, incurring some danger of an early breakdown. (It is perfectly possible to get 150 V across the capacitor with only 115 V on the motor. The inductance of the motor winding is in series with the capacitor; as the capacitance is reduced, series resonance* is approached, increasing the capacitor voltage.) Using a capacitor that is too large generally will not harm the capacitor, but the switching torque of the motor may be adversely affected. Therefore, the person servicing the motor should use the same value of capacitance that the manufacturer did. Electrolytic capacitors are all vented; for reasons of safety, this vent must never be sealed.

Capacitor testing boxes are sold for testing motors to determine, by means of a test on the motor, the size of capacitor to use. The locked-rotor torque is determined with different values of microfarads in series with the auxiliary winding, and the person providing service has often been advised to use that value of capacitance which gives the maximum locked-rotor torque. However, use of this amount of capacitance may cause a serious reduction in switching torque; hence the authors recommend always using fewer microfarads than are required for maximum locked-rotor torque; 80 percent of this value is probably a safer figure to use than 100 percent. A much safer procedure is to obtain the required *voltage and microfarad* rating from the motor manufacturer.

8-4 Testing Electrolytic Capacitors. When an electrolytic capacitor is suspected of being defective, it should be tested for

*See any elementary textbook for an explanation of the phenomenon of series resonance.

1. Short circuits
2. Open circuits
3. Capacitance in microfarads
4. Power factor
5. Grounds

The testing procedure to use is as follows: The capacitor should be connected in series with a suitable fuse across a 115-V 60-Hz line. (1) If the capacitor is *short-circuited,* the fuse will blow. (A lighting-out lamp cannot be used, for a capacitor will pass enough current to light an ordinary light bulb.) If the capacitor is not short-circuited, meters should be connected in the circuit as shown in Fig. 18-4. The voltage across the capacitor should be adjusted to the rated voltage of the capacitor. A reading of watts and amperes input and also applied voltage should be taken, as explained in Art. 18-12, making sure to correct for meter losses. (2) If no current can be measured, the capacitor is *open-circuited.* (3) If a readable value of current input is obtained, the *capacitance in microfarads* can be determined by comparing the reading with the currents given in Table 8-1. The tested value of microfarads should be checked against the rating marked on the can. Or the microfarads and power factor can be computed from the following formulas:

For 60 Hz:

$$\text{Microfarads} = \frac{2650 \times \text{amperes}}{\text{applied voltage}} \tag{8-1}$$

For 50 Hz:

$$\text{Microfarads} = \frac{3180 \times \text{amperes}}{\text{applied voltage}} \tag{8-1a}$$

$$\text{Power factor} = \frac{\text{watts}}{\text{volts} \times \text{amperes}} \tag{8-2}$$

(4) The power factor is computed from the test readings, using Eq. (8-2). It should not be greater than 10 percent and will generally be less if the capacitor is in good condition. (5) If the capacitor has a metal case, it should be tested to determine whether the internal element is grounded to the case; this ground test can be made at rated voltage. These readings should be taken as quickly as possible, for electrolytic capacitors are intermittently rated.

This method gives only an approximation of the power factor; most wattmeters are not accurate in the low range of power factor encountered in capacitors.

8-5 Causes of Failure. When an ac voltage is applied to the capacitor terminals, one anodic film becomes conductive during one-half cycle; the other becomes conductive during the other half cycle. The entire voltage stress appears across the film every half cycle. The current passing through the electrolyte causes considerable heating effect, which, if not controlled, will damage the anodic film.

(1) Excessive voltage. Voltage in excess of rated voltage causes localized heating which results in a puncture of the anodic film. The auxiliary phase current is then sufficient to cause a permanent short circuit.

(2) Excessive duty cycle. If the voltage is applied for a longer period than that recommended by the manufacturer, the heat generated will dry out the moisture content of the electrolyte, the paper will char, and the capacitor will cease to function. The capacitor will likely test as open.

(3) Excessive temperature. This causes an increase in the rate of evaporation of the moisture content of the electrolyte, causing the capacitor to dry up and fail. In addition, an excessive temperature lowers the puncture point of the anodic film, thereby increasing the possibility of voltage failure.

(4) Excessive vibration. The potting compound described in Art. 8-1 serves to restrict the radial and axial movement of the capacitor element. In some applications where the capacitor is subjected to excessive vibration, the capacitor should be mounted separately, or a high-vibration capacitor should be used. Normally, the failure due to excessive vibration results in rupturing of the ribbons (leads) between the capacitor element and the terminals.

(5) Regeneration. For a discussion of the causes and effects of regeneration the reader is referred to Art. 5-27.

8-6 Mounting Means. Most electrolytic motor starting capacitors are mounted on the top or side of the motor frame in a metal cover which serves to protect the plastic case from damage and also serves as an enclosure for the exposed terminals. Capacitor manufacturers have standardized case diameters; however, for a given rating the length may vary from one manufacturer to another. Table 8-2 gives the diameter and length of the case sizes referred to in Table 8-1.

When a capacitor is replaced, it is essential that sufficient electrical spacing be provided between the terminals and cover. Also, if the capacitor being replaced is quite old, wrapping corrugated paper around the case to prevent rattling may be necessary.

Capacitors may be mounted in an outlet box attached to the motor

TABLE 8-2 Capacitor Case Sizes, Diameter, and Length*

Case size	Diameter, in	Length, in
0	1 7/16	2
1	1 7/16	2 3/4
2	1 7/16	3 3/8
3	1 7/16	4 3/8
4	1 13/16	3 3/8
5	1 13/16	4 3/8
6	2 1/16	3 3/8
7	2 1/16	4 3/8
8	2 9/16	4 3/8

*Data from Mallory Capacitor Company.

frame, or inside the motor end shield. In some appliances the capacitor is mounted remote from the motor. Regardless of where the capacitor is located, it is necessary to take into account the ambient temperature around the capacitor.

AC FIXED-PAPER DIELECTRIC OIL CAPACITORS

8-7 Construction. Fixed-paper dielectric oil-filled capacitors, commonly called oil-filled capacitors, came into use in motors in the early 1930s. They are formed similarly to the dry electrolytic type, in that two or more sheets of paper are wound between aluminum foil electrodes. However, unlike in the electrolytic types, there is no anodic film on the aluminum electrodes. The number of sheets of paper varies with the voltage and temperature requirements. Tinned copper tabs are wound with the aluminum foil to make internal electrical connections. A typical element is shown in Fig. 8-2. The entire core is impregnated with oil, attached to terminals, and sealed in a metallic case.

Early designs had the core wound flat and placed into a flat can which appeared to be, and gave cause to be called, a "sardine can." Later, the cores were wound in a cylinder and placed into a round can with a cover attached to the can. Most often these cans had two leads instead of terminals and often had a bracket attached for mounting on the motor frame or end shield. The modern design is oval in shape, with standard case dimensions; the most common case for use with motors is a 1¼-in flat oval can.

8-8 Characteristics of Motor-Run Capacitors. These capacitors are designed for use on alternating current and for continuous service. The

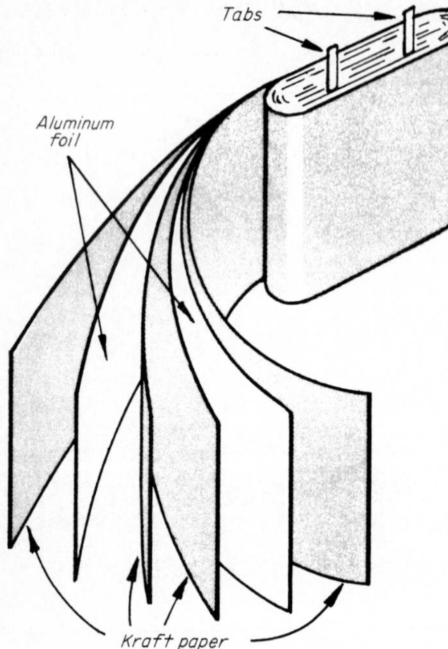

Fig. 8-2 Elements of a fixed-paper dielectric capacitor. The Kraft paper is the dielectric between the aluminum foil electrodes. Tinned copper tabs are wound with the aluminum to make internal connections. *(Aerovox Incorporated.)*

microfarad rating is much lower and the voltage rating much higher than an electrolytic capacitor of the same dimensions.

(1) Voltage rating. The voltage rating is stamped on the case. The most common rating for use with motors is 370 V, although 236 and 440 V are also used. The voltage rating is often two to three times or more the voltage rating of the motor with which it is used.

(2) Capacitance rating. The capacitance in microfarads (mfd) is stamped on the case as a single value. Table 8-3 gives the more common ratings used with motors. Only the 236-, 370-, and 440-V ratings are given. Sardine can style capacitors were almost always rated 1 or 2 mfd, and additional capacitance was accomplished by connecting a number in parallel. The standard tolerance of capacitance is plus or minus 10 percent.

(3) Temperature rating. The electrical characteristics are normally tested at 25°C, but the capacitance value remains fairly constant from −50 to +70°C case temperature. At 70°C there is only a 5 percent reduction of capacitance.

(4) Duty cycle. Oil-filled capacitors are rated for continuous duty at their rated voltage.

(5) Power factor. The power factor of oil-filled capacitors is normally about 1 to 2 percent.

TABLE 8-3 Ratings, Case Sizes, and Test Limits for AC Fixed-Paper Dielectric Oil Capacitors with Pressure Interrupters*

Rating, mfd (± 10%)	236 V				370 V				440 V			
	Amperes		Length, in		Amperes		Length, in		Amperes		Length, in	
	Min.	Max.	Case 32†	Case 37†	Min.	Max.	Case 32†	Case 37†	Min.	Max.	Case 32†	Case 37†
2	0.16	0.19			0.25	0.31	1.69		0.30	0.36	2.0	
3	0.24	0.29	1.69		0.37	0.46	2.0		0.45	0.55	2.5	
4	0.32	0.39	2.0		0.50	0.62	2.38		0.60	0.73	3.13	
5	0.40	0.49	2.13		0.63	0.77	2.75		0.75	0.91	3.63	
6	0.48	0.59	2.5		0.75	0.92	3.13		0.90	1.10	4.25	
7.5	0.60	0.73	2.88		0.94	1.15	3.75		1.12	1.37	5.0	3.13
8	0.64	0.78	3.0		1.01	1.23	3.88		1.20	1.46		3.25
10	0.81	0.98	3.5		1.25	1.53	4.63	2.88	1.5	1.82		3.88
12.5	1.0	1.22	4.25	2.75	1.57	1.91	5.63	3.38	1.86	2.28		4.63
15	1.2	1.47		3.13	1.88	2.3		3.88	2.24	2.74		5.38
17.5	1.4	1.72		3.5	2.2	2.68		4.38	2.61	3.2		6.0
20	1.6	1.96		3.88	2.51	3.07		4.88	3.0	3.65		6.75

Note: Voltages in this table are all mean-square-root values at 60 Hz.

*Data from Mallory Capacitor Company.

† Case 32 is 1 5/16 × 2 5/32 in, flat oval; case 37 is 1 29/32 × 2 29/32 in, flat oval. Add 1/2 in for terminals to obtain overall length.

8-9 When Replacing Oil-Filled Capacitors. When a defective capacitor is replaced, it is imperative that the new capacitor be of the same microfarad rating and of an equal or higher voltage rating. Many permanent-split capacitor motors built in the 1950s used capacitors rated 236 V; the common value increased to 330 V in the early 1960s, then to 370 V in the mid-1960s, where it remains today.

A common electrolyte Askeral, a polychlorinated byphenyl (PCB), was used until about 1975. This material was found to be nonbiodegradable and hazardous. The capacitor industry changed to other types of oil; none of the present-day capacitors contains PCB, and they are marked "Non PCB." However, the new types of oil were found to be flammable under certain conditions, as described in Art. 8-10.

8-10 Pressure-Sensitive Interrupter. When an internal short develops in a capacitor, there is local heating which increases the internal pressure. When sufficient pressure is attained, the terminal end of the case bulges out and ruptures the seam between the case and end plate. The internal pressure expels the oil through the rupture, and if the internal short ignites the oil, a fire hazard exists. Underwriters Laboratories (UL) and the capacitor industry developed requirements for internal current interrupters to open-circuit the capacitor prior to the case-rupture point. The cross section of a capacitor with a typical pres-

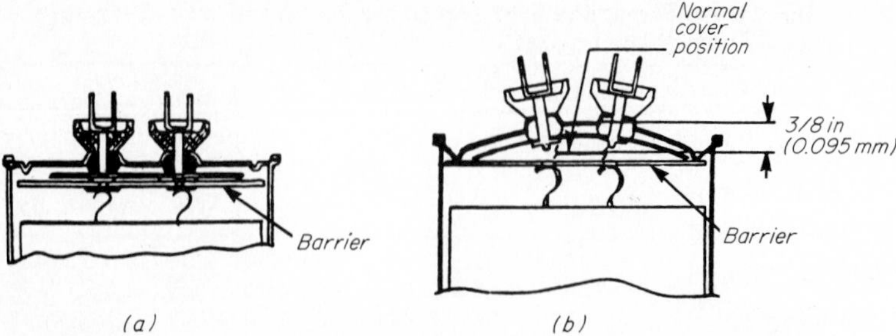

Fig. 8-3 Operation of a pressure-activated current interrupter. (*a*) The normal position of pressure-activated current interrupter with current flowing; (*b*) activated position of pressure-activated current interrupter in which excessive internal pressure causes the cover to bulge outward, thereby interrupting the current flow, to prevent rupture of the case. *(Cornell-Dubilier Electronics.)*

sure-sensitive interrupter is shown in Fig. 8-3. A capacitor that contains an interrupter is marked "protected" and includes the level of fault current protection, such as 5000 amperes fault current (AFC).

When a defective capacitor is replaced, the new capacitor should be one that has an interrupter and 5000-AFC protection. In addition, to assure operation of the interrupter and electrical spacings, at least ⅝-in clearance is required between the terminals and dead metal. Furthermore, extreme caution should be exercised in disposing of the defective capacitor containing PCB.

8-11 Polypropylene Dielectric. Capacitor manufacturers have developed oil-filled capacitors with a polypropylene dielectric. The electrodes are either paper with vacuum-deposited metal or a combination of vacuum-deposited metal on paper for one electrode and aluminum foil for the other. Figure 8-4 shows the construction of a typical polypropylene-film dielectric capacitor with both electrodes of metal deposited on paper. Figure 8-5 shows the construction of another design that consists of two sheets of polypropylene film, each with a thin layer of vacuum-deposited aluminum on one side. The entire edge of each electrode is bonded by end spray that has a high current-carrying capability. The end spray serves as the connection point for the internal electric tabs.

Polypropylene capacitors have a lower power factor than do fixed-paper oil-filled types, and, therefore, they have lower losses. Also, polypropylene capacitors have lower leakage current and are more suited for applications, such as hospital equipment, in which leakage current can be hazardous.

Dry-film polypropylene capacitors with even lower losses and leak-

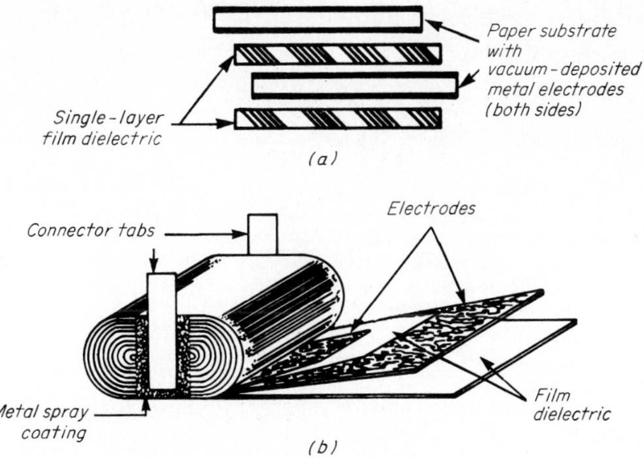

Fig. 8-4 A metallized capacitor. (*a*) Position of elements: paper with vacuum-deposited metal on both sides to form the electrodes, which are separated by a polypropylene film dielectric; (*b*) a formed capacitor with internal leads attached to end spray. *(Cornell-Dubilier Electronics.)*

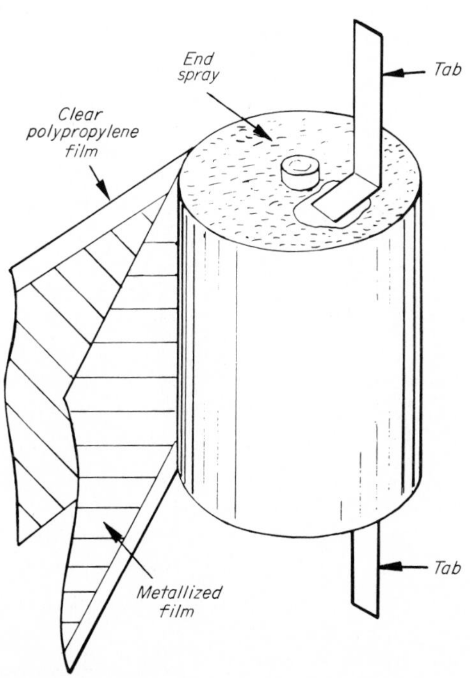

Fig. 8-5 Capacitor design of two sheets of polypropylene film each with a thin layer of vacuum-deposited aluminum on one side. Each aluminum layer serves as an electrode, separated by the polypropylene-film dielectric. The edges are bonded by a current-carrying end spray which also connects the electrical tabs. *(Aerovox Incorporated.)*

age currents have been developed by manufacturers. These have the added advantage of being smaller in size than previous types.

Bibliography

Deely, Paul McKnight: *Electrolytic Capacitors*, Cornell-Dubilier, New Bedford, Mass., 1938.

Electronic Industries Association: *Fixed Aluminum Electrolytic Capacitors for Alternating Current Motor Starting, Heavy Duty (Type 1) and for Light Duty (Type 2)*, Standard RS-463, EIA, 2001 I Street, NW, Washington, DC 20006, 1979.

Electronic Industries Association: *Fixed Paper and Film-Paper Dielectric Capacitors with Non-PCB Impregnants for Alternating Current Applications*, Standard RS-454, EIA, 2001 I Street, NW, Washington, DC 20006, 1978.

Electronic Industries Association: *Metallized Film Dielectric Capacitors for Alternating Current Application*, Standard RS-456, EIA, 2001 I Street, NW, Washington, DC 20006, 1978.

Georgiev, Alexander M.: *The Electrolytic Capacitor*, Murray Hill Books, Inc., Technical Division, New York, 1945.

Peck, G. V.: *Electrolytic Motor Starting Capacitors*, 5th ed., Mallory Capacitor Company, Indianapolis, 1976.

CHAPTER

9

SHADED-POLE MOTORS

Shaded-pole induction motors are used in a wide variety of applications requiring an induction motor of ¼ hp or less, even down to less than 1 mhp. In the subfractional range—horsepower ratings below 1/20—it is the standard general-purpose constant-speed ac motor. It is simple in construction, low in cost, and extremely rugged and reliable—like a polyphase induction motor—because it needs no commutator, switch, collector rings, brushes, governor, or contacts of any sort. Its torque characteristics and applications are similar to those of a permanent-split capacitor motor, except that it has a lower efficiency and a lower power factor. But efficiency and power factor are seldom of importance in these sizes. It is used in an extremely wide variety of applications: home appliances such as rotisseries, fans of all kinds, humidifiers, slide projectors; small business machines such as photocopy machines; vending machines and advertising displays. They are available with integral gear reducers to obtain almost any speed, even down to less than one revolution per month. Some manufacturers offer integral clutches or brakes. Although the simple shaded-pole motor is inherently nonreversible, reversible motors are built in different forms. Also, through special constructional arrangements, their use has been extended to various servosystem applications, as well as to use as permanent-split capacitor motors.

BASIC PRINCIPLES OF OPERATION

9-1 Shaded-Pole Motor Defined. A shaded-pole motor may be defined as a single-phase induction motor provided with an auxiliary short-circuited winding or windings displaced in magnetic position from the main winding.

Such a definition does not recognize synchronous shaded-pole motors, which are discussed in Chap. 10; the latter are widely used in clocks and timing devices. Only induction types are discussed in this chapter. In the usual form of construction, salient poles are used, and the auxiliary short-circuited winding consists of a single turn placed around a portion of the main pole. This coil is known as a shading coil because it causes the flux in that portion of the pole surrounded by it to lag behind the flux in the rest of the pole. Sometimes two or three shading coils are used on each pole, each coil surrounding a different percentage of the main-pole face.

9-2 Essential Parts of a Shaded-Pole Motor. Figure 9-1 gives a schematic representation of a simple shaded-pole motor. There is but a single winding, which is connected to the line, a second winding permanently short-circuited upon itself, and a squirrel-cage rotor. This short-circuited winding is displaced from the main winding by an angle which can never be as much as 90 electrical degrees. It has to be shifted from the axis of the main winding by a definite amount in order to set up a component field along an axis in space different from that of the main winding; furthermore, this shift has to be less than 90° so that a voltage can be induced in the short-circuited winding by transformer action of the main winding. Constructions may vary considerably in detail, but the preceding are the essential elements found in all shaded-pole induction motors. Additional elements are found in some shaded-

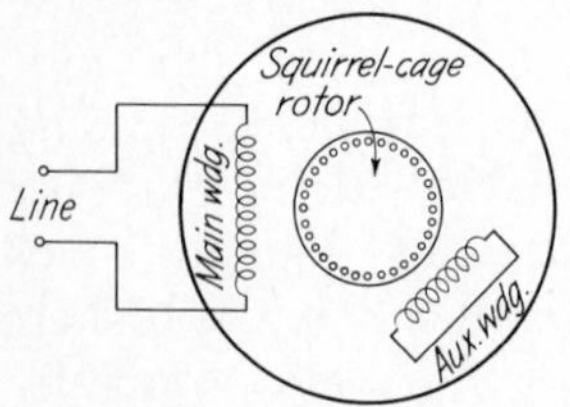

Fig. 9-1 Shaded-pole induction motor.

pole motors, developed for special purposes; some of these will be discussed later in this chapter. Figure 9-2 shows three stages in the manufacture of a four-pole shaded-pole stator. Figure 9-2*a* shows the stator core with shading coils in place, Fig. 9-2*b* the stator core with molded ground insulation and integral coil supports, and Fig. 9-2*c* the fully wound stator.

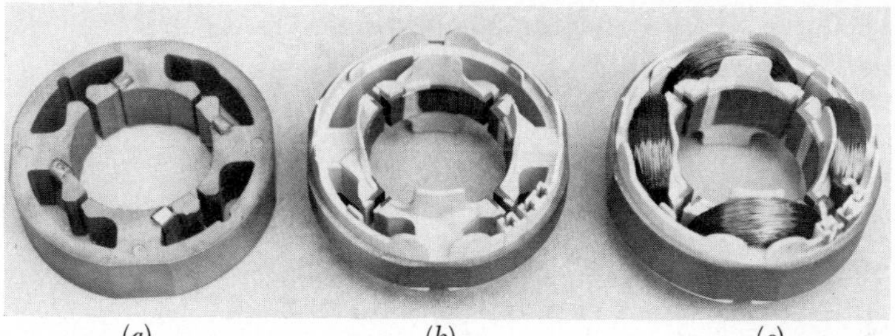

(*a*) (*b*) (*c*)

Fig. 9-2 Three stages in the manufacture of a four-pole shaded-pole stator. (*a*) Stator core with shading coils in place; (*b*) stator core with molded ground insulation and integral coil supports; (*c*) fully wound stator. *(Emerson Motor Division.)*

9-3 Principle of the Shaded-Pole Motor. Like any other induction motor, the shaded-pole motor is caused to run by the action of a revolving magnetic field set up by the primary windings. The method of obtaining the revolving field is slightly different from that of other types of induction motors. It was seen in Chap. 2 that the revolving field in a two-phase motor is the resultant of two stationary fields displaced 90° apart in space and 90° apart in time. In a split-phase motor at standstill, the fields are 90° apart in space, but they are displaced considerably less than that in time, as shown in Fig. 4-2; however, there is enough of a rotating field to produce considerable torque. Actually, it is not necessary that the fields be displaced in space by a full 90° in order to set up a sort of a rotating field; it is necessary only that there be *two* component fields, displaced both in time and in space. This con-

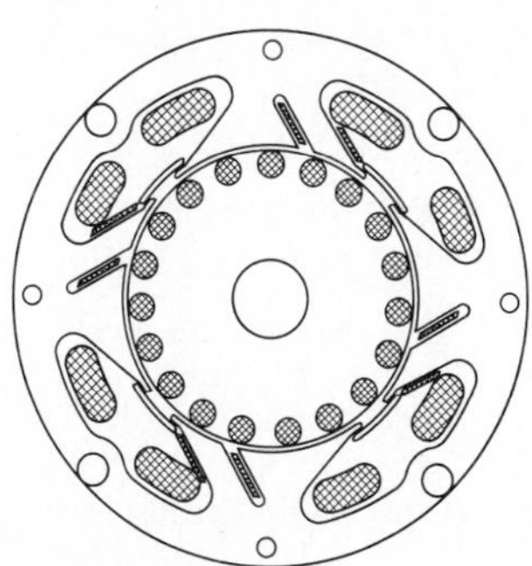

Fig. 9-3 A shaded-pole motor with tapered poles and magnetic wedges. *(Electrical Engineering.)*

dition is fulfilled in all shaded-pole motors, as will be shown. Each pole may be considered as split into two sections:

1. Main body of the pole
2. Shaded portion, i.e., the portion surrounded by the shading coil

An elementary understanding of the action of the shading coil may be gained from a study of the simple magnetic circuit of Fig. 9-4*a*,

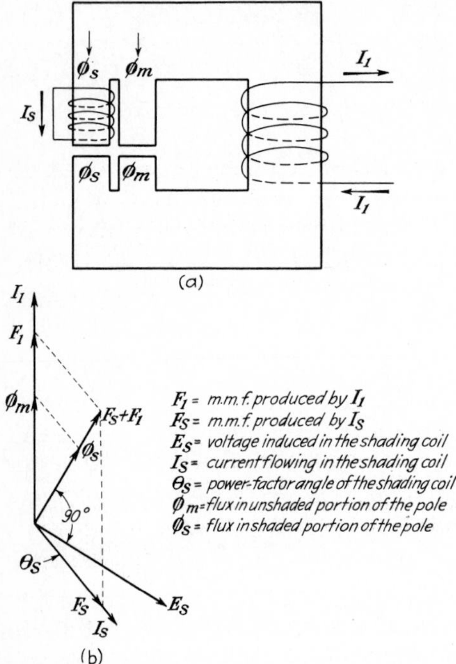

Fig. 9-4 How a shading coil reduces the flux threading it and delays it in time phase. (*a*) Simple magnetic circuit with a shading coil; (*b*) phasor diagram for part *a*.

which shows a single pole divided into two equal parts, one of which is shaded. Consider the instant of time when the current flowing in the primary coil is actually zero but is just starting to increase positively. In the main portion, the flux ϕ_m is just starting to build up in phase with the current. Likewise the flux ϕ_s in the shaded area is starting to build up, but this change in flux induces a voltage in the shading coil which causes current to flow. Since an induced current always flows in such a direction as to oppose the change in flux which induces it,* the current in the shading coil delays the building up of the flux ϕ_s. Further, when

*This is an application of Lenz's classical law, which is discussed in practically every elementary text on electricity.

the current in the primary coil starts to decrease, flux ϕ_m starts to decrease immediately. However, the current induced in the shading coil by a decreasing flux still tends to oppose the change in flux; i.e., in this case, the induced current tends to maintain the flux. Hence, flux ϕ_s continues to lag behind ϕ_m during the second part of the cycle. By similar considerations, it is evident that flux ϕ_s always lags behind flux ϕ_m in time. Hence, the net effect of this time and space displacement is to produce a shifting flux in the air gap which shifts always toward the shading coil. *Therefore, the direction of rotation of a shaded-pole motor is always from the unshaded to the shaded portion of the pole.*

A somewhat more rigorous and precise explanation of the action of a shading coil is given in the following article.

9-4 Action of a Simple Shading Coil Explained. Figure 9-4*b* is the phasor diagram for the simple shading coil illustrated. It was constructed so as to fulfill simultaneously the following conditions, all of which must be satisfied in accordance with elementary principles; these determine the diagram:

1. ϕ_m is proportional to and in phase with F_1.
2. ϕ_s is proportional to and in phase with $F_s + F_1$.
3. E_s is proportional to and 90° in time behind ϕ_s.
4. I_s is proportional to and θ_s° behind E_s.
5. F_s is proportional to and in phase with I_s.

After the diagram is constructed to fulfill simultaneously the conditions outlined previously, it becomes at once apparent, by comparing the magnitude and position of ϕ_s with that of ϕ_m, that the effect of the shading coil is to reduce the flux in the shaded portion of the pole and, at the same time, to shift it in time phase behind the flux in the unshaded portion.

Thus, Fig. 9-4 demonstrates qualitatively why the shading coil reduces the flux in the shaded portion of the pole and why it shifts the phase angle. In a practical motor, the effect is modified somewhat by the action of the rotor currents, but it still persists. The net effect of the shifting of the phase angle of the flux in the shaded portion of the pole is to produce the same effect as that of a gliding flux which is constantly shifting from the *main body* toward the *shaded portion*, tending to cause the rotor to move in that direction.

9-5 Use of Magnetic Wedges or Bridges. Figure 9-3 illustrates a construction that is in general similar to that of Fig. 9-2, except that magnetic wedges are used to hold the stator coils in place. Some engineers have found that magnetic wedges such as those shown often improve the performance of the motor, while other engineers disagree.

Figure 9-5 shows another construction, which uses a solid and permanent magnetic bridge. In this case, the stator core is made in two parts and the pole sides are straight and parallel. There are four shading coils, as shown. The four stator coils can be wound directly on the poles quite easily, or they can be wound on molds and readily slipped onto the pole projections, after which the internal part of the core and its windings are pressed into the outer ring of punchings. Sometimes, when this construction is used, two diametrically opposite coils are omitted for economy in winding; in this case, however, there are twice as many turns on each of the two poles as there would be in a single coil if there were four coils.

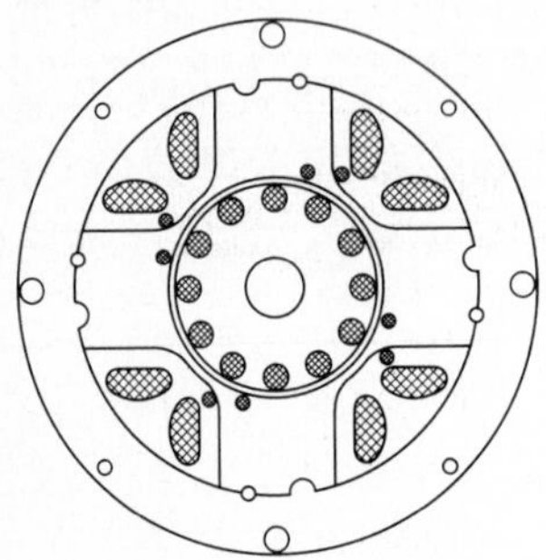

Fig. 9-5 A shaded-pole motor with a two-piece stator. *(Electrical Engineering.)*

9-6 Use of a Reluctance Effect in the Leading Pole Tip. A number of manufacturers report that they now use some form of additional reluctance effect in the leading pole tip to improve the performance of their shaded-pole motors, especially those in the larger sizes.

In 1937 Rall[1,*] described a reluctance-start motor which was built essentially like the motor of Fig. 9-3, except that there was no provision for a shading coil and the air gap for half of the pole face was approximately doubled. Such a motor, he reported, gave performance comparable to that of a shaded-pole motor. Why such a motor would develop locked-rotor torque is not easy to explain. (This motor is illustrated in Fig. 13-10 of an earlier work by Veinott,[2] as well as in Fig. 5-13 of another work.[3]). One theory advanced is that because there is relatively less iron in the magnetic circuit—compared with the amount of air—on the wide-gap side of the pole, there is less damping of the flux by the eddy currents in the iron; therefore, the flux in the wide-gap section leads the flux in the narrow-gap section, producing the

*For numbered references, see References at end of this chapter.

effect of flux shifting from the wide to the narrow gap, so that the motor runs from the wide to the narrow gap.

Baum[4] made an extensive mathematical analysis of this motor, and he ascribes the locked-rotor torque to two causes:

1. The difference in leakage inductance of the rotor bars on one side of the pole from that on the other causes a different current to flow on each side, producing a resultant torque as in a repulsion motor.

2. The difference in mutual reactance on the two sides of the poles gives the rotor current under the two sides some difference in phase, which also produces a resultant torque.

More recently, Poloujadoff[5] has studied this problem in greater depth.

One way of adding this reluctance effect to a shaded-pole motor is illustrated in Fig. 9-6; here, a portion of the leading pole tip is trimmed away. In trying to understand this effect, it may be helpful to note that, although the shading coil causes considerable damping of the flux within the area it embraces, there is some damping in the unshaded portion of the pole, caused by the eddy currents in the iron itself. Now, if some iron is trimmed away from the leading pole tip, the amount of damping that would have been produced by the iron in that pole tip is eliminated by the removal of the iron; hence the flux in the wide-gap section of the leading pole tip lags less than the flux in the middle of the pole. Hence we have three parts of the pole each carrying flux, slightly displaced from that of the other two.

Another way of achieving the reluctance effect in the leading pole tip is illustrated in Fig. 9-7. Here, the effect is achieved by piercing a rectangular hole in the leading pole tip.

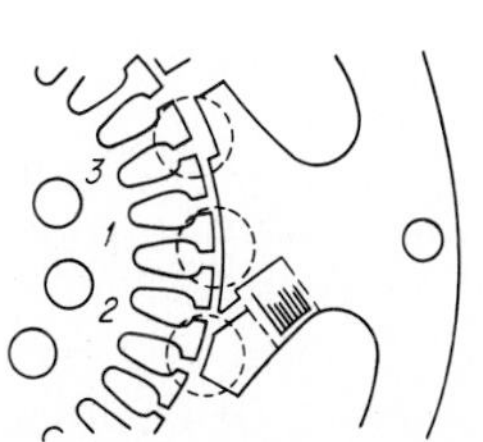

Fig. 9-6 The Tri-flux principle. Reluctance effect of leading tip 3 supplements shading-coil effect at 2. *(Controls Company of America.)*

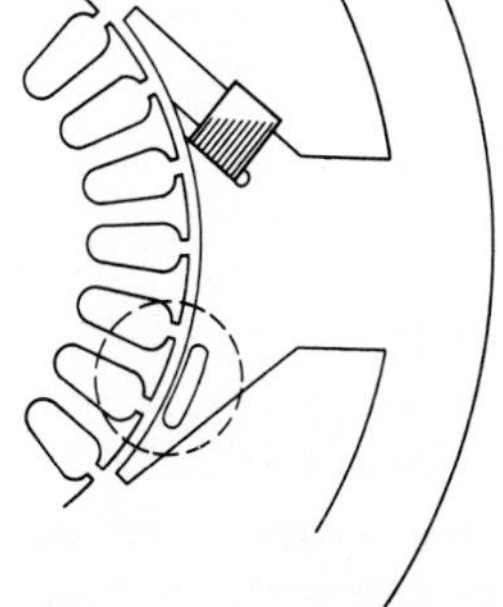

Fig. 9-7 A shaded-pole motor with reluctance slot in the leading pole tip. *(Universal Electric Company.)*

9-7 Ratings, Performance, and Torque Characteristics. Shaded-pole motors are built in a very wide variety of ratings and performance characteristics.

(1) Horsepower and speed ratings. Horsepower and speed ratings as specified by the National Electrical Manufacturers Association (NEMA) are given in Chap. 19, Table 19-4. Motors rated from 1/20 to 1/4 hp are generally built in a typical round construction, such as those shown in Figs. 9-2, 9-3, 9-5, and 9-11, for example; they may have four, six, or even eight poles. Motors rated 40 mhp or less are more often built in a skeleton construction, such as in Figs. 9-12, 9-17, or 9-24, for example; these have but two poles. Full-load speeds of such motors, for 60 Hz, may range from 2000 to 3000 rpm.

(2) Full-load inputs and efficiencies. Typical values of inputs and efficiencies for shaded-pole motors are given in Fig. 9-8. Since this curve was plotted, efficiencies have been raised somewhat, particularly for the motors at the upper end of the horsepower range.

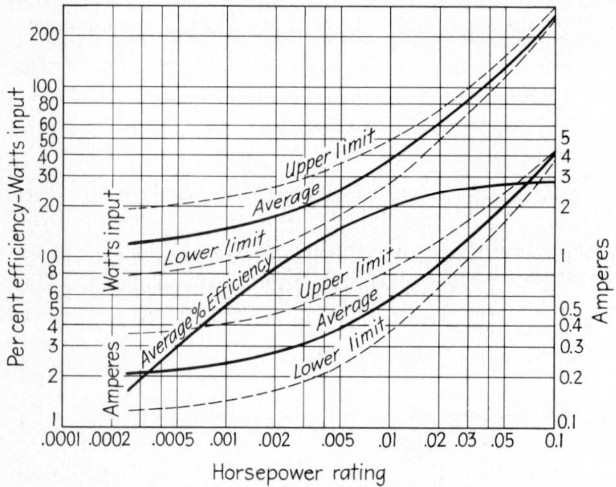

Fig. 9-8 Full-load inputs and efficiencies of shaded-pole motors.

(3) Speed-torque characteristics. Performance characteristics of an 8-mhp motor, taken from data published by Trickey,[6] are given in Fig. 9-9. This speed-torque curve is generally similar in shape to that of a polyphase induction motor, except that locked-rotor and breakdown torques are considerably lower, in terms of full-load torque. He shows the motor worked at close to breakdown torque, a practice still typical today. Perret and Poloujadoff[7] showed that the shape of the speed-torque curve is influenced by saturation.

Other typical speed-torque curves are given in Figs. 9-10 and 9-23; the former shows a slight cusp or saddle point at one-third synchronous

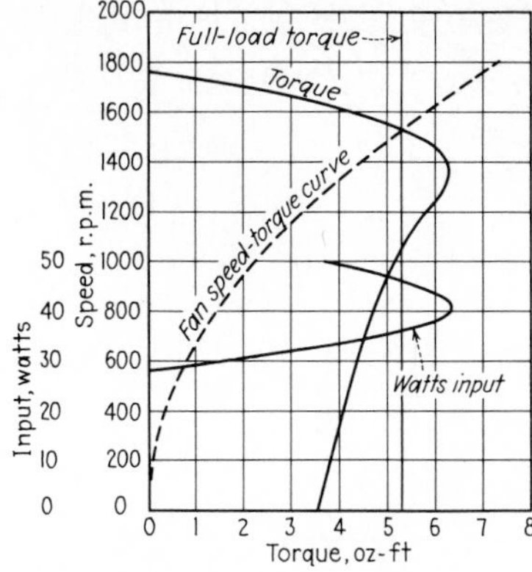

Fig. 9-9 Performance curves of a shaded-pole motor.

speed, due to the third space harmonic, and this cusp is sometimes greater than shown in this figure.

(4) Locked-rotor and breakdown torques. Locked-rotor torques are low, of the order of 35 to 90 percent of full load, with a tendency toward the higher torques in the lower horsepower ranges. Breakdown torques are also low, of the order of 120 to 150 percent of full load. NEMA values for breakdown torques are given in Chap. 15 in Table 15-4.

(5) Duty ratings. Many shaded-pole motors carry a continuous rating and do not depend upon supplemental cooling. Many are suitable for continuous duty only if the air from a fan is continuously flowing over them; this may be the fan they are driving or a fan mounted on the motor for only this purpose. Many carry only an intermittent or short-time rating, such as 5, 10, 15, or 30 min; this feature is present even in the low millihorsepower ranges and is designed to make the motors as compact and as economical in price as possible.

(6) Overheating protection. Underwriters Laboratories standards specify that when the inherent impedance of an open motor is sufficiently high to limit the total winding temperature to 150°C for Class A or 175°C for Class B motors under any condition of load including locked-rotor, the motor is said to be *impedance-protected* (against overheating).[8] Otherwise, thermal protectors are often installed in shaded-pole motors to protect against overheating.

9-8 Variations in the Operating Speed. The operating speed of a shaded-pole motor driving a fan may be subject to wide variations due either to manufacturing variations or to changes in applied voltage, or both. These variations are greater than are encountered when a split-phase or capacitor-start motor drives a fan. Figure 9-10 is one used by NEMA to explain why this is so.

In Fig. 9-10, a typical speed-torque curve is represented by the solid line. The curves drawn with dashed lines illustrate the range in char-

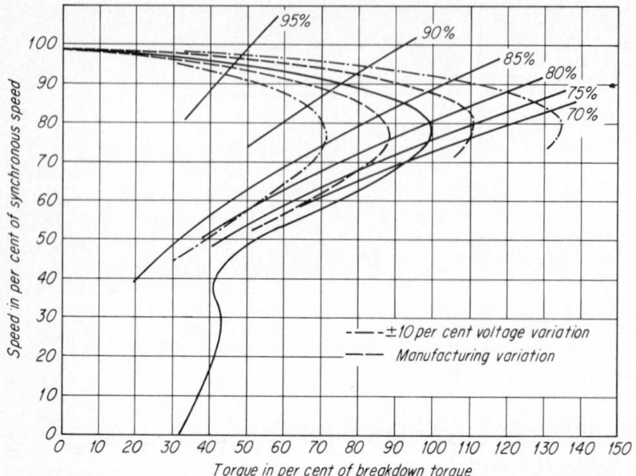

Fig. 9-10 Typical speed-torque curves for a shaded-pole motor showing speed variations expected because of manufacturing and voltage variations. *(From NEMA Motor and Generator Standards, Publ. no. MG1-1978, National Electrical Manufacturers Association, Washington, 1978.)*

acteristics due to manufacturing variations when the motor is operated at rated voltage and frequency. The curves drawn with dot-dash lines illustrate the total effect of adding a voltage variation of plus or minus 10 percent to the manufacturing variations. Now, a series of fan speed-torque curves are drawn with solid lines, and these are labeled 95%, 90%, etc. The 95% fan loads the typical motor at rated voltage, at 47 percent of breakdown torque, and runs at 95 percent of synchronous speed. (This corresponds to a breakdown torque of 212 percent.) Now, with this 95% fan, which loads the motor only lightly, the speed variations are small, from about 93 to 97 percent of synchronous speed. On the other hand, the 80% fan loads the motor right up to breakdown torque and, on average, runs at 80 percent of synchronous speed. Note that with this fan, the variation that may be expected in operating speed due to manufacturing and voltage variations is very great, being from 58 to 88 percent of synchronous speed, which represents a variation in operating rpm of minus 38 percent to plus 10 percent. Note that similar effects are encountered with multispeed permanent-split capacitor motors, as discussed in Arts. 7-9 to 7-11.

UNIDIRECTIONAL (FIXED-ROTATION) MOTORS

9-9 Single-Speed Motors. Motors in the fractional horsepower range are generally built with four or more poles and are of typical round-frame construction. One such construction is illustrated in Fig. 9-11. Motors in the subfractional horsepower range are usually rated in millihorsepower and built in some form of skeleton-type construction, sometimes referred to as "C-frame" motors. One such construction is illustrated in Fig. 9-12.

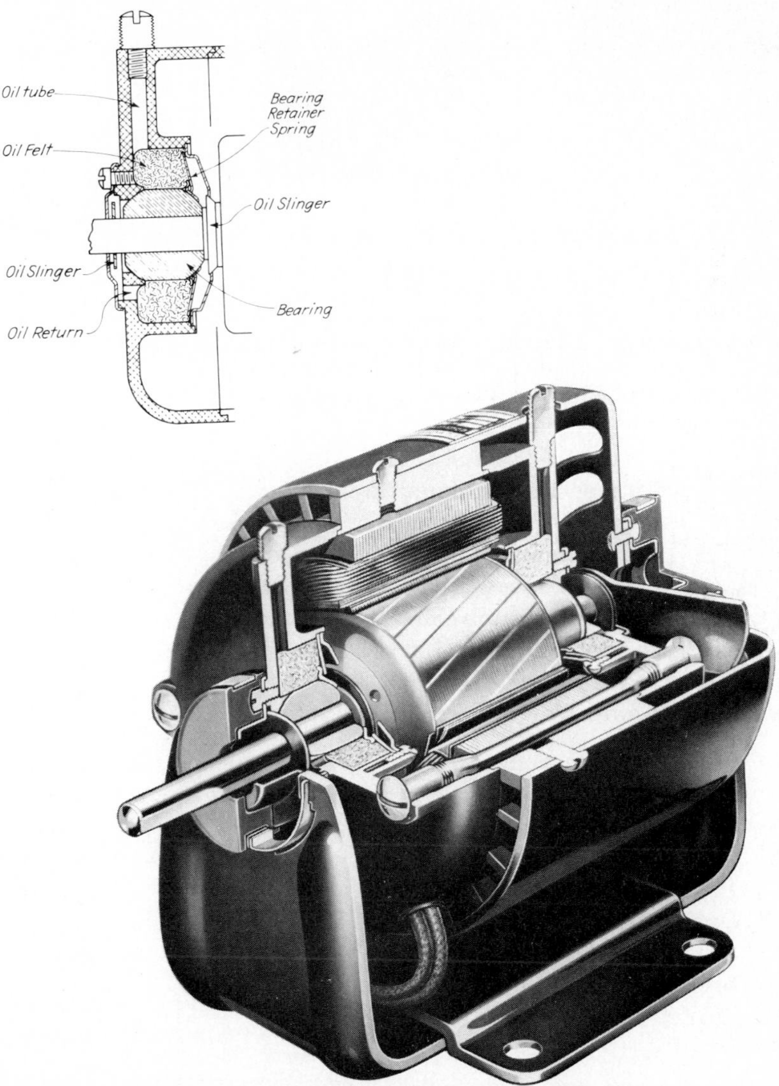

Fig. 9-11 Cutaway view of a totally enclosed fan-cooled shaded-pole motor with self-aligning sleeve bearings. Inset shows details of the bearing construction. *(Motor Division, Controls Company of America.)*

(*a*)

(*b*)

Fig. 9-12 Subfractional horsepower shaded-pole motors for solid or resilient mounting. Stator laminations are bonded together, and end shields, containing controlled-porosity bearings, are bonded to the stator core. Oil from the storage wicks, mounted on the end shields, is metered to the bearing-shaft clearance during operation. (*a*) Solid-mounted (stud-mounted) motor; (*b*) motor with resilient rings. *(General Electric Company.)*

9-10 Multispeed Motors. In general, multispeed motors are offered over the entire range of ratings, both fractional and subfractional. Usually multispeed operation is achieved with tapped windings. A line connection diagram for two-speed and also for three-speed motors, with leads tagged as established by NEMA, is given in Fig. 9-13. A three-speed motor has a blue lead added between black and red, so that the lines are connected to white and blue for medium speed.

A furnace-fan motor which can readily be connected by the user for

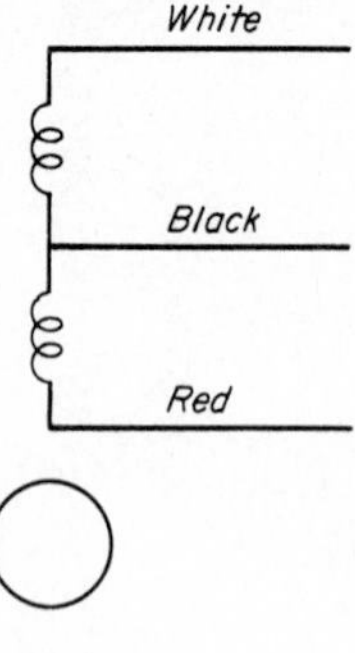

	L1	L2	Open
High speed	White	Black	Red
Low speed	White	Red	Black

Fig. 9-13 Line connections for two-speed motors. *(From NEMA Motor and Generator Standards, NEMA Publ. no. MG1-1978, National Electrical Manufacturers Association, Washington, 1978.)*

Fig. 9-14 Four-speed shaded-pole furnace fan motor with connection plug, which can be plugged into any one of four positions to give any one of four speeds. *(Westinghouse Electric Corporation.)*

any one of four speeds is illustrated in Fig. 9-14. This permits the furnace installer, or the user, to adjust the speed of the furnace blower to the best speed to provide a quiet uniform flow of heat and to change it later if conditions so require—without the need for an expensive controller. Internal wiring connections for this motor are given in Fig. 9-15.

A typical stator connection diagram for a six-pole four-speed tapped-winding shaded-pole motor is given in Fig. 9-16.

It should be pointed out, however, that shaded-pole motors are generally somewhat less satisfactory as multispeed motors than permanent-split capacitor motors. Part of the reason is that the characteristic cusp at one-third speed makes them more unstable when driving a fan, as can readily be seen by inspection of Fig. 9-10. Also, permanent-split motors can be arranged to have a higher ratio of locked-rotor to breakdown torque on the low-speed connection than they do on the high. These points were discussed in Arts. 7-8 to 7-13.

REVERSIBLE MOTORS

9-11 Use of Two Motors in Tandem. Since the ordinary shaded-pole motor is inherently unidirectional, because of the shading coil, reversibility poses a problem. Some manufacturers solve this problem by

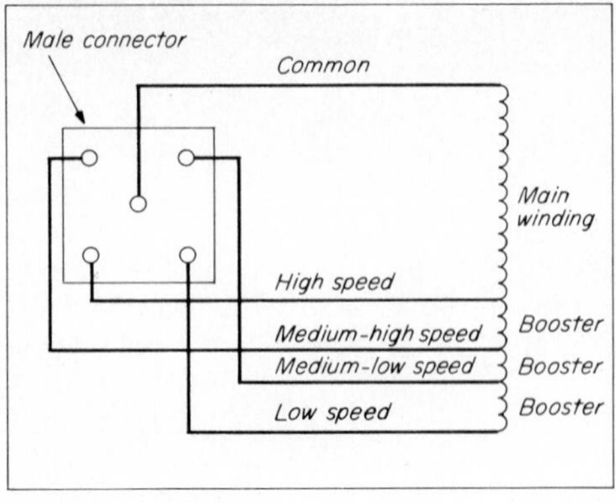

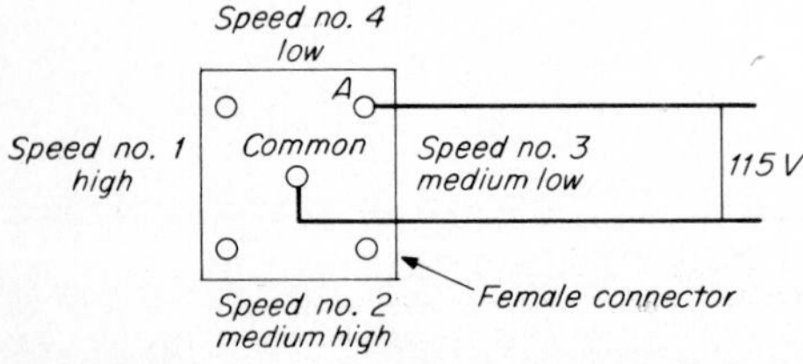

Fig. 9-15 Wiring and line connection diagrams for the four-speed motor of Fig. 9-14. The male portion of the connector is built into the motor end shield, and the connections to it shown in the top sketch are made at the factory. Line connections are made to the female portion as shown in the lower sketch. Line voltage can be applied between common and any one of the four speed taps by properly positioning the plug. *(Westinghouse Electric Company.)*

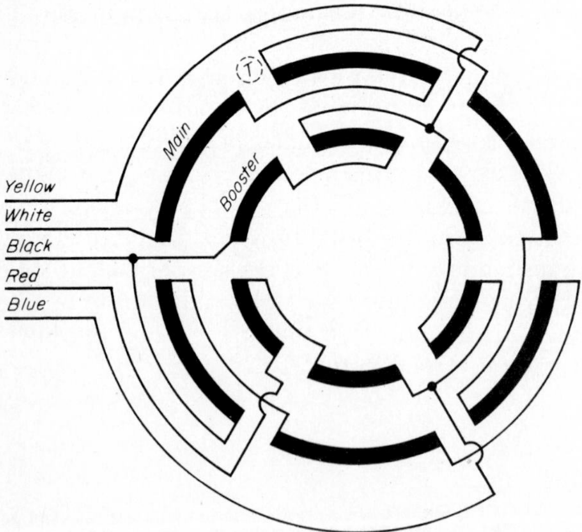

Fig. 9-16 Stator connection diagram for a four-speed six-pole shaded-pole motor with a tapped booster winding. *(Westinghouse Electric Corporation.)*

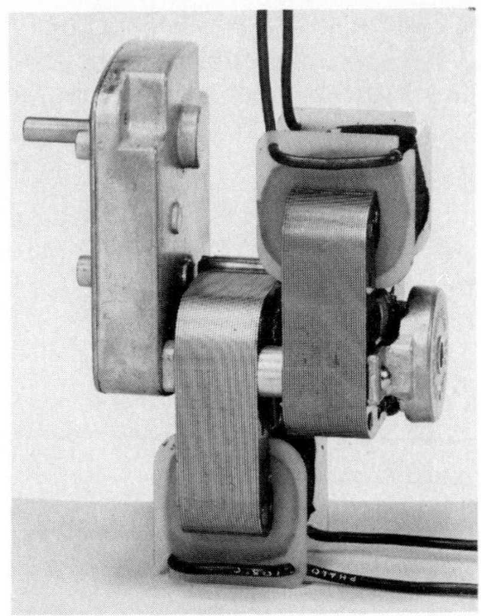

Fig. 9-17 A reversible shaded-pole geared motor that uses two motors in tandem. *(Brevel Products Corporation.)*

using two motors, one for each direction of rotation. Both rotors are mounted on the shaft, and one stator or the other is energized, depending upon the direction of rotation desired. One such motor is illustrated in Fig. 9-17.

9-12. Use of Two Wound Shading Coils per Pole. Figure 9-18 shows schematically how a shaded-pole motor can be reversed if it is provided with two shading coils per pole. Direction of rotation is determined by which shading coil is short-circuited. The figure shows a switch in only the shading coil circuits; it may or may not be necessary to deenergize the main coil when the shading coil circuits are open and the motor is not running.

It is obvious, of course, that the leading pole tips of a reversible motor cannot be trimmed to improve torque. So far as achieving reversibility is concerned, each of the two shading coils could embrace half of the pole. However, when the middle portion of the pole is not embraced by either shading coil, some remarkable effects can be achieved, as will be discussed in the articles that follow.

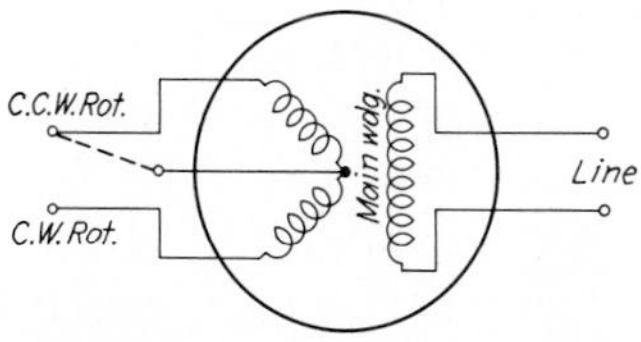

Fig. 9-18 A reversible shaded-pole motor with two auxiliary windings.

9-13 Production of a Space-Quadrature Field with Wound Shading Coils. Figure 9-19 shows the laminations for a two-pole skeleton-type shaded-pole reversible motor, using two shading coils per hole, or four wound shading coils in all. Note that there is a middle section of each pole which is not embraced by either shading coil; this is important, as will be shown. Positions of the four shading coils are represented in Fig. 9-20, and the coils are designated as A_1, A_2, B_1, and B_2. Either the A pair or the B pair of coils is short-circuited, depending upon the desired direction of rotation.

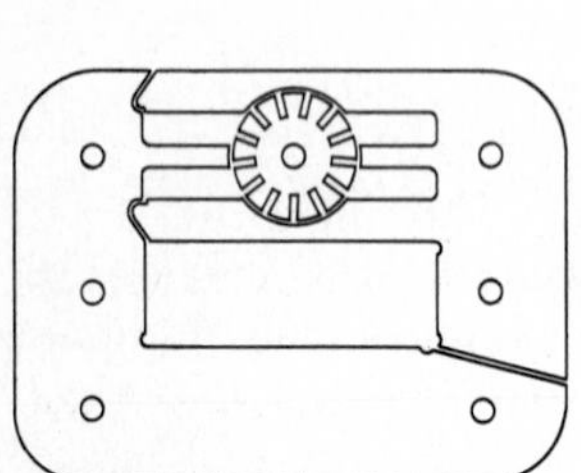

Fig. 9-19 Laminations for a motor with four shading coils. They are shown slightly separated to demonstrate the way they are assembled. *(Barber-Colman Company.)*

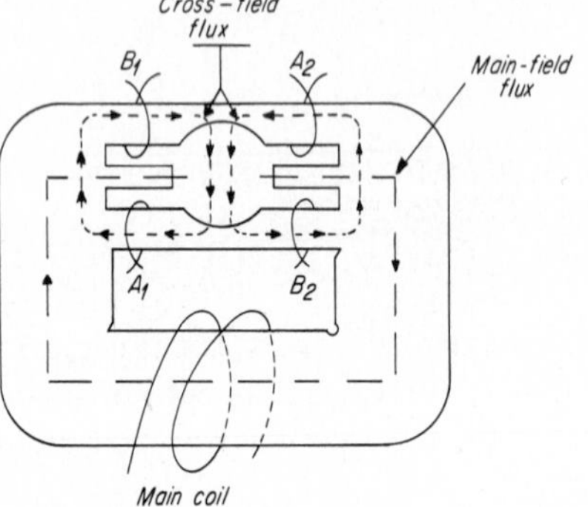

Fig. 9-20 Production of a cross-field flux in space quadrature with the main flux by four shading coils. A_1, A_2, B_1, and B_2 are the four shading coils.

Now, what happens if the A coils and the B coils are connected in series and a current is passed through them? It is obvious that for proper functioning as shading coils, the two A coils must be so connected as to set up a flux in the *same horizontal direction;* the same consideration applied to the two B coils. Now, let the B coils be connected so that their mmfs are directed to the right, and let the A coils be so connected that their mmfs are both directed to the left, as shown in Fig. 9-20. Clearly then, as can be seen from the figure, the resultant flux due to the energization of these four shading coils can cross the air gap only in the vertical direction, that is, in a direction at right angles to the main-axis flux which flows through the centers of the two poles and is horizontal. Thus, since these four shading coils were connected in series to produce a space-quadrature flux, the total circuit of the four coils so connected represents a space-quadrature winding, just like the auxiliary winding of a capacitor-start or permanent-split capacitor motor. It should be obvious to the reader that the only net voltage that can be induced in the four windings thus connected in series is by a cross-axis flux.

These considerations are represented schematically in Fig. 9-21. The two A coils are both displaced from the main winding but are in space

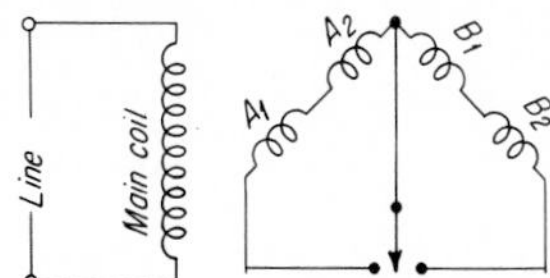

Fig. 9-21 Schematic diagram of connections for a reversible motor using four shading coils as in Fig. 9-20.

phase with each other. Likewise, the two B coils are in space phase with each other but displaced from both the main coil and the A coils. Either pair of coils can be short-circuited, depending upon the direction of rotation desired. The four coils in series are in space quadrature with the main winding. If the B_2 coil had been connected to the end of the A_2 coil, the effect would have been to put the four coils in series *in space phase* with the main winding, as shown schematically in Fig. 9-18. Hence when the four coils are to be used as a quadrature winding, it is important that they be connected properly.

9-14 Versatility of the Motor with Four Shading Coils. The ability to use the two sets of shading coils in series as a quadrature winding makes possible a variety of uses and arrangements in order to obtain special effects.

(1) Simple reversible shaded-pole motor. It has already been discussed how this motor can be used as a reversible motor, as shown in Fig. 9-21. In general, five-wire control is needed in order to deenergize the motor completely. From the standpoint of torque output from the same amount of material, such a motor suffers severely as compared with a unidirectional motor of the same physical size and weight. Because the wound shading coils require insulation space, less room is left for iron; also leading pole tips cannot have increased air gaps to boost the shading effect with a reluctance effect. The magnitude of these effects is illustrated in Fig. 9-23, by comparing curves 1 and 2.

(2) Permanent-split capacitor motors. Figure 9-22 shows two arrangements of permanent-split capacitor connections, which take advantage of the ability to connect the four coils as a space-quadrature winding. Figure 9-22*a* shows a conventional permanent-split capacitor arrange-

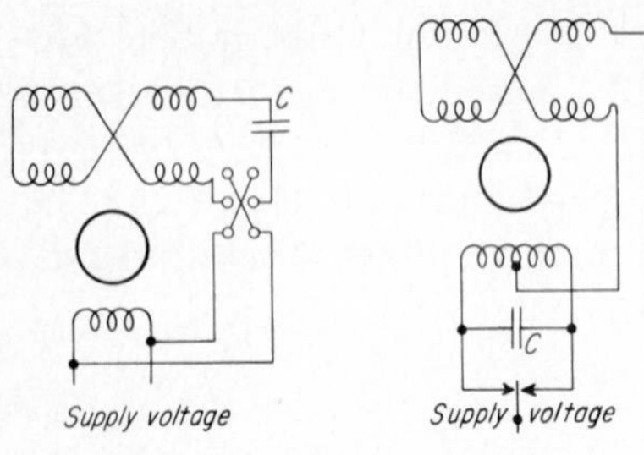

Fig. 9-22 A shaded-pole motor connected as a permanent-split capacitor motor. (*a*) Four-wire circuit; (*b*) three-wire circuit. *(Barber-Colman Company.)*

ment; rotation is reversed by reversing the capacitor phase. The very great improvement in torque output over the shaded-pole connection is shown in Fig. 9-23, curve 4. Four wires from the motor are needed.

A somewhat different arrangement, requiring only three motor leads, is shown in Fig. 9-22*b*. More capacitance is required, and the torque is slightly less, but it is still much more than with the shaded-pole connection, as can be seen in Fig. 9-23.

(3) Dynamic braking. Dynamic braking can be achieved by leaving the main coil excited and short-circuiting the four shading coils in series.

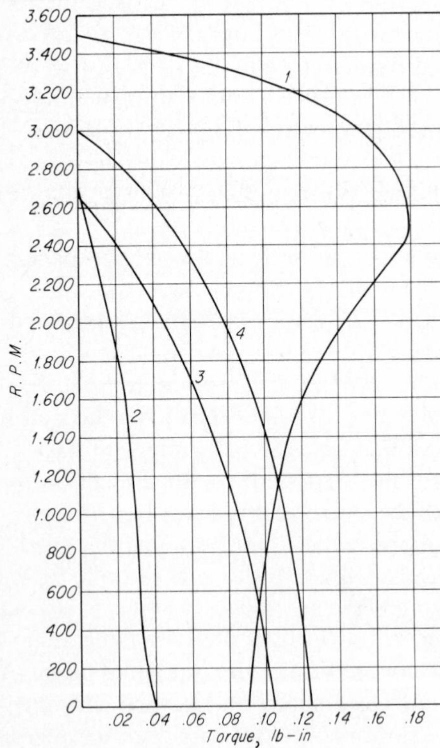

Fig. 9-23 Comparison of torque characteristics of skeleton-type shaded-pole motors weighing about 29 oz. Curve 1, conventional unidirectional motor with two shading coils. Reversible motor with four shading coils variously connected. Curve 2, two shading coils short-circuited. Curve 3, as a three-wire permanent-split capacitor motor with 3.0 mfd. Curve 4, as a three-wire permanent-split capacitor motor with 2.0 mfd. *(Plotted from data supplied by Barber-Colman Company.)*

This is a familiar method of braking: short-circuiting an auxiliary winding while the main is still excited.

(4) Servomotor. Since there are two windings in space quadrature, this motor can be used as a two-phase servomotor or by using the proper capacitor with one of the windings, as a single-phase servomotor. Speed-torque characteristics, as shown in Fig. 9-23, are quite suited to servo operation.

(5) Tachometer generator. With rated voltage applied to the main winding, an output voltage is generated in the four shading coils which is proportional to the speed, over quite a range of speeds.

MISCELLANEOUS

9-15 Geared Motors. Shaded-pole motors with an integral gear-head speed-reducer are offered by many manufacturers. One example of a reversible geared motor is shown in Fig. 9-17. Also, built-in clutches and brakes are offered by a number of shaded-pole motor producers.

9-16 Battery-Operated Shaded-Pole Motor System. Some manufacturers are offering a shaded-pole motor powered by a battery. By electronic means, the dc power is inverted to ac power to drive the motor. One such system is illustrated in Figs. 9-24, 9-25, and 9-26. This system uses the transistor oscillator circuit shown in Fig. 9-25.

9-17 Shaded-Pole Ceiling-Fan Motors. Many of the original multiple-pole motors used in ceiling fans were the shaded-pole type. Because of the number of poles required to obtain the slow speeds, these were inherently large in diameter and heavy. Unlike the permanent-split

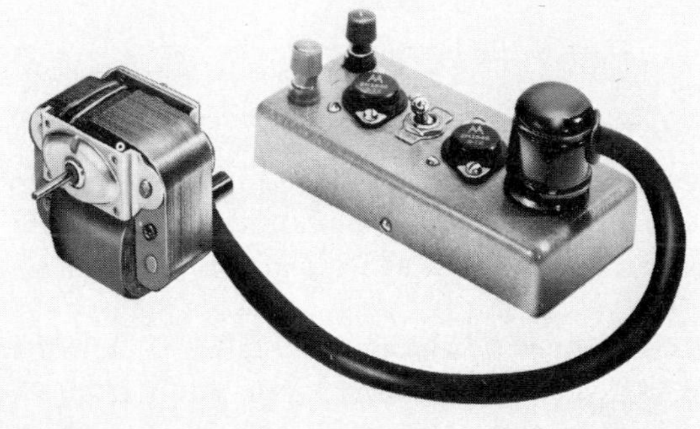

Fig. 9-24 A shaded pole motor with transistorized power supply. *(Barber-Colman Company.)*

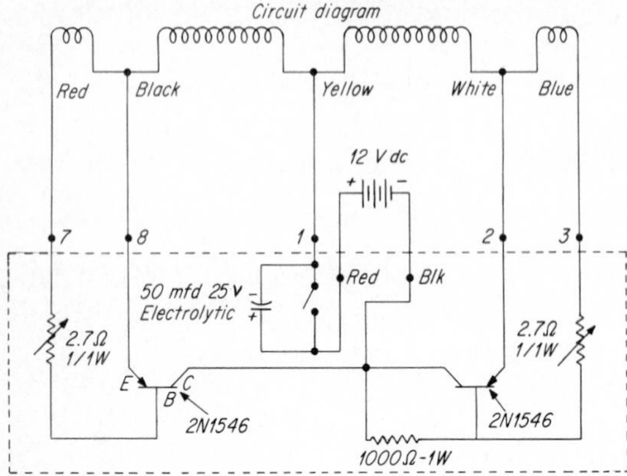

Fig. 9-25 Circuit diagram for the motor of Fig. 9-24.

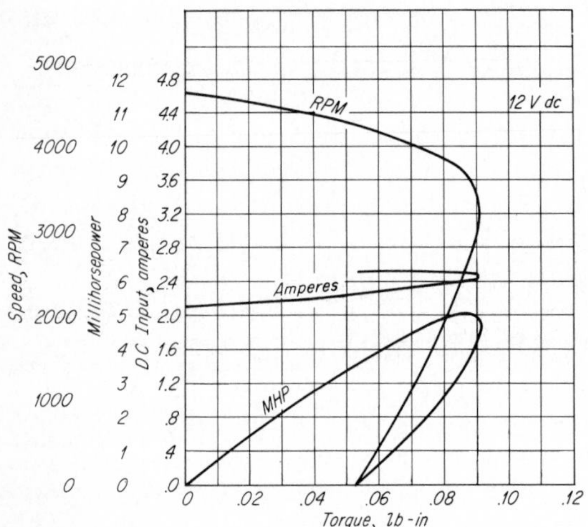

Fig. 9-26 Performance characteristics for the motor of Fig. 9-24 on a 12-V dc source.

capacitor motors discussed in Art. 7-18, the shaded-pole motors are nonreversible. Although a majority of ceiling fan motors today are permanent-split capacitor motors, shaded-pole motors are still used.

9-18 How To Determine Direction of Rotation. As stated in Art. 9-3, the direction of rotation of a shaded-pole motor is always from the unshaded to the shaded portion of the pole. If this principle is applied to the motor depicted in Fig. 9-3, the rotor will rotate in a clockwise direction. If a shaded-pole motor is constructed so that the stator area near the bore is visible, it is often possible to determine the direction of rotation by inspection.

By reversing the stator end-for-end with respect to the rotor, the direction of rotation can be reversed. However, not all motor designs can be reversed in this manner, since the stator and rotor may not be in axial alignment.

9-19 Rewinding or Reconnecting Motor for a Different Voltage and the Same Performance. If a shaded-pole motor is series-connected, it can be reconnected for half voltage by connecting the poles in parallel. Sometimes this is difficult because most shaded-pole motors are machine-wound and the connections between poles are too short to attach leads. If the poles are reconnected in parallel, every other pole must be connected in series with each other.

Many shaded-pole motors are wound with their poles in parallel. If it is desired to double the voltage rating of such a motor, the poles should be connected in series. However, each adjacent pole must be connected so that it is of opposite polarity.

If it is desired to rewind a 115-V motor for 230 V, and to keep the same type of connection, wire three sizes smaller with twice as many turns is required.

References

1. Rall, Carl A.: "This Motor Meets Many Product Needs," *Electr. Manuf.*, June 1937, pp. 31–35.
2. Veinott, C. G.: *Fractional Horsepower Electric Motors*, 2d ed., McGraw-Hill Book Company, New York, 1948.
3. ————: *Theory and Design of Small Induction Motors*, McGraw-Hill Book Company, New York, 1959. (Out of print, but available from University of Microfilms, Inc., P.O. Box 1647, Ann Arbor, Mich.)
4. Baum, John L.: "The Asymmetrical Stator as a Means of Starting Single-Phase Induction Motors," *AIEE Trans.*, vol. 63, 1944, pp. 245–250.
5. Poloujadoff, M.: "General Rotating MMF Theory of Squirrel-Case Induction Machines with Non Uniform Air Gap and Several Non Sinusoidally Distributed Windings," *IEEE Trans.*, vol. PAS-101, no. 3, March 1982, pp. 583–591.
6. Trickey, P. H.: "An Analysis of the Shaded Pole Motor," *Electr. Eng.*, September 1936, pp. 1007–1014.
7. Perret, R., and M. Poloujadoff: "Characteristics Analysis of Saturated Shaded-Pole Induction Motors," *IEEE Trans.*, vol. PAS-95, 1976, pp. 1347–1353.
8. *Impedance Protected Motors*, UL 519, Underwriters Laboratories Inc., 333 Pfingsten Rd., Northbrook, IL 60062, 1982.

Bibliography

Braymer, Daniel H., and A. C. Roe: *Rewinding Small Motors*, McGraw-Hill Book Company, New York, 1932, Chap. XX.

Chang, S. S. L.: "Equivalent Circuits and Their Application in Designing Shaded-Pole Motors," *AIEE Trans.*, vol. 70, pt. I, 1951, pp. 690–699.

Eastham, J. F., and S. Williamson: "Generalised Theory of Induction Motors with Asymmetrical Air-Gaps and Primary Windings," *Proc. IEE*, vol. 120, no. 7, July 1973.

Herzog, G. E., and G. H. Sherer: "The Calculation of Shaded-Pole Motor Performance by the Use of a Digital Computer," *AIEE Trans.*, vol. 78, pt. III, 1959, pp. 1607–1610.

Kimberly, E. E.: "The Field Fluxes of the Shaded-Pole Motor," *AIEE Trans.*, vol. 68, pt. I, 1949, pp. 273–277.

Kron, G.: "Equivalent Circuits of the Shaded-Pole Motor with Space Harmonics," *AIEE Trans.*, vol. 69, pt. II, 1950, pp. 735–740.

Suhr, F. W.: "A Theory for Shaded-Pole Induction Motors," *AIEE Trans.*, vol. 77, 1958, pp. 509–515.

Veinott, C. G.: "Discussion on Trickey's Paper," *Electr. Eng.*, May 1937, pp. 612–613.

CHAPTER

10

SYNCHRONOUS MOTORS

Exactness of speed is the outstanding characteristic that has made synchronous motors so popular in fractional and subfractional horsepower sizes. Speed of a synchronous motor is exactly proportional to the line frequency, which is regulated on all large 60-Hz power systems so closely that clocks are driven by synchronous motors; the frequency is rarely in error by more than a small fraction of 1 percent, and these small errors are averaged out over a period of time. Hence, synchronous motors are widely used for clocks and timing motors.

10-1 General. Synchronous motors are built over a horsepower and speed range greater than that of any other type of motor. Horsepower ratings range from 5000 or greater down to one-millionth of a horsepower; output speeds range from 24,000 rpm (400-Hz motor) down to as low as one revolution per month (with reduction gearing). In integral horsepower sizes, their popularity is due in large measure to their ability to correct system power factor but, in fractional and subfractional horsepower sizes, dc excitation is almost never used (except for dynamic braking), and the power factor is poorer than for induction motors of comparable rating. In sizes from ⅛ hp down to about 1 mhp, synchronous motors are used for such applications as teleprinters, facsimile picture transmission, instruments of all kinds, and sound-recording or sound-reproducing apparatus. Also, they are often used in textile applications in which several motors may be driven from the same variable-frequency source. In subfractional horsepower ratings (below 1 mhp), they are widely used for clocks and all kinds of timing devices. Almost invariably, they are self-starting, but some clock motors have been intentionally made so that manual starting is required.

10-2 Types of Synchronous Motors. Synchronous motors are built in a wide variety of types and constructions. However, most of them have one feature in common with induction motors: a stator structure with windings which, when properly energized with alternating current, set up a revolving magnetic field as explained in Chap. 2. The speed of this magnetic field, known as the *synchronous speed*, is exactly proportional to the frequency. Synchronous motors are so built that they lock into step with the rotating magnetic field and rotate at exactly the same speed as the latter. Induction motors, on the other hand, always rotate at a speed a little slower than synchronous. It is the construction of the rotor that determines whether it locks into step with the rotating field or rotates at a slightly slower speed. Hence, synchronous motors are usually classified (1) by the type of rotor construction and (2) by the stator winding arrangement. Large synchronous motors (that is, the larger integral horsepower sizes) use a salient-pole rotor with a dc-excited field coil on each pole and a squirrel-cage damper winding for starting and for preventing hunting when in synchronism. However, dc excitation is not used in fractional horsepower sizes, though permanent magnets are often used in one way or another. Popular constructions are reluctance, hysteresis, permanent magnet, and inductor. Reluctance, hysteresis, and permanent magnet types use a stator structure like that of an induction motor, but the inductor type uses a special construction for both stator and rotor.

RELUCTANCE MOTORS

10-3 Reluctance Motor Definitions. Types of reluctance motors include the following.

(1) Reluctance motor. A reluctance motor is a synchronous motor similar in construction to an induction motor, in which the member carrying the secondary circuit has salient poles, without dc excitation. It starts as an induction motor but operates normally at synchronous speed.

(2) Subsynchronous reluctance motor. A subsynchronous reluctance motor is a form of reluctance motor which has a number of salient poles greater than the number of electrical poles of the primary winding, thus causing the motor to operate at a constant average speed which is a submultiple of its apparent synchronous speed.

10-4 Construction and Principles of Operation. As implied in its definition, the reluctance motor is really a special case of a dc-excited synchronous motor with an open field circuit. However, the small reluctance motor is usually built from induction-motor parts, except that some teeth are cut out of the squirrel-cage rotor; in addition, flux barriers are often added in the yoke section of the rotor to accentuate the saliency effect.

(1) Construction. Large synchronous motors use a salient-pole rotating field which is excited with direct current. The salient poles are provided with squirrel-cage damper windings which are used to start the motor as an induction motor and then to provide stability against hunting once the motor is in synchronism. Rotors of reluctance motors have a squirrel-cage damper winding and salient poles, but no winding for dc excitation; one form of rotor punching is shown in Fig. 10-1, made for a six-pole motor. In each of six places, teeth have been removed. If, as in the example, the sizes or locations of these cutouts are slightly unsymmetrical, better locked-rotor torque is obtained than if the cut-

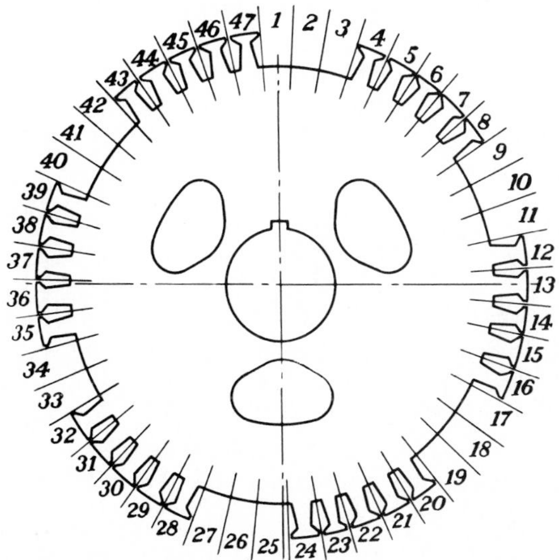

Fig. 10-1 Rotor punching used in a reluctance motor.

outs were uniform in size and spacing, because cogging effects are thereby reduced. Although the teeth were removed, the end rings were left intact, and the bars were left in place. These projecting groups of teeth provide something that can be "grabbed" by the rotating field and "held" so that the rotor stays in step with the rotating field, thereby giving synchronous speed. Other rotor laminations are illustrated in Fig. 10-2. These, too, show the projecting or salient poles, as well as flux-barrier slots.

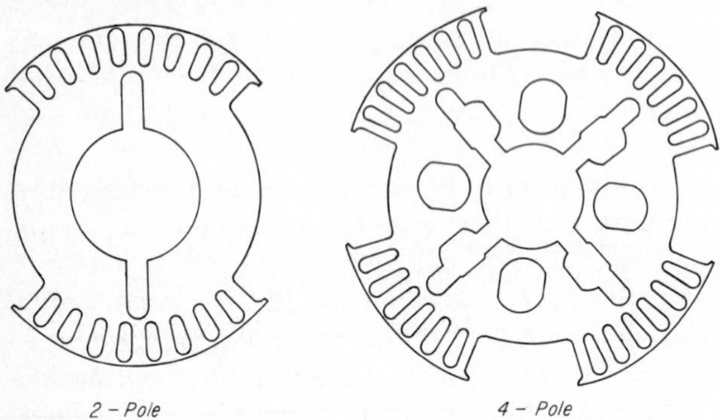

Fig. 10-2 Rotor laminations for polyphase reluctance motors, showing flux-barrier slots. *(Reliance Electric Company.)*

A somewhat more accurate way of explaining the synchronizing action of the salient poles than the preceding one follows. In any simple magnetic circuit, consisting of a fixed and a movable flux-carrying element, the movable element will tend to assume a position such that the reluctance of the magnetic circuit is a minimum, and a mechanical force is exerted on the movable piece to try to make it assume this position of minimum reluctance. When the centers of the salient groups of teeth in Fig. 10-1 line up with the centers of the poles of the revolving field, the reluctance of the magnetic path is at a minimum, and hence the rotor punchings tend to assume and hold this position.

(2) Effect of the flux-barrier slots. Figure 10-2 shows the use of flux-barrier slots to improve performance. That they do improve performance is based upon the fundamental principle that the force exerted upon a movable flux-carrying element in a magnetic field of constant strength is proportional to the rate of change of reluctance caused by the movement. In the reluctance motor, the strength of the magnetic force tending to align the salient projections of the rotor with the rotating magnetic field depends upon how much the reluctance of the magnetic circuit changes with change in rotor position with respect to the rotating magnetic field. When centers of the revolving magnetic field poles line up with the cutout slots, the reluctance is greatest; it is least when the flux poles line up with the salient groups of teeth. The synchronous torque depends upon the difference in reluctance of the magnetic circuit for these two extreme positions. Each of the two rotor punchings illustrated in Fig. 10-2 uses a long radial slot in the center of each main pole. This barrier slot interposes no appreciable additional reluctance when the salient poles are lined up with the centers of the revolving field poles, because the barrier slots are parallel to this flux and, therefore, the flux does not have to cross the slots. However, when the poles of the revolving field are aligned with the centers of the cutouts, these barrier slots do interpose a substantial increase in reluctance because the flux in the yoke (core section) has to flow *across these slots.* Thus, the effect of the barrier slots is to increase the synchronous torque because they accentuate the change in reluctance that occurs when the flux passes from the cross axis of the rotor to the main axis.

In textbooks and other theoretical treatises on synchronous motors, the reactance of the stator winding when the rotor is in the position of *minimum* reluctance is called the *direct-axis reactance;* the reactance of the stator winding when the rotor is in the position of *maximum* reluctance is called the *quadrature-axis reactance.* It is shown analytically in such treatises that synchronizing torque is a function of the difference between these two reactances. Expressed in these terms, the effect of the barrier slots is to decrease the quadrature-axis reactance, thereby increasing the synchronizing torque.

(3) How a reluctance motor pulls into step. In the discussion that follows, let us assume, for purposes of simplicity, that we are talking about a polyphase reluctance motor, which has a true rotating field. Although we shall be thinking about the polyphase motor, the remarks that follow apply qualitatively to reluctance motors with other types of stator excitation.

Starting from rest, the motor accelerates as an induction motor. As it approaches synchronous speed, the salient poles of the rotor slip by the poles of the rotating magnetic field which are set up by the stator currents, but the nearer the rotor gets to synchronous speed, the slower the poles slip. As a rotor pole passes and begins to lag behind a pole of the rotating field, a synchronizing torque is developed, tending to make the rotor catch up with the rotating field. This torque tends to accelerate the rotor above its induction-motor speed, because it is pulling the rotor in the direction of rotation. If the rotor has not pulled into step by the time its salient pole slips back halfway to the next salient pole, the synchronizing torque disappears. When the rotor has slipped back just over halfway to the next pole of the rotating magnetic field, the latter will try to pull the rotor pole *back* toward it. The effect of this is to decelerate the rotor so long as the salient pole of the rotor is ahead of the nearest pole of the rotating magnetic field; that is, the synchronizing torque in this case is negative. The average effect of these successive pulls in opposite directions is zero; therefore, unless the rotor can accelerate from the induction speed to synchronous speed while a rotor pole passes a pole of the stator field, the motor will not synchronize at all. In other words, synchronization starts just as the centers of the poles of the rotating field line up with the centers of the salient groups of teeth; synchronization has to be completed before the rotor can slip back as much as half a pole pitch (90 electrical degrees).

(4) Effects of inertia and rotor resistance. *Pull-in torque* of a synchronous motor is defined as the maximum constant torque against which the motor will pull its connected load into synchronism. When a motor is driving a load at a steady speed, all of the internal torque (torque developed by electromagnetic action) is available to carry the load. However, synchronization involves a sudden change in speed, as we have seen, and a torque is required to overcome the inertia of the rotating parts in order to effect this change in speed. Hence the torque available to carry the load is reduced by the amount consumed in overcoming inertia. Now, the inertia torque required to effect the speed change necessary for synchronization is proportional both to the speed change and to the total rotational inertia. Since a low rotor resistance means that, as an induction motor, the motor will come up close to synchronous speed, the speed change required is small, and so is the inertia torque. Hence a low rotor resistance helps the pull-in torque, although it reduces the starting torque. Load inertia increases the inertia torque, so that load inertia decreases the pull-in torque.

(5) Effect of frequency. The difficulty of designing and building a reluctance motor to operate on a high-frequency circuit increases very rapidly with frequency. To understand why this is so, consider a 60-Hz reluctance motor, and let us try to operate it on 120 Hz. If the voltage is doubled, the synchronizing torque will be about the same, but because the frequency is doubled, the synchronization has to take place in half the time. Moreover, the change in speed during synchronization will be somewhat greater because the synchronous speed is now doubled, and the slip, in rpm, will normally be increased in order to obtain locked-rotor torque. In other words, the difficulty of building a reluctance synchronous motor increases approximately as the square of the frequency.

10-5 Polyphase Reluctance Motors. Reluctance motors were discussed a long time ago by Trickey,[1,*] who later published methods for

*For numbered references, see References at end of this chapter.

making performance calculations on them.[2] Polyphase reluctance motors use regular polyphase windings in the stator, and a rotor with cutouts, of the general form of Fig. 10-2. The motor starts and comes up to speed as an induction motor, pulls into step, and operates as a synchronous motor. A typical speed-torque curve is shown in Fig. 10-3. The induction-motor breakdown torque and the average locked-rotor torque are little affected by the rotor cutouts, although the minimum locked-rotor torque is appreciably reduced because of the greater cogging of the synchronous rotor. The synchronous pull-out torque is from one-third to one-half the induction-motor breakdown torque. Because of this, the pull-out torque of reluctance motors is usually made lower than the breakdown torque of an induction motor of the same rating; also, reluctance motors are worked a little harder and often use more material, even if in the same frame, than induction motors of comparable rating.

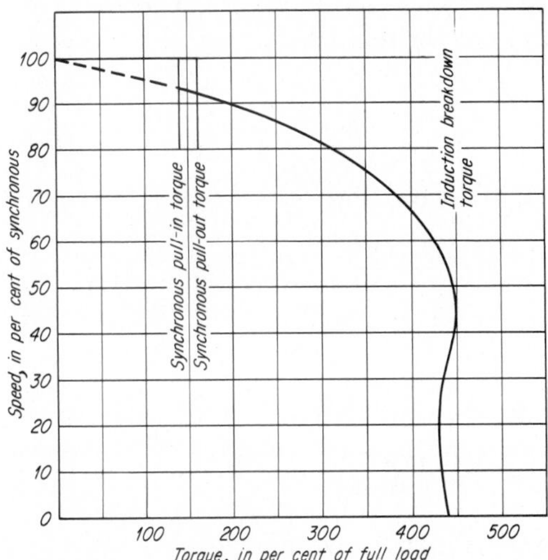

Fig. 10-3 Speed-torque curve of a polyphase reluctance motor.

10-6 Split-Phase Reluctance Motors. Split-phase reluctance motors use a rotor construction as described in Art. 10-4, but the stator is wound as a split-phase motor. In other words, a split-phase reluctance motor is a split-phase motor with a reluctance rotor.

10-7 Capacitor-Type Reluctance Motors. Capacitor-type reluctance motors are capacitor-type motors with a reluctance rotor. The windings could be any one of the four capacitor types: capacitor-start, two-value

Fig. 10-4 Small capacitor-type synchronous or nonsynchronous gearmotor with built-in helical gear reduction. Standard 60-Hz motors are from 0.7 to 300 rpm, and output torques are from 110 to 1.3 oz-in, respectively. Motor has three leads and is electrically reversible. *(Bodine Electric Company.)*

capacitor, permanent-split capacitor, or split-phase capacitor. Reluctance motors with a permanent-split capacitor winding are widely used in instrument applications. Figure 10-4 illustrates one such motor. National Electrical Manufacturers Association (NEMA) connection diagrams, given in Figs. 7-3 and 7-4, apply to synchronous capacitor motors as well as to asynchronous capacitor motors. Table 10-1 gives some typical performance data.

10-8 Shaded-Pole Reluctance Motors. Shaded-pole reluctance motors are shaded-pole motors with reluctance rotors. Performance curves for two shaded-pole reluctance motors are given in Fig. 10-5. In general,

TABLE 10-1 Performance Specifications of Small 60-Hz Reluctance-Type Single-Phase Capacitor Motors*

		Torque, oz-in			
Horsepower	rpm	Full-load	Locked-rotor	Pull-in	Pull-out
1/100	1800	5.6	8.0	6.0	8.0
1/50	1800	11.2	11.2		
1/30	1800	18.6	14.0	18.6	19.5
1/20	1800	28.0	18.0	27.0	28.0
1/10	1800	56.0	22.0	57.0	70.0
1/25	3600	11.2	10	14	16
1/75	900	15	13	15	27

*Data supplied by Ashland Electric Products Div.

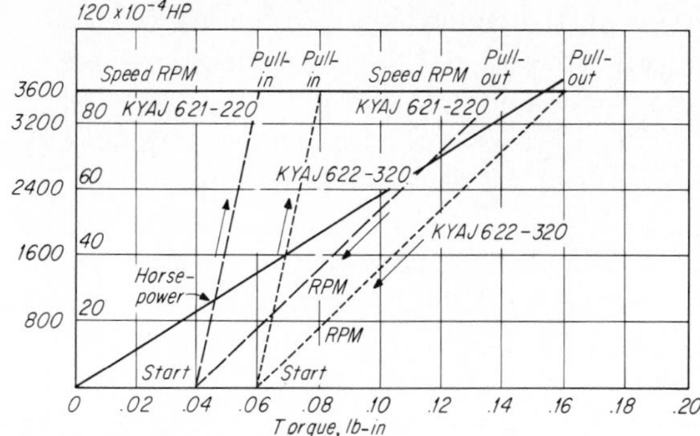

Fig. 10-5 Performance curves of two different shaded-pole reluctance motors, each the size of the motor shown in curve 1 in Fig. 9-23. *(Barber-Colman Company.)*

these motors are of the skeleton-type construction shown in Fig. 9-24, and they are of approximately the same physical size as the motor for which the characteristics are given in Fig. 9-23. Note that the pull-out torque compares quite favorably with curve 1 of Fig. 9-23, but the pull-in torque is very much less. This is probably due to the inherently high slip of the shaded-pole motor. We saw, in Art. 10-4(4), that a high-slip induction motor has reduced pull-in torque.

Some shaded-pole synchronous motors employ a permanent magnet on either end of the rotor and are called *polar synchronous motors.* Such a motor has, with respect to the line voltage, only one rotor position per pair of poles, into which it pulls into step, whereas the conventional reluctance motor has two positions per pair of poles into which it can pull into step.

10-9 Making a Synchronous Motor in the Repair Shop. A reluctance synchronous motor can easily be made in the repair shop from almost any induction motor, whether it be polyphase or single-phase. One should not expect to get much, if any, more than about one-third the rating of the induction motor from which it is made. To make such a synchronous motor, it is recommended that the easiest way is probably to mill flats on the rotor surface, equally spaced, with as many flats as there are poles in the stator winding. The total area of the flats should be of the order of 40 percent of the total cylindrical area of the rotor before the flats were milled; this means that the unmilled cylindrical area will be almost half again as large as the milled flats. In such a case, the rotor resistance rings and the conductors in the flat part will be partially, but not entirely, removed. One must not mill through the resistance rings! Usually this partial removal of conductors and resistance rings will do little appreciable harm. The motor should be able to carry roughly about one-third its nameplate rating.

An alternative, and probably a better, although a more tedious, way is to chisel out 35 to 40 percent of the rotor teeth, without disturbing the conductors or resistance rings. The teeth should be cut out at approximately equal intervals, there being as many cutouts as there are poles, as shown in Fig. 10-1. A motor of this type is particularly useful for checking tachometers, as well as for numerous other applications.

A synchronous motor can be made out of any wound-rotor polyphase induction motor by connecting two of the three-phase secondary leads to one side of a dc source and the third secondary lead to the other side of the dc source. The motor is first brought up to speed as an induction motor with secondary leads shorted; then the dc excitation is applied. Such a motor is known as a *synchronous induction motor.*

HYSTERESIS MOTORS

Unlike reluctance motors, hysteresis motors pull into or out of step smoothly, because there are no physical "poles" or projections on the rotor. Hence, hysteresis motors can pull into step just about any load they can carry, regardless of the connected inertia of that load. They do not pull into step in any predetermined position with respect to the applied voltage wave. In this respect they differ from reluctance motors which always pull into step with the rotor in one or more positions with respect to the applied voltage; with ordinary reluctance motors, there is one position per pole, but with polarized[3] synchronous motors, there is one position per pair of poles. Although hysteresis motors pull into step smoothly, they may produce a "flutter"[4] (variation in speed during a revolution) which can be objectionable in certain applications.

10-10 Hysteresis Motor Defined. A hysteresis motor is a synchronous motor without salient poles and without dc excitation which starts by virtue of the hysteresis losses induced in its hardened steel secondary member by the revolving field of the primary and operates normally at synchronous speed because of the retentivity of the secondary core.

10-11 Construction Rotor construction for a typical hysteresis motor is illustrated in Fig. 10-6*a*. Hysteresis rings of a special magnetic material, such as an alloy of steel, containing 3.5 percent chromium; or 3 to 36 percent cobalt; or alnico, are carried on a supporting arbor made of

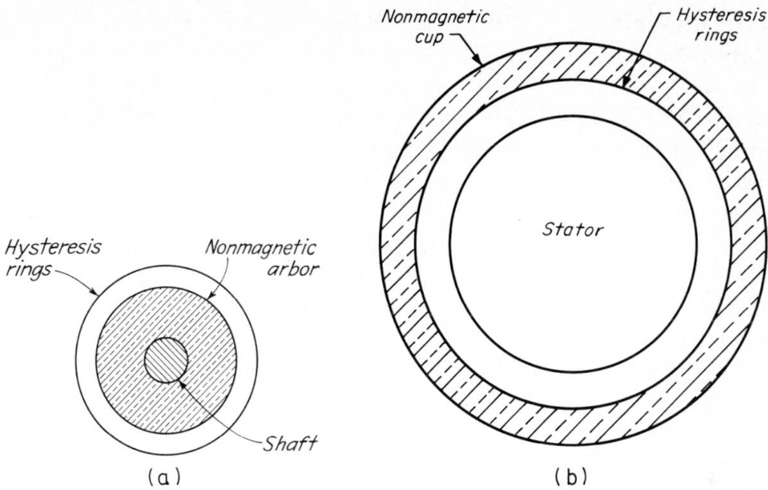

Fig. 10-6 Rotor constructions for a hysteresis motor. (*a*) Conventional construction: (*b*) inverted-rotor construction.

a nonmagnetic material such as brass; the assembly is carried on the shaft. Hysteresis rings are usually from thin stock, and several of them are assembled to give a built-up laminated rotor. In the smallest sizes, the rotor may consist of a single solid ring, or cylinder.

In the *inverted* or *umbrella* construction the hysteresis rings are pressed into a cup, as shown in Fig. 10-6*b*, and the whole revolves around the stator, which is inside the hysteresis rings. With such a rotor design, the flywheel effect of the increased rotor inertia provides a more uniform angular velocity and smoother operation. The inverted construction is ideal for gyroscopes, for the hysteresis rings can be pressed directly into the flywheel. Figure 10-7 illustrates a motor with

Fig. 10-7 Hysteresis motor with inverted-rotor construction used for tape recorders and video recorders. *(Hurst Manufacturing Corporation.)*

an inverted rotor which is used for tape recorders and video recorders.

Almost any form of stator construction that sets up a rotating magnetic field can be used; polyphase stators set up more uniform rotating fields than single-phase induction motors, but the latter are quite successful and widely used. Mechanical arrangements of the stator vary widely. Principles of operation of both polyphase and single-phase motors are discussed in the following paragraphs.

10-12 Polyphase Hysteresis Motors. Assume that the rotor of Fig. 10-6*a* is placed inside a conventional polyphase stator which, when energized, sets up a rotating field such as is shown in Fig. 2-4. In the stator iron, the field is an alternating or pulsating field. This point can be more easily understood by referring to Fig. 2-4 and noting what happens to the strength and direction of the flux in the iron back of slot 7 at each of the four times illustrated.

Time	Strength	Direction
1	4	Clockwise
2	1	Clockwise
3	2	Counterclockwise
4	4	Counterclockwise

Similarly, it can be shown that the iron back of every stator slot carries an alternating flux. Likewise, the hysteresis rings on the rotor carry an alternating flux when the rotor is stationary. (It is to be noted that the flux back of each slot is slightly out of time phase with the flux back of each of the other slots, but it is an *alternating flux* back of all slots.) It is well known that an alternating flux in an iron circuit causes losses due to eddy currents and to hysteresis. In a polyphase hysteresis motor, the torque is directly proportional to the hysteresis loss in the rotor. In fact, the hysteresis torque, expressed in synchronous watts, is equal to the hysteresis loss in the rotor at standstill. This is another way of saying that a hysteresis loss of, say, 5 W, under locked-rotor conditions, produces a locked-rotor torque numerically equal to the running torque required at synchronous speed to develop an output of 5 W. An interesting comparison is afforded by the polyphase induction motor, in which the locked-rotor torque, expressed in synchronous watts, is numerically equal to the locked-rotor secondary I^2r loss.

In other words, if the rotor of Fig. 10-6*a* is placed inside the stator of a polyphase motor that is excited to produce a rotating magnetic field, two kinds of iron losses will be produced in the hysteresis rings: eddy-current losses, which produce induction-motor torque, and hysteresis losses which produce hysteresis motor torque. There is an important distinction between these two kinds of torque. Induction-motor torque decreases to zero at synchronous speed, whereas, in an "ideal" hysteresis motor, the hysteresis torque is constant at all speeds from standstill to synchronism; i.e., *locked-rotor, pull-in,* and *pull-out torques* are all mutually equal to one another. Such a speed-torque curve is shown in Fig. 10-8. In a practical motor, the shape of the speed-torque curve is modified somewhat from the theoretical curve shown by harmonics in the rotating field and by other irregularities. For a more extended and mathematical treatment of the theory of hysteresis motor torque, the reader is referred to a paper on the subject by Teare.[5]

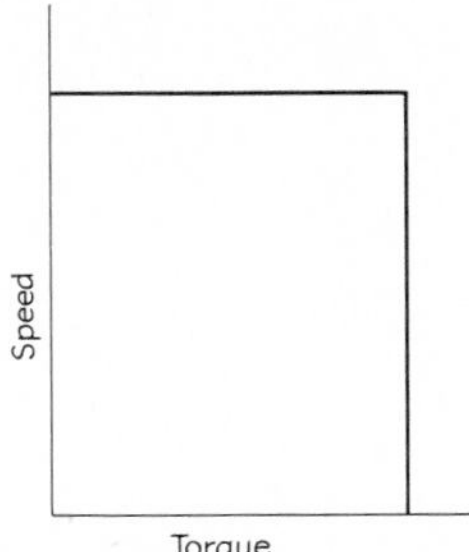

Fig. 10-8 Speed-torque curve of an "ideal" polyphase hysteresis motor.

For military applications, polyphase hysteresis motors are built in sizes down to as small as 21/32-in diameter by 1.335-in length and weighing only 2 oz. One such motor built by Globe Industries develops 0.08-oz in torque at 24,000 rpm, which amounts to 1.9 mhp. It can be used on a single-phase 400-Hz circuit by using a phase-splitting capacitor. The Singer Company, Kearfott Division, builds a 400-Hz motor which has a ¾-in diameter by a 1.25-in length and which develops about 2 mhp at 8000 rpm. Larger motors that develop more power are also available.

10-13 Capacitor-Type Hysteresis Motors. Capacitor-type hysteresis motors are essentially the same as polyphase hysteresis motors except that the rotating field is set up by a capacitor-motor stator. Rotating fields set up by a capacitor motor are generally elliptical in shape (see Art. 2-12) rather than circular, as in a polyphase motor. Also, the shape of the ellipse varies with the load because the capacitor cannot maintain balanced conditions at more than one load. Consequently, the actual speed-torque curve of a capacitor hysteresis motor departs more from the "ideal" curve of Fig. 10-8 than does that of a polyphase hysteresis motor. In fractional horsepower sizes, a capacitor-start stator may often be used, but in sizes below about 1/20, permanent-split motors have found great acceptance.

Performance specifications for one line of permanent-split capacitor hysteresis motors, with horsepower and speed ratings from 5 to 125 mhp, are summarized in Table 10-2. Figure 10-9 shows a photograph of one of these motors. In horsepower ratings smaller than these, shaded-pole hysteresis motors are frequently used.

TABLE 10-2 Performance Specifications of Small 60-Hz Hysteresis Synchronous Permanent-Split Capacitor Single-Speed Motors*

Horsepower	rpm	Full-load torque, oz-in	Locked-rotor torque, oz-in
1/200	600	8.4	8.4
1/200	900	5.6	5.6
1/75	900	15.0	15.0
1/100	1800	5.6	5.6
1/75	1800	7.5	7.5
1/50	1800	11.2	11.2
1/20	1800	28.0	28.0
1/15	1800	37.4	37.4
1/12	1800	46.6	46.6
1/20	3600	14.0	14.1
1/8	3600	35.0	35.0

* Data supplied by Ashland Electric Products Div.

Fig. 10-9 Hysteresis synchronous permanent-split capacitor motor. It may have a single-speed winding with performance as given in Table 10-2, or it may have dual-speed windings with performance as given in Table 10-3. *(Ashland Electric Products.)*

10-14 Multispeed Hysteresis Motors. Since hysteresis motors usually have a smooth cylindrical rotor, it is a simple matter to obtain more than one speed of operation by putting more than one set of windings on the stator. Polyphase windings can be wound two-speed by using consequent-pole windings. In general, however, it is not practicable to make the windings of a permanent-split capacitor motor consequent-pole, because a consequent-pole connection doubles the electrical space angle between the main and auxiliary windings. It is much better to use two separate sets of windings for a two-speed capacitor motor.

Performance on one manufacturer's line of two-speed hysteresis motors is summarized in Table 10-3. A number of manufacturers offer two-speed hysteresis motors; some offer three-speed motors.

TABLE 10-3 Performance Specifications of Dual-Speed 60-Hz Hysteresis Synchronous Permanent-Split Capacitor Motors*

Horsepower		rpm		Full-load torque, oz-in		Locked-rotor torque, oz-in	
Low-speed	High-speed	Low-speed	High-speed	Low-speed	High-speed	Low-speed	High-speed
1/300	1/150	600	1200	5.6	5.6	5.6	5.6
1/150	1/75	600	1200	11.2	11.2	11.25	11.25
1/37	1/17	600	1200	45	49	38	28
1/300	1/100	600	1800	5.6	5.6	5.6	5.6
1/300	1/150	900	1800	3.7	3.7	3.7	3.7
1/100	1/50	900	1800	11.2	11.2	11.2	11.2
1/60	1/30	900	1800	18.7	18.7	18.7	18.7
1/25	1/12	900	1800	45	46	30	28
1/60	1/40	1200	1800	14	14	14	14
1/15	1/12	1200	1800	56	46	40	20

* Data supplied by Ashland Electric Products Div.

PERMANENT-MAGNET MOTORS

Ceramic magnet materials that have been developed over the last few years have made possible the development of small permanent-magnet synchronous motors which have magnetic salient poles but not physical salient poles. It is the extremely high coercivity of the ceramic magnetic materials that has made this development possible.

10-15 Permanent-Magnet Synchronous Motor Defined. A permanent-magnet synchronous motor is one with a rotor so constructed as to have functional premagnetized magnetic poles in a magnetic material of very high coercivity.

10-16 Construction. Typical construction of the rotor of a permanent-magnet synchronous motor is illustrated in Fig. 10-10. It consists of a hollow cylinder of a very high coercivity ceramic magnet material. The cylinder has been premagnetized to set up permanent magnetic poles, alternating north and south in polarity, as shown in the figure. With such a rotor, the stator can be of a wide variety of constructions and designs and can be wound for single-phase or polyphase excitation. Of course, the stator has to be wound for the same number of poles as the rotor.

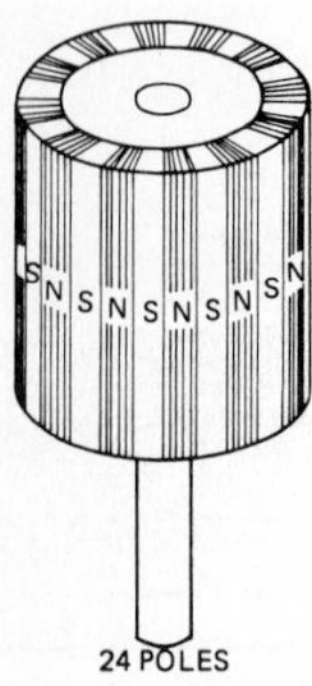

Fig. 10-10 Rotor for a permanent-magnet synchronous motor. *(Hurst Manufacturing Corporation.)*

10-17 Performance Characteristics. In operation the magnetic poles of the rotor must lock in with the poles of the rotating magnetic field produced by the stator. Because the rotor poles are permanently fixed in position, the rotor and connected load have to accelerate from standstill to synchronous speed in approximately one-half cycle of the supply frequency. This makes the permanent-magnet motor very sensitive to inertial loads and limits the amount of shaft-connected inertia that can be brought up to synchronous speed. However, if the load is directly coupled to the shaft, some torsional flexibility in the coupling can alleviate the problem somewhat. If the motor is geared to its load, the inertia of the load "seen" by the motor is reduced by the square of the gear ratio.

Starting in the wrong direction could be very disconcerting, if not disastrous in some cases. Because this can happen with permanent-magnet motors, directional reliability can be a critical item in their design and application, for in nearly all applications the motor must always start in the right direction. Hence, directional reliability is usually more important than efficiency or pull-out torque. With a permanent-split capacitor-type stator, two conditions can detract from directional reliability: overvoltage or undervoltage and improper capacitor value. These factors are illustrated in Fig. 10-11. This figure also illustrates the effect of operating this particular motor on 50 Hz.

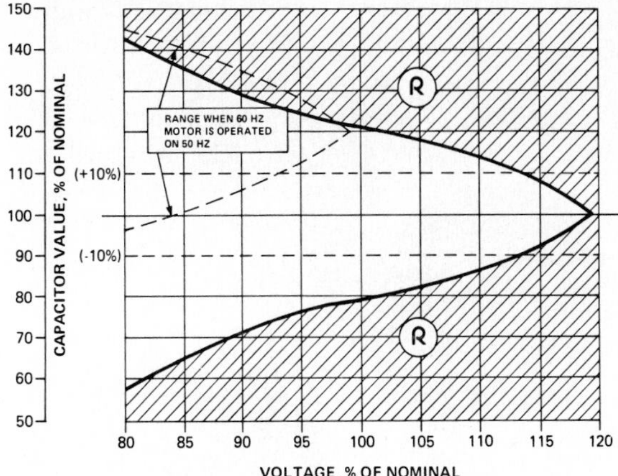

Fig. 10-11 Effect of voltage and capacitance on the directional reliability of a permanent-magnet synchronous motor. *(Hurst Manufacturing Corporation.)*

INDUCTOR MOTORS

Inductor-type motors are essentially inductor-type synchronous generators operated as motors. They have the interesting characteristic that they will run only at synchronous speed, or not at all; they do not operate at subsynchronous speeds as can reluctance and hysteresis motors. However, the small inductor-type motors described in the following paragraphs use permanent-magnet excitation, instead of dc excitation as used in conventional inductor alternators. Many small ac timing motors are inductor-type.

10-18 Construction: "50 Frame" Motors. Parts and assemblies of a synchronous inductor motor are illustrated in Fig. 10-12. Rotor punchings are mounted on an alnico disk-shaped hub. Two rotors are mounted on the shaft, as shown. Figure 10-13 shows the path of the permanent-magnet flux; all the teeth of one rotor are of the same magnetic polarity, and all the teeth of the second rotor are of the opposite magnetic polarity. There are eight polar projections to the stator punchings, and each of these has a toothed pole face with teeth having the same pitch as the rotor teeth. The stator winding consists of eight coils, each one wound around one tooth; these coils are connected as in a conventional two-phase winding.

10-19 Principles of Operation. Principles of operation are best explained with the help of Fig. 10-14, which shows eight stator teeth—

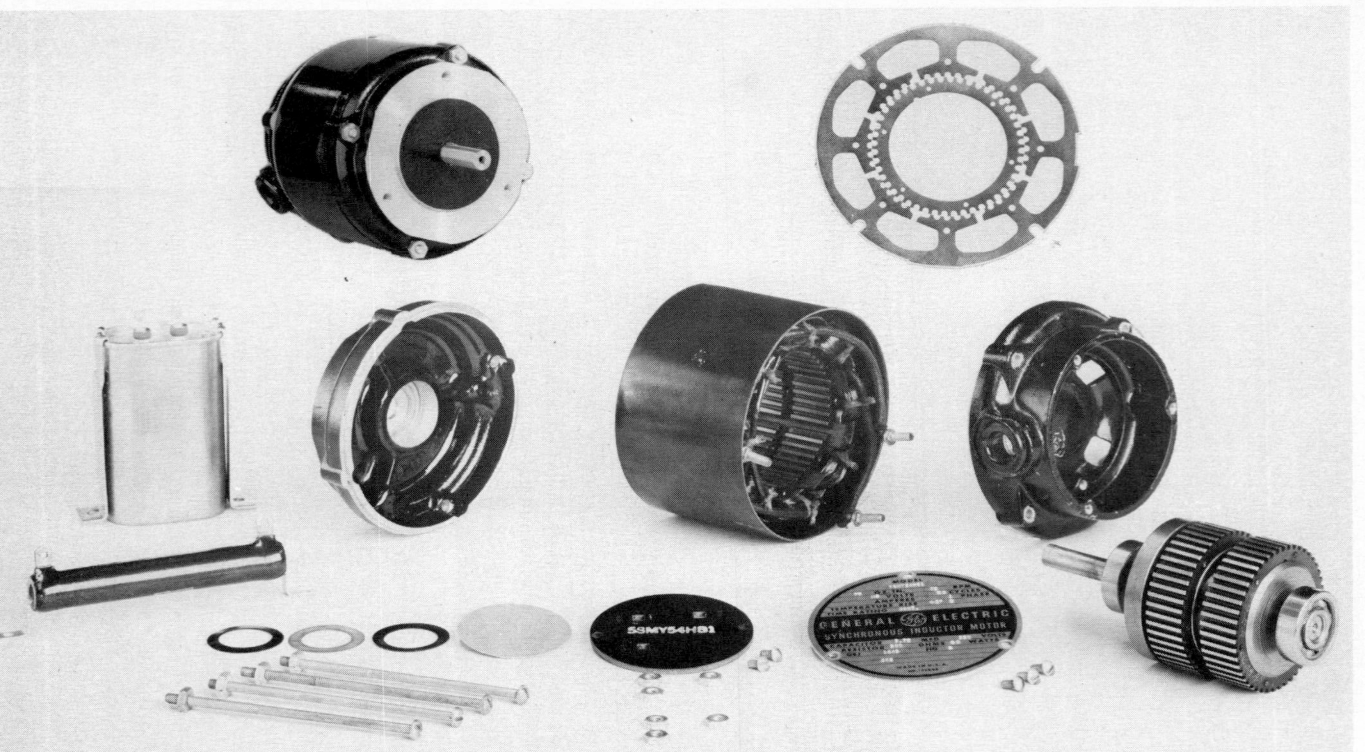

Fig. 10-12 Disassembled view of a type SMY 50-frame synchronous inductor motor. At the top left is a view of the complete motor, and at the top right are shown stator and rotor laminations in detail. Available torques are 75 to 250 oz-in, in steps of ratings. All run at 75 rpm on 60 Hz. *(General Electric Company.)*

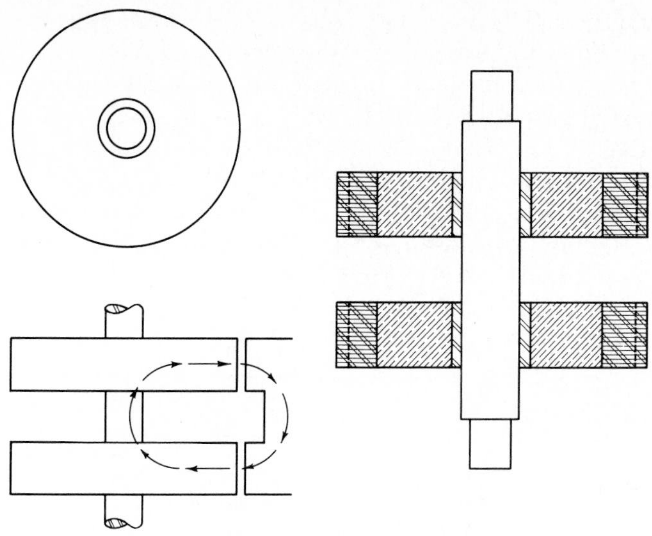

Fig. 10-13 Schematic sectional view of double-rotor inductor motor showing path of permanent-magnet flux. *(General Electric Company.)*

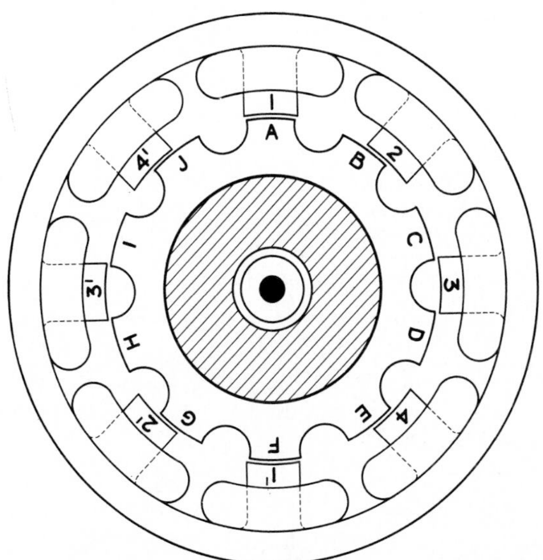

Fig. 10-14 Schematic sectional view of inductor motor with one stator tooth for each stator coil. Actual motor has several teeth per coil (see stator punching in Fig. 10-12), but principles of operation are the same. *(General Electric Company.)*

one per coil—and ten rotor teeth. Stator coils are connected as a four-pole two-phase winding, and excited, say, so as to produce a clockwise-rotating field. Assume, for purposes of discussion, that all the rotor teeth are magnetized as south poles as a result of the alnico disk. Now consider the instant of time when coils 1 and 3 are energized so that 1 is a north pole and 3 a south pole. It is obvious that the rotor locks in the position shown because pole 1 attracts A and pole 3 repels teeth C and D equally. (At this same instant, coils 2 and 4 are not excited. Although teeth B and E are attracted to poles 2 and 4 because of the permanent-magnet excitation, these two forces balance each other and produce no net torque.) If the rotor is displaced slightly from the position shown in Fig. 10-14, the magnetic forces tend to return it to that position. One-quarter cycle later, pole 2 becomes a north pole, and poles 1 and 3 are not excited. Since pole 2 is now a north pole, the rotor will turn clockwise until B and 2 are in alignment, with C displaced only one-quarter of a tooth pitch from 3; thus, the rotor has turned a quarter tooth pitch in a quarter cycle of the applied voltage. Poles 3, 4, and 1 successively become north poles at one-quarter-cycle intervals, and at each interval the rotor moves a quarter of a tooth pitch. Thus, in one cycle of power frequency, the rotor moves *one rotor tooth pitch,* and not one stator pole pitch. Therefore, when the speed of an inductor-type motor is figured, each rotor tooth must be counted as *two* poles. In Fig. 10-12, the rotor has 48 teeth; hence, from the equation in Art. 2-17 the synchronous speed for 60-Hz excitation becomes

$$\frac{120 \times 60}{2 \times 48} = 75 \text{ rpm}$$

Motors of the double-disk construction are built with two rotor-stator units spaced apart as shown in Fig. 10-12 in order to prevent excessive magnetic leakage in the axial direction. However, common stator coils thread like poles of both units. Rotors, of course, are magnetized so that opposite polarities appear at the peripheries. Stator slots are in alignment, but the teeth in two rotors are displaced circumferentially by 180 electrical degrees. Windings, instead of being two-phase, are usually of the permanent-split capacitor type. Bifilar-type motors have two windings in each of the two phases.

10-20 Characteristics and Applications. Inductor motors start almost instantaneously; they attain synchronous speed in a quarter to a half cycle of the power supply, or in 4 to 8 ms. If they did not, they would never synchronize at all! They stop equally fast, for the magnetic rotor provides a built-in braking action, as well as a holding torque—of as much as 10 oz-in—after the motor has stopped. When necessary, the holding torque can be increased by passing a small amount of direct current through the windings. Current consumption is practically independent of the load, and the motor may be stalled indefinitely without harm.

(1) Servomotor and stepper-motor applications. The characteristics just enumerated make this motor well suited to certain off-on servo applications. A circuit as simple as that of Fig. 7-3 can be used for such a purpose. Load inertia should be kept low, not over 2.5 lb-in^2, unless 5° of angular freedom is provided by the coupling, in which case more inertia can be handled. By use of proper circuitry, the motor can be used as a stepper motor, making 96 or 192 steps per revolution, depending upon the circuitry used. (See also Chap. 15.)

(2) Bifilar-type motors. These motors are supplied with two windings in each phase to simplify external circuitry when a transistorized switching circuit is used; pulsing the two coils alternatively avoids the necessity for external polarity reversals.

10-21 A Small Synchronous Inductor Motor

(1) Construction. A smaller inductor motor has been developed, using the same basic principles of operation but differing considerably in the arrangement of the active elements, which are shown in Fig. 10-15. The rotor consists of four cup-shaped stampings, each having 36 teeth. These punchings are arranged in two pairs, back to back, and adjacent to the opposite ends of a cylindrical alnico magnet with a large clearance hole for the shaft. The entire assembly is held in a fixture and die-cast as a single unit; since the die-cast material is nonmagnetic, it does not affect axial magnetization of the magnet. The outside diameter is then ground. The stator likewise consists of four cup-shaped metal stampings, each having 36 teeth. These punchings are arranged in two

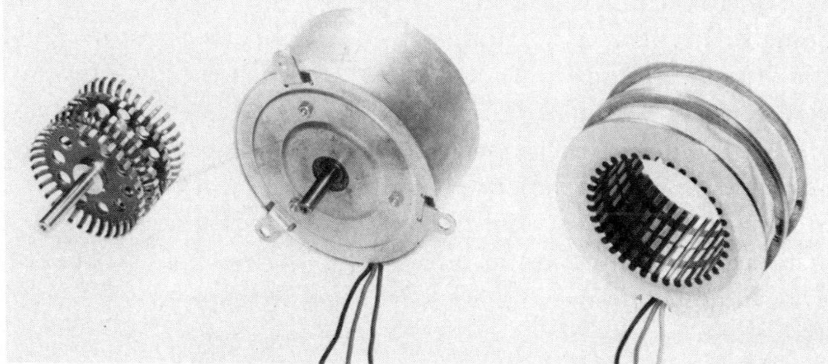

Fig. 10-15 Small synchronous inductor motor (20-frame) of different construction than the motor of Fig. 10-12. The motor has a rated torque of 2 oz-in and runs at 100 rpm on 60 Hz. *(General Electric Company.)*

pairs adjacent to each other axially, with the two middle punchings back to back. The entire assembly is encased in a single plastic-molded unit resembling a double spool. The inside bore is ground to expose four complete rings of uniformly spaced teeth in axial alignment.

(2) Principles of operation. Principles of operation are similar to those of the larger motor. In this motor, the stator teeth are axially aligned, and the two rotor punchings comprising one pair are displaced from each other circumferentially by one-half tooth pitch. The second pair of rotor punchings, identical with the first, is displaced circumferentially from the first pair by one-quarter tooth pitch. Consider the moment when the current in one winding is maximum and the current in the other zero; consider only the two stator punchings and the two rotor punchings associated with the excited coil. All 36 teeth of one of the stator punchings are magnetized as north poles; all 36 teeth of the other as 36 south poles. Now, all 72 rotor teeth have the same polarity since they are all on the same end of the alnico magnet; say that they are all south poles. Then, the rotor will assume a position such that 36 rotor teeth line up with the 36 stator teeth magnetized as north poles, like A and 1 in Fig. 10-14; the other 36 rotor teeth will fall halfway between the other 36 stator teeth magnetized as south poles, just as 3 falls between C and D in Fig. 10-14. One-half cycle later, the rotor will have moved one-half a tooth pitch; it may be in either direction without the second phase, which acts similarly but which imparts a definite predetermined direction of rotation to the rotor.

(3) Characteristics and applications. This small inductor motor has many of the same characteristics as its larger counterpart: quick starting and stopping, automatic holding, etc. It is used chiefly as a timer motor. It is supplied with three external leads to be used as a reversible motor. The torque at synchronous speed is a series of pulsations, with peak values occurring four times per cycle. Because of this torque pulsation, ball-bearing models can run in the direction opposite to what is intended, though this seldom happens with sleeve-bearing motors. This can be avoided by keeping the motor loaded at all times, preferably near its rated load; it should definitely not be used when the load may overhaul the motor. For a fast, smooth start, the manufacturer recommends allowing 5° of free play between the motor and its load.

TIMING MOTORS

AC timing motors are small synchronous motors, with power outputs generally only a fraction of a millihorsepower, often as little as 0.001 mhp. They are used in industrial and military control systems for timing any number of functions. Less costly versions are widely produced for driving small clocks or for other timing applications. Usually they are supplied with a built-in gear reduction to give an output-shaft speed of 1 rpm. A popular construction is to put these gears into a pear-shaped cup as shown in Figs. 12-13 and 12-14. Commonly used electrical types are shaded-pole hysteresis, capacitor hysteresis, and inductor. Manufacturers of timing motors are listed in Table 15-1.

10-22 Multipolar Shaded-Pole Hysteresis Motor with a Single Stator Coil. A shaded-pole hysteresis motor of novel construction, used for demand meters, two-rate meters, recording instruments, and time switches, is illustrated in Fig. 10-16. It has but a single cylindrical stator coil which excites 12 main coils. Two washers serve as shading coils for all 12 poles. Synchronous speed of a 12-pole motor is 600 rpm for a 60-Hz circuit.

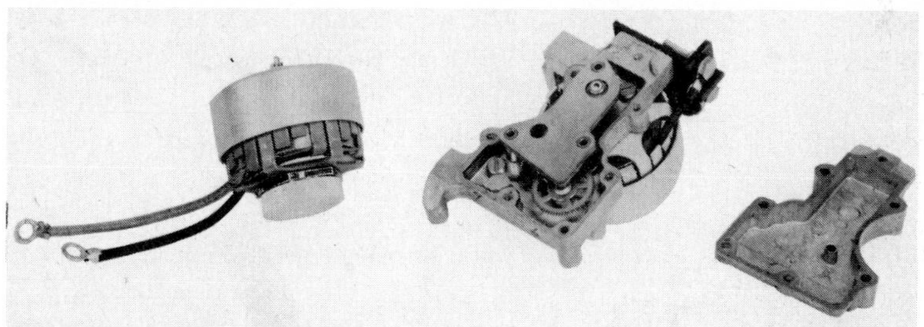

Fig. 10-16 Multipolar shaded-pole hysteresis motor used for demand meters, timing switches, etc. *(Westinghouse Electric Corporation.)*

10-23 Multipolar Capacitor Hysteresis Motor with Only Two Stator Coils. A 16-pole capacitor hysteresis motor of essentially the same construction discussed in the preceding article is shown in Fig. 10-17.

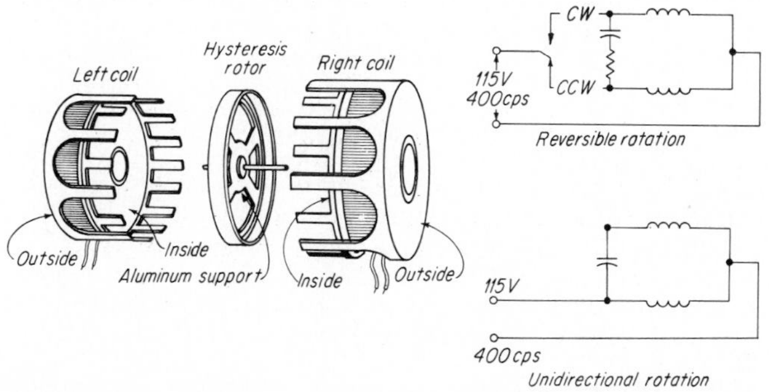

Fig. 10-17 Typical capacitor-type 400-Hz timing motor. *(Reprinted from Richard J. Kavanaugh and Edwin Ellner, "Timing Motors," Mach. Des., The Electric Motor Book, 1961, pp. 138–147.)*

Schematic wiring diagrams are given in the figure for a three-lead reversible arrangement or for a two-lead nonreversible arrangement. When the two stator coils are identical, the motor can be operated from a two-phase supply. Motors of this construction are available for 400-Hz circuits and also for 60-Hz circuits. On the former, they operate at 3000 rpm; on the latter, at 450 rpm.

10-24 Inductor-Type Timing Motors. Inductor-type timing motors may be of the type described in Arts. 10-18 to 10-21, or they may use a shaded-pole primary to determine direction of rotation. As pointed out earlier, such motors run either synchronously or not at all; they have to synchronize in a quarter to a half cycle or less, or they will not synchronize. Also, they are limited in the amount of inertia they can pull into step.

References

1. Trickey, P. H.: "The Non-Excited Synchronous Motor." *Electr. J.*, April 1933.
2. ———: "Performance Calculations on Polyphase Reluctance Motors," *AIEE Trans.*, 1946, pp. 191–193.
3. Winston, Joseph: "The Polarized Synchronous Motor," *Electro-Technology*, July 1963, p. 92.
4. Miller, Norman A.: "Hysteresis Motor Flutter," *Electro-Technology*, July 1964, pp. 123–127.
5. Teare, B. R., Jr.: "Theory of Hysteresis Motor Torque," *AIEE Trans.*, vol. 59, 1940, p. 907.

Bibliography

Chang, S. S. L.: "An Analysis of Unexcited Synchronous Capacitor Motors," *AIEE Trans.*, vol. 70, pt. II, 1951, pp. 1978–1982.

Chi Yung Lin: "Characteristics of Reluctance Machines," *AIEE Trans.*, vol. 70, pt. II, 1951, pp. 1071–1077.

———: "Equivalent Circuits of Reluctance Machines," *AIEE Trans.*, vol. 71, pt. III, 1952, pp. 1–9.

Douglas, J. F. H.: "Synchronous Performance of Reluctance Motors by Improved Circle Diagrams." *AIEE Trans.*, vol. 77, 1958, pp. 1038–1041.

———: "Current Loci of Permanent-Magnet Synchronous Motors," *AIEE Trans.*, vol. 78, pt. III, 1959, pp. 76–78.

———: Pull-In Criterion for Reluctance Motors, *AIEE Trans.*, vol. 79, 1960, pp. 139–142.

Lee, C. H.: "The Theory and Design of a Very Slow-Speed Reluctance Motor," *AIEE Trans*, vol. 78, 1959, pp. 1683–1699.

Ohkawa, M., and S. Nakamura: "Characteristics and Design of Permanent-Magnet Synchronous Motors," *Electr. Eng. (Japan)*, vol. 90, no. 6, 1970, pp. 125–136.

Talaat, M. E.: Steady-State and Transient Synthesis of 3-Phase Reluctance Motors, *AIEE Trans.*, vol. 70, pt. II, 1951, pp. 1963–1970.

CHAPTER

11

UNIVERSAL MOTORS

The universal motor is characterized by its ability to operate, with substantially the same performance, on direct as well as alternating current of frequencies up to 60 Hz. It develops more horsepower per pound than other ac motors, principally because of its high speed. These motors are series-wound and have series characteristics on both alternating and direct current, except when governors or other means are used to control their speed. No-load speeds are high, sometimes well over 20,000 rpm; but the armatures are designed so that they will not be damaged at these speeds. Power ratings vary from 10 mhp to 1 hp, for continuous-rated motors, and even higher for intermittent-rated motors. They are usually designed for full-load operating speeds of 4000 to 16,000 rpm in the larger horsepower ratings, and up to 20,000 or more in the smaller power ratings. At the higher speeds, better universal characteristics (i.e., more nearly the same performance characteristics on both alternating and direct current) can be obtained, as well as more output per pound. Universal motors are generally custom-built for a specific application and are very often sold as parts rather than as complete motors. Usually they are not off-the-shelf items, though a number of companies do stock them.

11-1 Universal Motor Defined. A universal motor is a series-wound or a compensated series-wound motor designed to operate at approximately the same speed and output on either direct current or single-phase alternating current of a frequency not greater than 60 Hz and of approximately the same root-mean-square (rms) voltage.

11-2 Application and Torque Characteristics. Very popular applications for universal motors include portable drills, saws, routers, vacuum cleaners, sewing machines, food mixers, blenders, and many others. There used to be two major types of universal motors, *noncompensated* and *compensated*, but the latter has now all but disappeared. The noncompensated motor usually is built with concentrated or salient poles. The speed-torque curve of a noncompensated concentrated-pole universal motor, for both ac and dc operation, is given in Fig. 11-1. Similar speed-torque curves for a compensated motor are given in Fig. 11-2. It is to be noted that the compensated universal motor has better universal characteristics (i.e., more nearly the same speed on both alternating and direct current) than the noncompensated motor. It is also to be noted that the superiority of the compensated motor in terms of universal characteristics is more marked at low speeds than at high speeds.

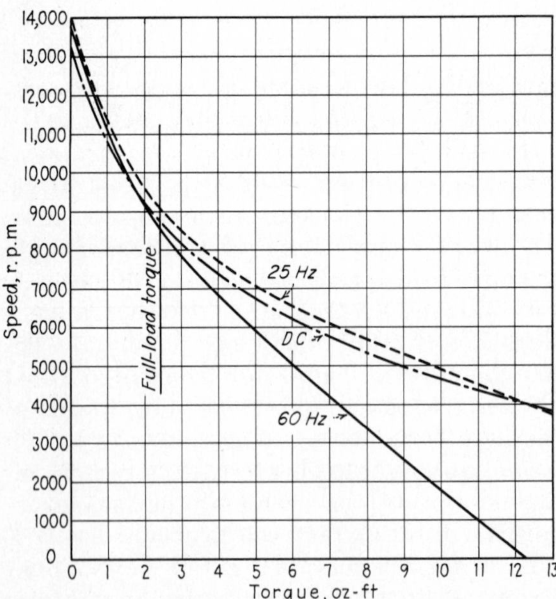

Fig. 11-1 Speed-torque curves of a noncompensated universal motor rated at ¼ hp and 8000 rpm.

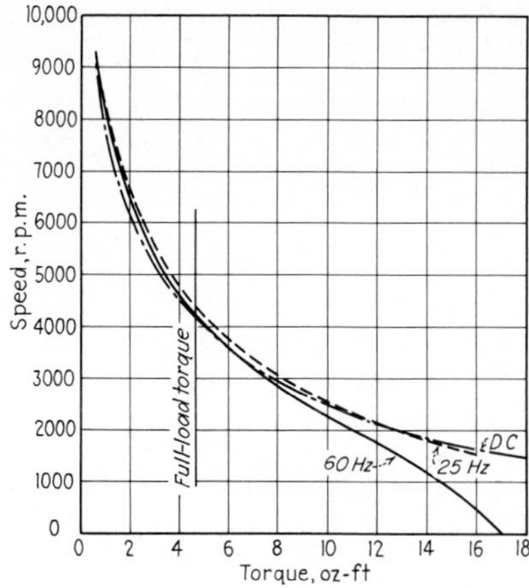

Fig. 11-2 Speed-torque curves of a compensated universal motor rated at ¼ hp and 4000 rpm.

The noncompensated motor is less expensive and simpler in construction and is more generally used for those reasons.

It is to be noted that with either type, the speed drops off rapidly with an increase in load and increases with a decrease in load. This characteristic is most desirable in vacuum-cleaner service, for if the cleaner is used under conditions which decrease the volume of air handled, the load on the motor decreases. This decrease in motor load is accompanied by increased motor speed and increased vacuum, so that the cleaner will actually handle more air than it would if a constant-speed motor were used. Likewise, this characteristic of speeding up on light loads is very desirable in the case of portable drills, for the motor will drive small drills at high speed and larger drills at a lower speed.

When dc distribution networks were commonly used in the downtown areas of large cities, universal motors were popular for driving business machines of all kinds, because of their ability to operate on either direct or alternating current. Constant speed, when needed, was obtained by means of a mechanical governor. Now that such dc distribution networks have all but disappeared, the universal motor in business machines has been replaced by some form of induction motor which is inherently constant-speed, quieter, and less expensive; moreover, the induction motor does not produce the radio interference of a commutator-type motor. Also, when the application requires an adjustable-speed motor, electronic speed-control means are now feasible.

No industrywide standards of horsepower and speed ratings have been established for universal motors because these motors are often sold as sets of parts, because the choice of possible full-load operating speeds is virtually unlimited, and because most of their applications are highly specialized. However, the dimension standards in Table 11-1

TABLE 11-1 Dimensions for Universal Motor Parts*

BH, in +0.000 −0.002	CL, in	CO, in (to fit)
2.125	1.83	No. 6 screw
2.437	2.12	No. 8 screw
2.625	2.22	No. 8 screw
2.875	2.44	No. 8 screw
3.187	2.75	No. 10 screw
3.687	3.25	No. 10 screw
4.375	4.12	No. 10 screw

*For explanation of symbols, see Fig. 11-3.
Source: From *NEMA Motor and Generator Standards*, NEMA Publ. no. MG1-1978, National Electrical Manufacturers Association, Washington, 1978.

have been established by the National Electrical Manufacturers Association (NEMA).[1,*]

The standard direction of rotation is counterclockwise, facing the end opposite the shaft extension.

11-3 Essential Parts of the Universal Motor

(1) The concentrated-pole noncompensated motor. The parts for a concentrated-pole noncompensated universal motor for a built-in application are illustrated in Fig. 11-3. The shape of the stator punching and that of the complete wound stator are shown. The armature has a commutator and is wound like the armature of a dc motor or like the rotor of a repulsion-start induction motor. Figure 11-3 also serves to explain the lettered dimensions given in Table 11-1.

Since universal motors are so generally supplied as parts, NEMA has prepared Authorized Engineering Information (MG1-18.061)[1] on common practice in motor frame designs for universal motor parts, in order to assist the user in making proper use of them. The following is abstracted from this source.

> Parts sets which are intended for reversible motors should have armatures with neutral internal connections. For nonreversible motors, the armature should be constructed to have the commutator leads shifted to give the best brush life and commutation.
>
> Armature internal connections are normally based on the motor's brush mechanism being located midway between the poles of the stator.

The way to connect armature leads to the commutator to achieve the preceding effect is described in considerable detail in the section on

* For numbered references, see References at end of this chapter.

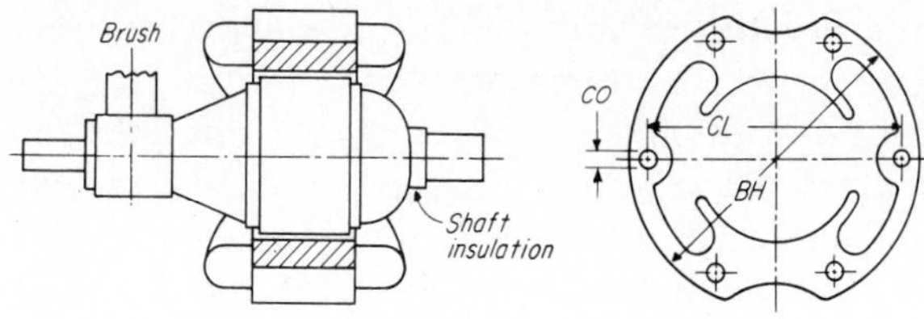

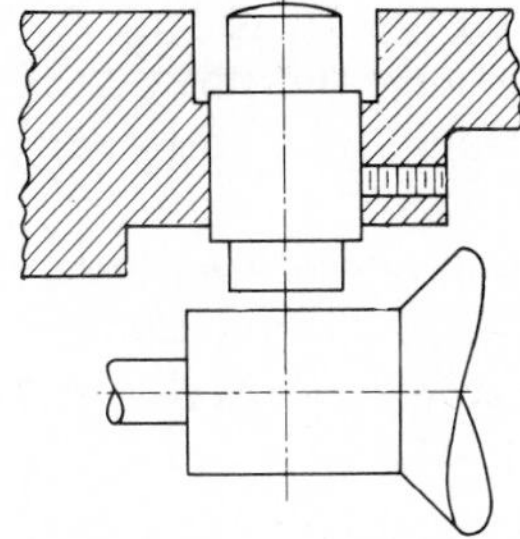

Fig. 11-3 Outline drawing showing arrangement of parts. The upper-right-hand view of the stator punching identifies the dimensions of Table 11-1. *(From NEMA Motor and Generator Standards, NEMA Publ. no. MG1-1978, National Electrical Manufacturers Association, Washington, 1978.)*

armature windings in Chaps. X and XII of an earlier edition of this book[2] and, to a lesser extent, later in this chapter.

(2) Distributed-field compensated universal motors. The armature of a distributed-field compensated universal motor is generally the same in appearance as that of noncompensated universal motors. However, the stator punching looks very much like the stator punching for a two-pole induction motor. Some compensated universal motors use two windings: a *field winding* and a *compensating winding,* displaced 90° from the main winding. Other compensated universal motors use a single winding and shift the brushes so that the single winding serves two purposes, as a single winding does in the repulsion-start motor. These winding arrangements and the reason for the compensating winding are discussed more fully in Art. 11-6.

PRINCIPLES OF OPERATION

11-4 Operation on Direct Current. The theory of the operation of the dc series motor is discussed in Arts. 12-3 and 12-6 and will not be repeated here.

11-5 Operation on Alternating Current. If alternating current is applied to a series motor, it will start and run. The current in the armature circuit, of course, reverses 120 times per second (for 60 Hz), but the field excitation and stator flux likewise reverse 120 times per second, and these reversals take place *in time phase with the armature current.* On alternating current, the torque varies instantaneously 120 times per second, but the torque developed is always in one direction. (It is, perhaps, superfluous to say that the motor operates in the same direction of rotation on alternating current that it does on direct current.) However, there are some effects present on ac operation that are not present on dc operation.

(1) Laminated-field construction. Because the stator flux alternates, it is necessary to use a laminated-field structure in order to reduce hysteresis and eddy-current losses.

(2) Reactance voltage. In a simple dc circuit, the current is limited by the resistance. In a simple ac circuit, the current is limited by the impedance and not solely by the ohmic resistance. The impedance is made up of two components, resistance and reactance. Reactance is present in an ac circuit whenever a magnetic circuit is set up by the current flowing in the electric circuit. Reactance is, therefore, present to a marked degree in the case of a universal motor. This reactance voltage, which is present during ac but not dc operation, absorbs some of the line voltage, reducing the voltage applied to the armature, so that the speed of the motor, for any given current, tends to be lower on alternating than on direct current.* In other words, the effective voltage on the armature for any given current is less on ac than on dc operation.

(3) Saturation effect. In the preceding paragraph, it was shown that the tendency of the reactance voltage is to make the speed lower on alternating than on direct current. There is another effect, which gives the opposite tendency. This effect is simply that a given *root-mean-square* (rms) *value* of alternating current will produce less rms alternating flux than will a direct current of the same value because of saturation effects in the iron.† At low currents and high speeds, the

*The *IR* drop and the reactance voltage of both the armature and field are all subtracted by phasors from the voltage applied to the motor to obtain the counter electromotive force (emf). The counter emf, hence speed, is reduced by the reactance voltage regardless of whether the latter appears in the field or in the armature.

†When the instantaneous value of the alternating current is exactly equal to the rms value, the magnetic flux will be substantially the same as it would be with the same amount of direct current flowing through the windings. Forty-five degrees later in time, the instantaneous value of the alternating current has risen to its *peak* value, which is 1.414 × its rms value. As a result of saturation of the iron parts of the magnetic circuit, the flux does not increase 41.4 percent; therefore, the effective, or rms, value of the flux is less for a given effective value of alternating current than for the same value of direct current.

reactance voltage is relatively unimportant, and this saturation effect usually causes the motor to operate at a higher no-load speed on alternating than on direct current. Likewise, under 25-Hz operation, the saturation effect is as pronounced as on 60 Hz, but the effect of the reactance voltage is appreciably less, in the ratio of 25:60. The net result is that the motor may sometimes operate at a higher speed on 25 Hz than it does on direct current. (See Fig. 11-1.)

(4) Commutation and brush life. The commutation on alternating current is substantially poorer than on direct current, and the brush life is likewise less. The principal reason for the poorer commutation on alternating current is because of the voltage induced in the short-circuited coils undergoing commutation by the transformer action of the alternating main field. No such transformer voltage exists when the motor is operated on direct current.

11-6 Compensated Universal Motors

(1) Purpose of compensating winding. It was mentioned in the preceding article that the principal reason that the speed is lower on alternating than on direct current under normal load conditions is the presence of the reactance voltage. The reactance voltage, or simply reactance, occurs both in the armature and in the field. The reactance voltage due to the field cannot be eliminated—unless the working field is eliminated—but the reactance voltage due to the armature can be practically eliminated by the use of a compensating winding. The compensating winding is a winding connected in series with the armature and so arranged that the ampere-turns of this winding oppose and neutralize the ampere-turns of the armature. To obtain this compensation, the compensating winding is displaced 90 electrical degrees from the field winding. When properly proportioned, the compensating winding virtually nullifies the effect of the reactance voltage due to the armature. There is another equally important effect, namely, that the compensating winding eliminates the distortion of the main-field flux and improves the commutation; a slight overcompensation further assists commutation.

One way to keep the field reactance down is to keep the product of field turns and flux as low as possible. The weaker the field is made, the stronger must the armature be (the more turns must it have) to develop the necessary torque and output. A strong armature will distort a weak field, and there are definite practical limits as to how low the ratio

$$\frac{\text{Field ampere-turns}}{\text{Armature ampere-turns}}$$

can be made. When a compensating winding is used, this ratio can be made lower than with a noncompensated motor.

To summarize: (1) The compensating winding neutralizes the reactance voltage that would be present in the armature, (2) it practically eliminates field distortion due to the armature, and for this reason, it permits the designer to take reactance out of the field and put it into the armature, where it can be neutralized, and (3) it is a pronounced boon to commutation. As a matter of fact, it is really because of the assistance to commutation that the compensated motor can be operated at weak fields. The compensated universal motor is in reality a miniature edition of an ac series motor such as is used in railway service.

(2) Two-field compensated motor. As mentioned previously, the compensating winding is displaced 90 electrical degrees from the main winding. Such an arrangement is shown in the distribution chart of Fig. 11-4*a*. In Fig. 11-4*b*, the direction of current flow in each of the stator slots and in each of the windings is shown. Likewise, the neutral position of the brushes is shown as opposite slots 18, 9, and 18. (The actual physical position of the brushes depends on the way the connections are made to the commutator, but these positions are shown as the reference axes for the brushes.)

(3) Single-field compensated motor. A little study will reveal how the two windings of Fig. 11-4*a* and 11-4*b* can be replaced by a single winding. For example, it will be noted that, in slots 7, 8, 16, and 17, the effect of one winding is to cancel the effect of the second winding. Therefore, since these conductors mutually cancel one another, the

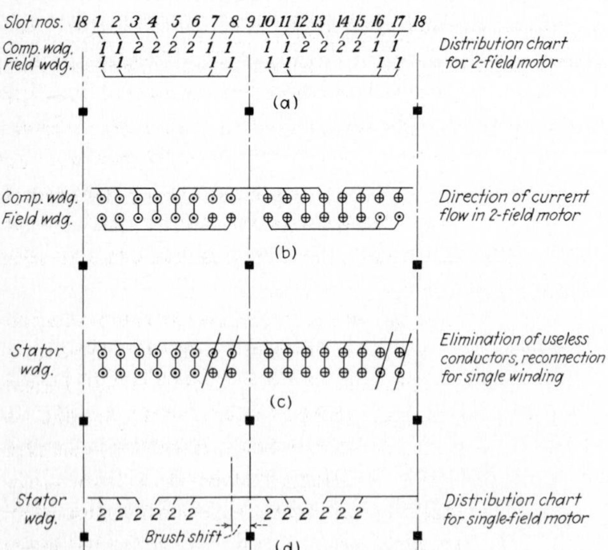

Fig. 11-4 The equivalence of a single-field winding with shifted brushes to two separate windings.

conductors themselves *could be omitted without affecting the operation of the motor.* These conductors are shown canceled in Fig. 11-4*c*. Now then, we can reconnect the conductors in the various slots shown in Fig. 11-4*c*, and we have here a single winding, the distribution chart of which is shown in Fig. 11-4*d*. Now the brushes were not moved, but the flux conditions for the same current in the single winding shown in Fig. 11-4*d* are exactly the same as the flux conditions for that same current in the two windings given in Fig. 11-4*a*. It is to be noted, however, that the center of this new single stator winding is displaced one slot from the brushes. In other words, the one stator winding with shifted brushes is exactly equivalent* to the two windings.

Thus, the net result of all the foregoing is to show that a single winding, as in Fig. 11-4*d*, *can give the same magnetic fields as the two windings represented in Fig.* 11-4*a*, *provided that the brushes are shifted one slot from the neutral axis of the stator winding.* The two-field motor, however, can be reversed by changing the connections of the main winding; but, in order to reverse the single-field motor, it is necessary to shift the brushes, just as in the case of a repulsion motor. The single-field motor uses less copper, and has fewer I^2R losses than its two-field equivalent.

STATOR WINDINGS

11-7 Concentrated-Pole Noncompensated Motors. The stator and brush-holder connections of a simple two-lead nonreversible concentrated-pole noncompensated universal motor are given in Fig. 11-5. It will be noted that the brushes are located halfway between the centers of the stator coils. When the brushes are so located, the connections from the armature coils to the commutator cannot be connected *on center*,† but they must be either *thrown left*† or *thrown right*;† i.e., the coil connections are brought straight out to commutator bars approximately opposite one of the two slots in which one of the two sides of the coil lies. In Fig. 11-5, arrows are drawn to assist in tracing the polarity of the stator coils. If the current is assumed to flow in through the upper leads and through the stator coils as indicated by the plus signs and dots and if, further, the armature connections are known, the direction of rotation can be predetermined as indicated in the tabulation of Fig. 11-5.

CAUTION: In Fig. 11-5, it is shown that the armature could be caused

*For an explanation of this term, see Art. 3-1.
†See Fig. 11-10.

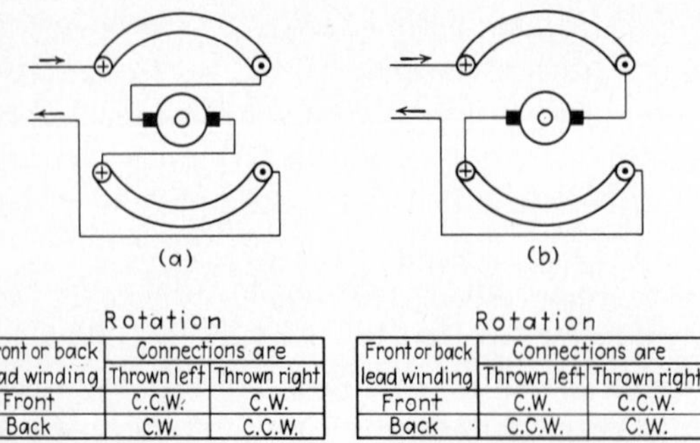

Rotation

Front or back lead winding	Connections are	
	Thrown left	Thrown right
Front	C.C.W.	C.W.
Back	C.W.	C.C.W.

Rotation

Front or back lead winding	Connections are	
	Thrown left	Thrown right
Front	C.W.	C.C.W.
Back	C.C.W.	C.W.

Fig. 11-5 Motor wiring and brush-holder connection diagrams for a series motor. From these diagrams and the table, the direction of rotation can be predetermined. Progressive windings are assumed in all cases, and rotation is viewed from the commutator end. (*a*) Connection diagram showing brush-holder connections; (*b*) brush-holder connections reversed.

to rotate in either direction simply by interchanging the leads on the brush holders. From this, it should *not* be inferred that any two-lead noncompensated motor can be reversed satisfactorily simply by interchanging the brush-holder leads, for it is a fairly common practice to connect the armature one or two bars or more "off neutral" in order to obtain better communication, better torque characteristics, and better universal characteristics. Such a motor can be reversed by interchanging brush-holder connections, but the commutation will be much poorer and the torque at any given normal operating speed may be reduced by as much as 50 percent. If a universal motor shows such symptoms, it may be that its brush-holder leads were improperly interchanged!

11-8 Split-Series Electrically Reversible (Reversing) Motors. The armature and field connections of a split-series three-wire reversible concentrated-pole universal motor are shown in Fig. 11-6. With this arrangement, one stator coil is used to obtain one direction of rotation, and the other stator coil to obtain the other direction of rotation, only one stator coil being in the circuit at a time. Needless to say, the armature connections must be on neutral in order to obtain satisfactory operation in both directions of rotation. It is interesting to note that the external reversing control of such a motor is identical with the control of a reversible high-torque capacitor motor described in Art. 7-19; the split-series motor is electrically reversible and may be used for plugging service.

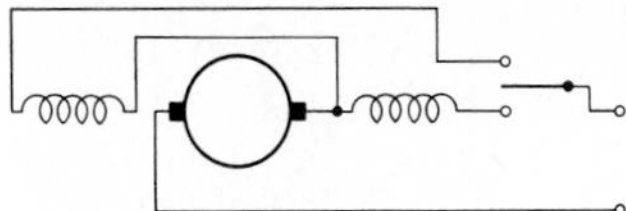

Fig. 11-6 Connections for concentrated-pole noncompensated split-series three-wire reversible universal motors.

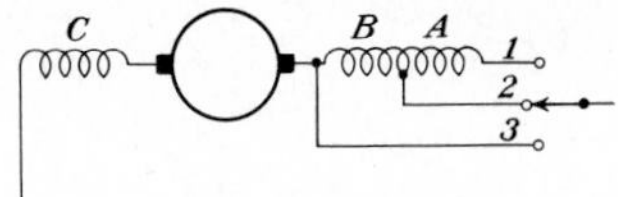

Fig. 11-7 Connections for concentrated-pole noncompensated tapped-field two- or three-speed nonreversible universal motors.

11-9 Tapped-Field Winding. Speed control of concentrated-pole noncompensated universal motors is sometimes effected by using tapped-field windings. Such an arrangement is represented in Fig. 11-7. There may be three taps as shown there, or simply the two taps 1 and 2. Taps 1, 2, and 3 give *low, medium,* and *high* speed, respectively. Sections A and B may comprise one field coil, and section C the other; it is to be noted, however, that the wire sizes in these three sections may all differ from one another.

It should be pointed out that the tapped-field universal motor still has varying-speed characteristics, regardless of which tap is used; but for fans, blowers, and many other applications, satisfactory control of the speed can be obtained.

Sometimes a tapped field is employed merely to obtain the same speed operating on direct current as on alternating current; for operation on alternating current, tap 2 is used, and for operation on direct current, tap 1. With such a tap properly located, it is possible to make the motor operate at the same speed on 60 Hz as it does on direct current for any *one* given torque load, but the speeds on direct and on alternating current may not be the same at any other load, although they will be closer together than if no tap were used.

11-10 Two-Field Compensated Windings. The stator and brush-holder connections for a reversible two-winding compensated universal motor are given in Fig. 11-8. It is to be noted that the compensating winding and armature must be reversed as a unit; it is not sufficient to reverse only the armature.

11-11 Two-Field Compensated Split-Series Windings. Two-field compensated universal motors can be connected as split-series motors. Here again it is to be noted that the compensating winding and armature are permanently connected in series as a single unit, regardless of the direction of rotation.

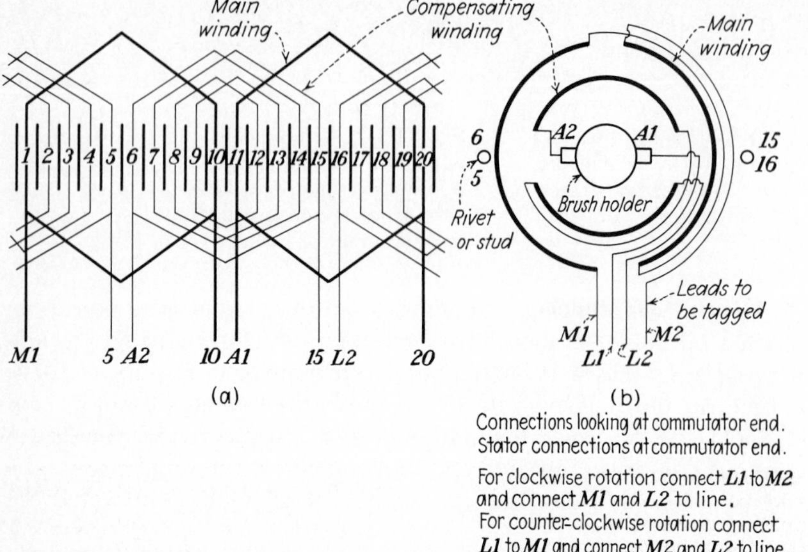

Fig. 11-8 Stator and brush-holder connections for a reversible two-winding compensated universal motor. (*a*) Stator windings; (*b*) connection diagram.

11-12 Single-Field Compensated Motor. In Art. 11-6, it was explained that a single distributed field winding could accomplish the same results as two windings displaced 90° apart, provided that the brush axis were shifted slightly from the axis of the stator winding. The stator and brush-holder connections of such a motor are given in Fig. 11-9. In this diagram, the brushes are shown shifted *with rotation.* The armature connections for this motor are on center. The single stator winding corresponds to the compensating winding on a two-field motor, and the normal brush position would be between the coils; however, in this fig-

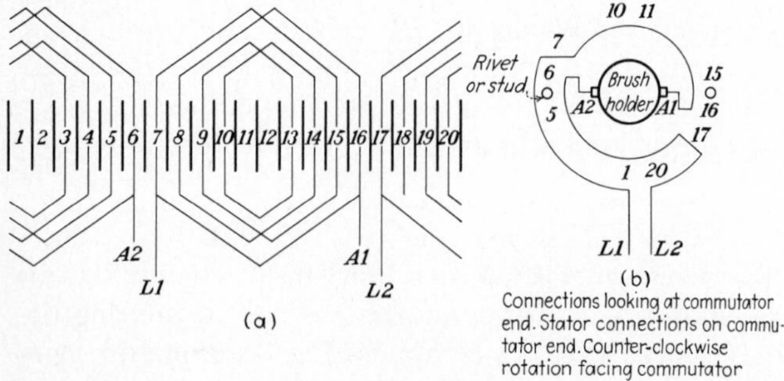

Fig. 11-9 Stator and brush-holder connections for a single-field compensated universal motor with counterclockwise rotation. (*a*) Stator windings; (*b*) connection diagram.

ure, they are shown displaced slightly in a counterclockwise direction because the armature is to rotate counterclockwise. It should be pointed out, however, that the brushes might be located between the centers of the stator coils, and the brush shift obtained by making the armature connections two, three, or four bars off neutral.

The single-field compensated motor can be reversed not by changing the connections of the winding but only by shifting the brushes.

ARMATURE WINDINGS

Armature windings are discussed briefly in this section. They were treated in greater detail in an earlier work.[2]

11-13 General. Universal motors almost invariably are wound for two poles. In principle, the windings are the same as the windings used in a dc or in a repulsion-start motor. But in universal motors (and also in small dc motors) another type of winding, known as the back-lead winding, is often used. (See Art. 11-15.)

11-14. Front-Lead Windings. In front-lead windings, the commutator connections from the armature coils are brought out in front of the armature, i.e., on the commutator side of the laminations. Three methods of making these armature connections to the commutator are illustrated in Fig. 11-10. The winding progresses to the left around the armature. For a more detailed description refer to previous works by Veinott.[2,3]

(1) Three main types of connections. Roughly speaking, there are the three main types of connections shown in Fig. 11-10: on-center, thrown-left, and thrown-right. When the connections are on center, the brushes will fall opposite the centers of the poles. When the connections are either thrown left or thrown right, the neutral position for the brushes falls between the poles. Direction of rotation depends upon which way the connections are thrown, as shown in tabular form in Fig. 11-5.

(2) "On-neutral" and "off-neutral" connections. When the armature connections are on neutral, the armature is suitable for rotation in either direction. However, an off-neutral connection is often used, for reasons given in Art. 11-7. The effect may be obtained either by shifting the brushes themselves against rotation, or by moving the armature connections to the commutator with rotation, as illustrated in Fig. 11-10.

(3) Single-field compensated motors. The neutral position of the brushes with respect to the stator can be determined by use of the diagrams in

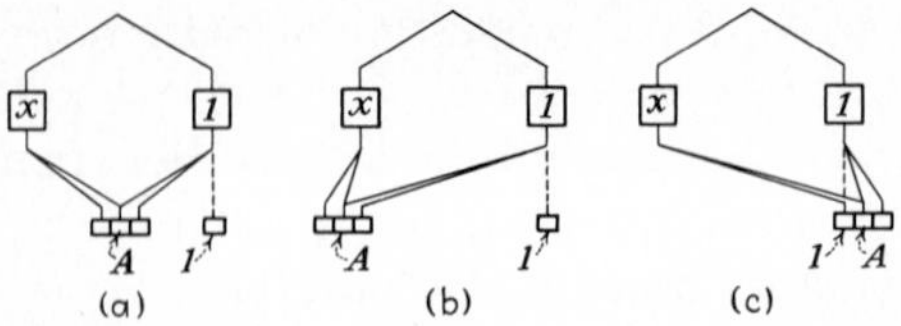

(a) (b) (c)

If the number of slots is even:

Brush position	Opposite centers of poles or opposite main coils	Between poles or opposite centers of compensating coils	Between poles or opposite centers of compensating coils
On neutral for CW or CCW rotation	$A = x$	$A = x + \frac{b}{4}$	$A = x + \frac{3}{4}b$
1 bar off neutral for CCW rotation	$A = x + 1$	$A = x + \frac{b}{4} + 1$	$A = x + \frac{3}{4}b + 1$
1 bar off neutral for CW rotation	$A = x - 1$	$A = x + \frac{b}{4} - 1$	$A = x + \frac{3}{4}b - 1$

If the number of slots is odd: $b/4$ comes out as 5½—a fraction. In such a case, drop the ½ and locate bar no. 1 to the left of the center line of slot no. 1 so that mica to right of bar no. 1 is opposite the center line of slot no. 1. Otherwise follow the formulas above.

Note: If from formula, $A > b$, subtract b from result to find A.

Fig. 11-10 Armature connections to the commutator. The armature has front-lead windings and two bars per slot. This figure shows how to lay out an armature connection diagram for any lap-wound armature with two bars per slot and front-lead windings. (*a*) On center; (*b*) thrown left; (*c*) thrown right.

Fig. 11-10, if the stator winding is considered—for purposes of location of neutral—as the compensating winding. This neutral corresponds to the hard neutral of a repulsion motor. The brushes must then be shifted a proper amount with rotation.

(4) Connection diagrams for three bars per slot. The diagrams of Fig. 11-11*a* and 11-11*b* show how to lay out the connections on center for three bars per slot. The connections can then be shifted one, two, or three commutator bars to the left or the right for off-neutral operation, as explained in detail in Fig. 11-10.

11-15 Back-Lead Windings. In back-lead windings, the wire is started at the back (the end opposite the commutator) of the armature, and all the commutator connections are brought out at that end. The progress of a typical winding around the armature is illustrated in Fig. 11-12. When there are two bars per slot, a short and a long loop are brought out of each slot. Short and long loops are for connections between coils, and the two different lengths identify the first and second coils in each

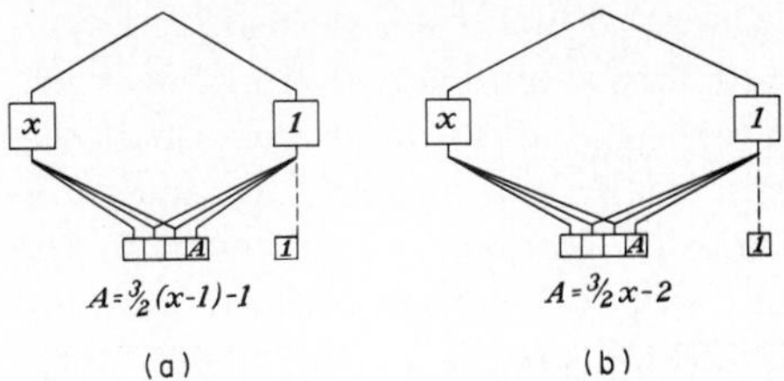

Fig. 11-11 Armature connection diagrams for lap windings. (*a*) For any simplex winding with three bars per slot, which has an even coil throw (x is an odd number); (*b*) for any simplex winding with three bars per slot, which has an odd coil throw (x is an even number).

slot. When the winding is complete, the *start* and *finish* wires form another loop. After the armature is completely wound and before wedging, the loops are pulled through the slots and connected to the commutator.

11-16 Notes on Rewinding and Reconnecting. When armature leads are soldered to the commutator bars, it is important to use a good solder, preferably pure tin. Likewise, it is essential that the slots in the commutator necks, and also the ends of the wires, be clean and free from

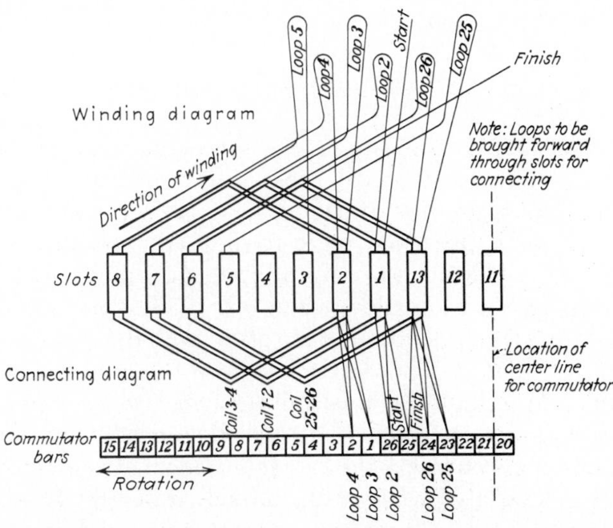

Fig. 11-12 Winding and connection diagram of a back-lead winding with 13 slots and 26 bars. The connections are thrown right and on neutral and are reversible.

dirt, varnish, or other insulation. A neutral flux, such as rosin and alcohol, should always be used; an acid flux should never be used. Care should be taken during the soldering operation not to overheat the commutator. Commutators can be loosened and ruined if the soldering operation gets them too hot. Moreover, excessive heat can ruin the insulation that is close to the commutator. When the proper equipment and know-how are available, welding the leads to the commutator necks is preferable to soldering.

Because universal motor armatures usually operate at high speeds, special care must be taken to wind and shape the coils so as to obtain a good balance. It is highly desirable to balance the complete armature dynamically after refinishing the commutator surface.

11-17 Refinishing the Commutator. After the dipping and baking process, the outside of the commutator should be turned on a lathe to true it up, to remove excess solder, to clear short circuits between adjacent bars, and to prepare it for the final finish on the brush surface. Mica insulation between bars should be undercut up to the commutator necks, or to within approximately ⅛ in (3 mm) from the rear (winding) end of the commutator. This mica can be removed with a hand scraper, but preferably by means of a circular saw. The saw used for this purpose should be about 0.005 in (0.1 mm) thicker than the mica; a standard thickness is 0.025 in (0.6 mm). It should be from ⅜ to ¾ in (10 to 20 mm) in diameter, with 12 to 16 teeth per inch (0.5 to 0.6 teeth per millimeter) of circumference, and hollow-ground on both sides. A speed of 7500 rpm or greater is recommended. The cuts should be made to a depth of 0.032 in (0.8 mm) from the final finished surface of the commutator.

The final finish to the commutator surface should be cut with either a diamond-point or a carboloy-point cutting tool. This tool should have a tip rounded off to a radius of approximately 0.005 in (0.1 mm) and ground for a rake of 5° and a clearance angle of 18°. The tool must be firmly clamped and rigidly held for the turning process, which should be done at a high speed—of the order of 6000 to 7500 rpm. Feed should be adjusted so that the tool does not advance more than 0.0005 in (0.01 mm) per revolution, nor should the depth of the final cut be more than 0.0005 in (0.01 mm). If the finish cut is taken in this manner, the burrs in the slots will be few and easily removed with a stiff brush.

For good commutation and brush life, it is important that the commutator surface be concentric with the shaft journals. After the finish cut, set the armature up on V blocks and indicate the commutator surface with a dial indicator, preferably one reading 0.0001 in per division. Rotate the armature slowly and note maximum and minimum readings over a revolution. For high-speed armatures, the total variation in reading should not exceed 0.0005 in (0.01 mm) [eccentricity of 0.00025 (0.006 mm)]. Moreover, the maximum variation from one bar to an adjacent one should not exceed 0.0002 in (0.005 mm).

11-18 Checking the Rewound Armature. Short-circuited coils usually can be detected by means of a growler. The armature is placed into a growler and rotated slowly, while an iron knife, or other light piece of iron or steel, is held against the top of the slowly revolving armature. If a coil is short-circuited, the two slots in which the two sides of the short-circuited coil lie will attract the metal piece. Open-circuited coils can also be found by using the growler. Every pair of adjacent bars in the commutator can be successively short-circuited by means of a knife kept in approximately the same position while the armature is slowly rotated by hand. The same amount of sparking should occur between every pair of bars; if a coil is open-circuited, no spark can be obtained. The armature is tested for grounds by means of a megger, a dielectric withstand tester, or other source of at least 500 V, impressing the voltage between the commutator and the shaft.

SPEED CONTROL OF UNIVERSAL MOTORS

Speed of a universal motor may be controlled by use of tapped field windings (see Art. 11-9), by a mechanical governor, by a series-connected external impedance, or by electronic means. The last three methods are discussed in this section.

11-19 By a Mechanical Governor. Mechanical governors to control the speed of universal motors were widely popular at one time, and a great many of them are undoubtedly still in use. However, for new applications, they have virtually disappeared. Of the manufacturers listed in Table 1-1, only one, Globe Motor Division, TRW Inc., offers governor-controlled universal motors.

The most popular form of governor was the Lee governor, which used a phenolic disk on which were mounted two contact arms. When the contacts were opened by centrifugal force, a resistance was inserted in series with the motor. Some of the governors operated only at a fixed speed; others were designed so that the speed could be adjusted even while the motor was running. In general, the latter was accomplished by means of a linkage mechanism that adjusted the position of the inner contact, so that more or less centrifugal force would be required to separate the contacts, thereby causing the governor to operate at a higher or lower speed, as desired. Governors for universal motors are discussed in the literature.[2,4] If the resistance, bridged across the contacts of the governor, is not sufficiently high, the no-load speed may be considerably higher than the governed speed.

11-20 By an External Impedance. The speed of a universal motor is often varied by means of an external resistor connected either in series with the motor line or across the brushes. This resistor may be a variable resistor, as in the case of a speed controller for a sewing-machine motor, or it may be a fixed resistor. When an external resistor is connected in series with the motor, the speed torque of the motor is more drooping than it would be without the resistor, and the locked-rotor torque is reduced substantially. The effect is illustrated in Fig. 11-13. With such a drooping torque characteristic, the motor may sometimes fail to start, and there will be wider-than-normal variations of speed with changes in load. A choke connected in series with the motor can also be used to reduce the speed of a universal motor.

11-21 By a Solid-State Electronic Controller. A number of electronic controllers using solid-state circuitry have been developed to control the speed of universal motors. Such circuits are used in electric drills, blenders, and other applications. Seven circuits for this purpose are described by Adem;[5] the remainder of this article was taken largely from Adem's article and is devoted to a discussion of one of his circuits.

Adem's circuit is reproduced in Fig. 11-14. It uses half-wave rectification, with feedback. Speed is adjusted by the potentiometer P_1. The way the circuit works will be described briefly.

The circuit arm, consisting of the resistor R_1, potentiometer P_1, supplied through the diode CR_2, provides an adjustable reference voltage V_c which is smoothed out by capacitor C_1. The motor is supplied with half-wave current through the silicon-controlled rectifier CR_1. The silicon-controlled rectifier (SCR) is fired when the IR drop in resistor R_2

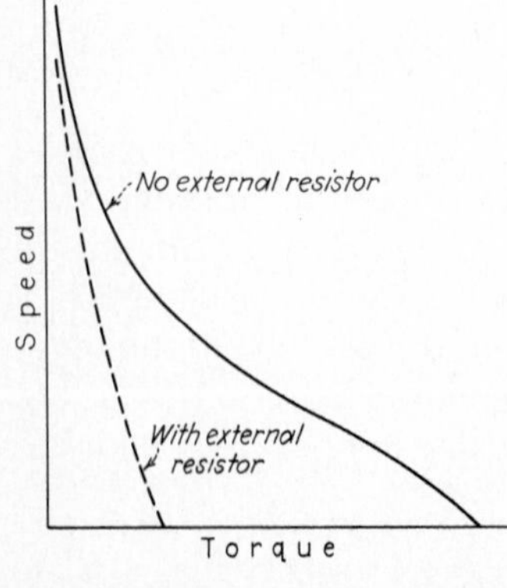

Fig. 11-13 How the speed-torque curve of a universal motor is affected by an external resistor connected in series with the motor windings.

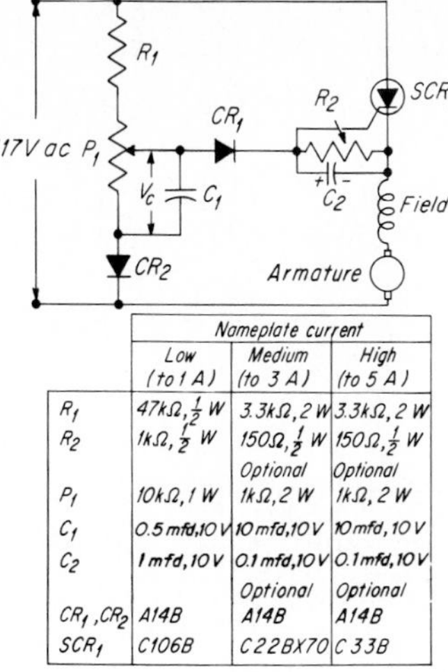

	Nameplate current		
	Low (to 1 A)	Medium (to 3 A)	High (to 5 A)
R_1	47kΩ, ½ W	3.3kΩ, 2 W	3.3kΩ, 2 W
R_2	1kΩ, ½ W	150Ω, ½ W Optional	150Ω, ½ W Optional
P_1	10kΩ, 1 W	1kΩ, 2 W	1kΩ, 2 W
C_1	0.5 mfd, 10 V	10 mfd, 10 V	10 mfd, 10 V
C_2	1 mfd, 10 V	0.1 mfd, 10 V Optional	0.1 mfd, 10 V Optional
CR_1, CR_2	A14B	A14B	A14B
SCR_1	C106B	C22BX70	C33B

Fig. 11-14 A solid-state electronic feedback control circuit for controlling the speed of a universal motor. Half-wave rectification is used. *(From A. A. Adem, "Seven Solid-State Circuits for Controlling Universal-Motor Speed," Mach. Des., Jan. 5, 1967, pp. 118–123.)*

becomes sufficiently great. During the no-current half of the cycle, the voltage induced in the motor, due to the motor's residual flux, is compared with the reference voltage V_c, to determine when to fire the SCR. When the motor is at standstill, no residual voltage is induced, so the SCR fires early in the cycle, causing maximum current to flow through the motor, which causes the motor to accelerate. As the motor attains speed, the voltage induced in the armature by residual flux increases, which delays the firing of the SCR and thereby reduces the current and torque until the motor ceases accelerating. A heavy load on the motor will cause the motor to slow down, which reduces the residual voltage and thereby causes the SCR to fire earlier in the cycle, so that increased current will flow, developing increased torque to take care of the heavier load.

The capacitor C_1 plays an important role in the low-speed performance of this circuit. Too much capacitance may cause an unstable operation (hunting) at low speeds because the overall gain is such that the SCR may skip-fire, delivering power to the motor for one cycle and skipping for several successive cycles. On the other hand, too little capacitance may not give enough phase shift to fire the SCR late in the cycle to obtain slow-speed operation. The capacitor must have sufficient capacitance to provide a sufficient phase shift for low-speed operation, but not enough to cause skip-firing.

If difficulty is found at low speed by hunting, the minimum value of V_c can be limited, either by inserting a fixed resistor between P_1 and CR_2 (with C_1 across both this resistor and P_1) or by means of a mechanical stop on the moving arm of P_1.

Even at the highest speed setting of P_1, only about 85 V is impressed across the motor because of half-wave rectification, and therefore full power cannot be developed. A simpler way to get around this is to bypass the SCR with a mechanical switch, which can be mechanically ganged with the potentiometer P_1 so that at the highest speed setting the SCR is bypassed.

Not all motors will work successfully with the circuit just described, for not all motors have enough residual flux. Moreover, different motors of the same design will not all run at the same speed for the same setting of the potentiometer. Since this is a feedback control circuit, the characteristics of the motor and circuit have to be carefully matched to ensure satisfactory operation. The constants in Fig. 11-14 were given by Adem as representative.

Also, it should be noted that the circuit of Fig. 11-14 will work only on an ac supply; it will not work with direct current. Electronic control of shunt-wound dc motors, operated from an ac supply, is discussed in Chapter 13.

MISCELLANEOUS SERVICE PROBLEMS

11-22 Setting or Checking Neutral. Most universal motors are operated with the brushes shifted slightly from the true neutral position. If the correct amount of this shift is known, and if the mechanical construction permits adjustment of the brushes, the true neutral position may be found experimentally and the brushes then shifted the *proper amount in the right direction.* For salient-pole motors, the "kick-neutral" method of Art. 12-23 may be used. The same method is also applicable to two-field compensated motors if the field winding is used; perhaps a better method is to use the compensating winding and follow the procedure for finding the hard neutral of a repulsion motor, as described in a previous work by Veinott.[3] The neutral position of a single-field compensated motor is the hard neutral of a repulsion motor, and the brushes are shifted with rotation from this position.

11-23 Rewinding for Different Characteristics. The rewinding of a universal motor for characteristics different from those for which it was designed almost invariably necessitates rewinding both stator and armature windings. In the case of a two-field compensated motor, the compensating winding must always be rewound if the armature is rewound. When a motor is rewound for different characteristics, care must be observed that the resulting armature current does not exceed brush capacity.

(1) Rewinding for a different voltage. To rewind to obtain the same performance on a circuit of a voltage higher or lower than rated voltage, rewind the stator (including both windings if there are two) with more or fewer turns in direct proportion to the increase or decrease in voltage. Likewise, rewind the armature with more or fewer turns per coil in direct proportion to the increase or decrease in voltage. In general, in all windings use wire of the largest size that will permit the correct number of turns to be put into the slots.

(2) For different torques or speed. Before attempting to rewind a universal motor to obtain a different speed with any given device or for a different torque, a careful application test should be made with the original winding. The applied voltage should be adjusted until the desired torque or speed characteristics are obtained. A temperature run should then be taken to determine whether or not the motor can safely operate on the newly found circuit voltage. By this procedure, a new operating voltage will be found; say this new voltage is 150 V for purposes of illustration. Consider now the motor as if it had been originally rated for a 150-V circuit, and, on this basis, rewind for a circuit of the voltage on which it must be operated, say 115, following the rules laid down in the preceding paragraph.

References

1. *NEMA Motor and Generator Standards,* NEMA Publ. no. MG1-1978, National Electrical Manufacturers Association, 2101 L Street N.W., Washington, DC 20037, 1978.
2. Veinott, C. G.: *Fractional Horsepower Electric Motors,* 2d ed., McGraw-Hill Book Company, New York, 1948.
3. ————: *Fractional- and Subfractional-Horsepower Electric Motors,* 3d ed., McGraw-Hill Book Company, New York, 1970.
4. Grott, R. W.: "Governor Controlled Fractional Horsepower Motors," *Prod. Eng.,* June 1936, pp. 216–218.
5. Adem, A. A.: "Seven Solid-State Circuits for Controlling Universal-Motor Speed," *Mach. Des.,* Jan. 5, 1967, pp. 118–123.

Bibliography

"A Brush Guide for Fractional h.p. Motors," *Appliance Eng.,* December 1982, p. 68.

Dickin-Zanger, C. W.: "Commutation in Fractional Horsepower Universal Motors," *Electro-Technology,* June 1962, pp. 139–140.

————: "Performance Analysis of Universal Motors," *Electro-Technology,* December 1962, p. 127.

Kawada, T.: "Analysis of Design Factors of Universal Motors," *Electrical Engineering in Japan,* vol. 85, June 1965, pp. 1–8.

Packer, L. C.: "Universal Type of Motors," *AIEE Trans.,* vol. 44, 1925, p. 587.

Puchstein, A. F.: "Universal Electric Motors," *Ohio State Univ. Eng. Exp. Sta. Bull.* 53, 1930.

Wier, W. E.: "Iron Loss in Universal Motors," *AIEE Trans.,* vol. 73, pt. III-B, 1954, pp. 1546–1552.

CHAPTER

12

DC MOTORS

In fractional and subfractional horsepower sizes, dc motors were initially used, for the most part, only where ac power was not available, and they represented only a small part of the small-motor industry. More recently, the use of dc motors has been growing due in part to the growing demands of automation for adjustable-speed drives and in part to the availability of low-cost controllable silicon rectifiers. Today, most fractional horsepower dc motors operate on rectified power; in subfractional horsepower sizes, a larger proportion of them, especially permanent-magnet, operate directly from batteries.

It should be pointed out that dc motors, unless regulating means of some sort are used, cannot be depended upon to operate so closely to their rated full-load speed as do ac motors. At full load, fractional horsepower single-phase induction motors will generally operate within 2 percent of their rated speed, whereas similar dc motors may vary as much as 7½ percent above or below their rated speed.

12-1 Types of DC Motors. Major types of dc motors with wound fields are illustrated schematically in Fig. 12-1. Permanent-magnet motors, which are now very popular, have no field windings of any kind.

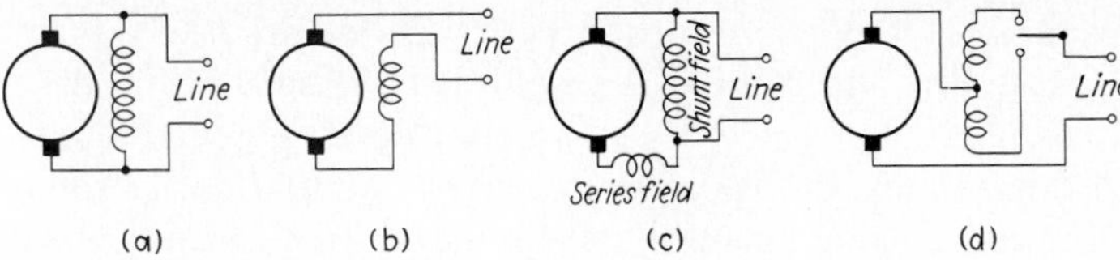

Fig. 12-1 The major types of wound-field dc motors. (*a*) Shunt motor; (*b*) series motor; (*c*) compound motor; (*d*) split-series motor.

(1) Shunt motors. Shunt-wound motors have the field connected in shunt with the armature and have essentially constant-speed characteristics similar to the speed characteristics of induction motors. *Stabilized shunt-wound motors* are shunt-wound motors with a light series winding added to stabilize the speed characteristics, i.e., to prevent any rise in speed when load is added. Tendency to rise in speed is due to the field-weakening effect of armature reaction, as explained in Art. 12-3(4). Fractional horsepower dc motors generally do not require or use a stabilizing winding.

(2) Series motors. Series-wound motors have the field connected in series with the armature and have the varying-speed characteristics of universal series motors, representative curves for which are shown in Figs. 11-1 and 11-2. A modified form of this type is the *split-series motor* shown in Fig. 12-1*d*. Its chief advantage is ease of reversibility, requiring only a single-pole double-throw switch for control; it is discussed further in Art. 12-10.

(3) Compound motors. Compound-wound motors employ both a series and a shunt field and have speed characteristics intermediate between those of shunt and series motors, depending upon the amount of the compounding. The series winding is connected in such a direction as to assist the shunt field, thereby increasing the flux. Compound-wound motors are normally known as varying-speed motors. ("Differentially compounded motors," wherein the series field is connected so as to oppose the shunt field, are seldom used because of their unstable speed characteristics.)

(4) Permanent-magnet motors. Permanent-magnet motors have armature windings, but no field windings; a permanent-magnet material is used in the field structure to set up the required flux. These motors are now available in a wide variety of sizes and ratings, and from a number of manufacturers, as shown in Table 1-1. They are discussed in greater detail later in this chapter (Arts. 12-15 to 12-19).

12-2 Essential Parts of a DC Motor. A fractional horsepower dc motor has essentially the same parts as a universal motor, as illustrated in Fig. 11-3.

(1) Field structure. A field structure is necessary to provide a magnetic circuit for the flux and to hold the field coils. Basically, the field structure consists of a yoke, pole bodies (one per pole), pole faces (one per pole), and, sometimes, interpole(s) to assist commutation. In some constructions the entire field assembly is laminated. In other constructions only the pole bodies and pole faces are laminated; they are bolted to a solid yoke.

(2) Field coils. The field coils may consist of (1) shunt coils wound with many turns of small wire which are connected in series with each other and across the line or in parallel with the armature, (2) series coils with fewer turns and larger wire, which are connected in series with the armature and line, (3) both shunt and series coils for a compound motor, and (4) sometimes an interpole.

(3) Armature. The armature is the rotating element, which consists of slotted punchings pressed onto the shaft. These punchings perform the dual purpose of providing a path for the magnetic flux and of carrying the armature winding, which is usually a lap or parallel winding of a closed-circuit type. (Such windings are discussed in some detail in an earlier edition of this book.[1])* Ordinarily, fractional horsepower dc motors for industrial applications have only two poles. However, aircraft dc motors often have four or more poles. Sometimes the armature is skewed to reduce noise.

(4) Commutator and brushes. The commutator and brushes are necessary to feed the current to the armature. Both cartridge-type and rocker-ring constructions are common. The rocker-ring construction, though more expensive, is often used to permit adjustment of the brush position for best commutation.

(5) Commutating poles. These are also called *interpoles* and are occasionally used in fractional horsepower dc motors to improve commutation. These are small auxiliary poles placed halfway between the main poles and connected in series with the armature. There may be as many interpoles as main poles, or only half as many.

PRINCIPLES OF OPERATION

12-3 Basic Principles Common to All Types. In the discussion that follows, it is assumed that the motors are all supplied from a source of steady dc voltage, such as a battery, or from a generator that has a very low ripple content. Special problems imposed by use of a voltage with

*For numbered references, see References at end of this chapter.

a high ripple content, as is obtained from a rectifier, for example, are discussed in Art. 13-5.

(1) How torque is developed. A simple dc motor is represented schematically in Fig. 12-2. Armature windings, together with commutator and brushes, are so arranged that the flow of current is in one direction

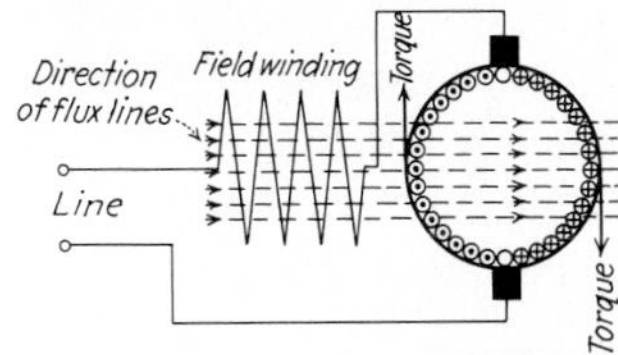

Fig. 12-2 A simple series motor.

in all the conductors on one side of the armature, and in the opposite direction in all the conductors on the opposite side of the armature. This condition is represented in the figure by the use of dots to indicate current flowing toward the observer, and by plus signs (representing the tail of an arrow) to indicate current flowing away from the observer, perpendicularly to the plane of the paper. The field winding sets up a magnetic field as shown in the figure. It is a simple fundamental law of motor action that, if current is passed through a conductor which is perpendicular to a magnetic field, a mechanical force will be exerted on the conductor, mutually perpendicular to both the conductor and the direction of the field. A rule for the direction of this force is given in the left-hand rule.* Application of this rule to Fig. 12-2 gives an upward force on all conductors to the left, and a downward force on all conductors to the right, so that a torque is developed in a clockwise direction.

(2) Counter emf. An understanding of the nature and role of counter electromotive force (emf) is absolutely essential to any comprehension of the way any type of dc motor works. Again let us refer to Fig. 12-2 and let us assume that the armature is revolving in a clockwise direction, the same direction as the torque developed by the current. Because of rotation, the armature conductors are cutting the lines of force of the field, thus generating voltages in the armature conductors.

*There is a left-hand rule for motor action similar to the Fleming right-hand rule for generator action explained in the footnote to Art. 2-10. Extend the thumb and first and second fingers of the left hand mutually perpendicular. The thumb points in the direction of motion if the first finger is pointed in the direction of the lines of force and if the second finger points in the direction of current flow.

Application of Fleming's right-hand rule (see footnote to Art. 2-10) shows that the direction of this induced voltage is away from the observer on the left, and toward the observer on the right; that is, *the direction of the induced voltage is opposed to the direction of current flow.* In sum then, in any dc motor, rotation of the armature induces a voltage in the armature circuit which opposes the flow of the current that causes the rotation. Because the direction of the voltage is counter to the applied voltage, it is called counter electromotive force. Its role will be discussed further in ensuing paragraphs.

(3) Basic performance equations. Performance characteristics may be succinctly expressed in the form of a few simple equations. Let

E = counter emf generated in the armature

I_a = armature current

N = rpm of the armature

R_a = total resistance in the armature circuit, including all windings in the circuit, and brush-contact resistance

T = torque developed by the armature

V = voltage impressed on the armature circuit

ϕ = flux per pole

K_1, K_2, K_3 = proportionality constants

Now the torque is proportional to both flux and current, or

$$T = K_1 \phi I_a \tag{12-1}$$

Voltage induced is proportional to rate of cutting flux, that is, to the product of speed and flux, giving

$$E = K_2 N \phi \tag{12-2}$$

The counter emf has to be the applied voltage less the *IR* drop, or

$$E = V - I_a R_a \tag{12-3}$$

Solving the above equation for armature current gives

$$I_a = \frac{V - E}{R_a} \tag{12-4}$$

Speed can be determined from counter emf and flux by rewriting Eq. (12-2)

$$N = \frac{K_3 E}{\phi} \tag{12-5}$$

The ways these equations are interpreted and used for the various types will be discussed separately for each type.

(4) Armature reaction. Consider again the schematic representation of a dc motor in Fig. 12-2. In all the conductors to the left of the brushes the current is flowing out of the paper; in all the conductors to the right, the current is flowing into the paper. This is exactly the effect that would be produced by a coil of wire, wrapped around the armature, with its axis vertical. This coil tends to produce a magnetic field along the brush axis and perpendicular to the main-field axis. In general, that is exactly what the armature current does: it sets up, or tends to set up, a magnetic field which is displaced 90 electrical degrees from the main-field flux. Such an effect is known as *armature reaction.* Armature reaction has at least two harmful effects: it distorts the shape of the main field, usually weakening it in the process; it causes a voltage to be induced in the coils undergoing commutation, tending to produce sparking at the brushes. Commutation can be improved by shifting the brushes, if the motor construction permits, *against rotation;* this procedure is limited to motors that operate in only one direction of rotation. If the brush-holder positions are fixed, the effect of shifting the brushes can be obtained by shifting the coil connections to the commutator as illustrated in Fig. 11-10; this procedure is practicable only if the direction of rotation is known before the armature coils are connected to the commutator.

(5) Interpoles. The effects of armature reaction may be overcome by use of compensating windings, such as are used in some universal motors; in fractional horsepower sizes, it is done more commonly by the use of one or more interpoles [see Art. 12-2(5)]. By selecting the proper number of interpole turns the field strength of the interpole is such as to cancel the armature-reaction magnetomotive force (mmf) plus counteract the inductive voltage of commutation, thus eliminating sparking at the brushes.

Interpole windings of motors must be connected so that any armature conductor passes under a main pole and then an interpole of the *same* polarity. Interpole windings of generators are connected so that any armature conductor passes under a main pole and then an interpole of the *opposite* polarity. If a motor is converted to a generator with the same direction of rotation, the interpole connections need not be changed, since the armature, and thus the interpole, current flows in the opposite direction.

12-4 How the Shunt Motor Works. The shunt motor, as shown in Fig. 12-1*a*, has both armature and field connected across the line. Now, let us use the equations in the previous article to deduce the operating characteristics. At no load, armature current I_a is negligible so that the counter emf E is substantially equal to the applied voltage V (Eq. 12-3). Then, the no-load speed is fixed since E is fixed (Eq. 12-5). As the motor is loaded, the speed N drops a little, lowering E, allowing more current I_a to flow, thereby increasing the torque. Since the torque is proportional to the current, the IR drop is also proportional to torque, and the speed falls off linearly with an increase in load. In practice, armature reaction may weaken the field so that the speed will not fall off quite so rapidly as the load, or as suggested by these equations. It is clear that locked-rotor torque will be very high, for E is zero at standstill; hence current drawn and torque developed will both be high.

The operating speed can be adjusted by adjusting the strength of the shunt field, either by insertion of a rheostat in series with it, as shown in Fig. 12-3, or by adjusting the voltage across the field. Weakening the field increases the speed (Eq. 12-5), and strengthening it decreases the speed. Without provisions for field control, a shunt motor is a *constant-speed motor.* With field control, a shunt motor is an *adjustable-speed motor.*

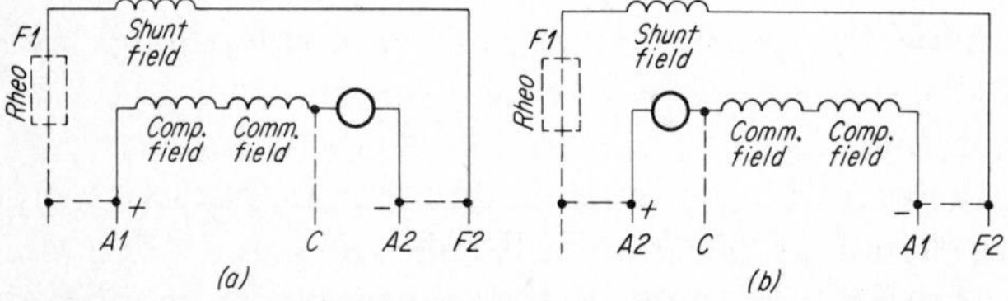

Fig. 12-3 Terminal markings and connections for dc shunt motors. (*a*) Counterclockwise rotation; (*b*) clockwise rotation. *(From NEMA, Motor and Generator Standards, NEMA Publ. no. MG1-1978, National Electrical Manufacturers Association, Washington, 1978.)*

12-5 How the Compound Motor Works. A compound motor is like a shunt motor except that it has a series field as well as a shunt field. The series field is wound on the same poles as the shunt field. As shown in Figs. 12-1 and 12-4, the shunt field is connected across the line, and the series field is connected in series with the armature. The series field is so connected that it strengthens the shunt field when armature current is drawn from the line; such a motor is a *cumulative-compound motor.* If the series field is connected so as to oppose the shunt field, it is known as a *differential-compound motor;* the latter is rarely used and is not recommended, for its speed characteristics tend to be unstable. The compound motor works in essentially the same way as, and has operating characteristics similar to those of, a shunt motor except that

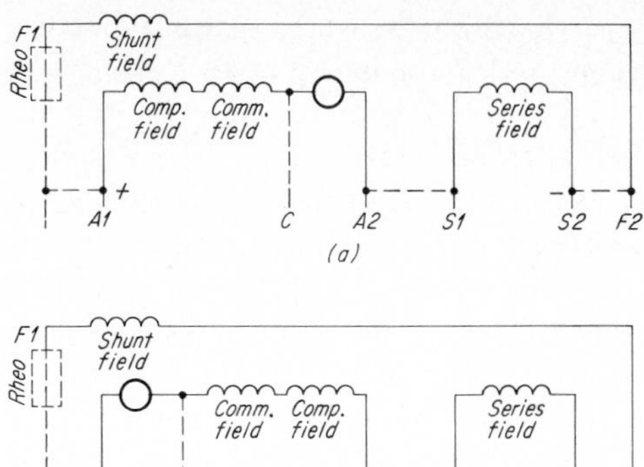

Fig. 12-4 Terminal markings and connections for dc compound or stabilized shunt motors. (*a*) Counterclockwise rotation; (*b*) clockwise rotation. *(From NEMA Motor and Generator Standards, NEMA Publ. no. MG1-1978, National Electrical Manufacturers Association, Washington, 1978.)*

the speed regulation (change in speed from no load to full load) and locked-rotor torque of the compound motor are greater.

12-6 How the Series Motor Works. The series motor is known as a varying-speed motor and has a very high no-load speed, as can be seen in Figs. 11-1 and 11-2. The speed regulation is very much higher than for a shunt or compound motor. At standstill, the counter emf is zero, and the current is high—as is the field flux—so that the motor accelerates rapidly. As the motor accelerates, the counter emf increases with the speed. Now, in the case of a shunt motor, the counter emf increases directly in proportion to the speed because the flux remains constant. However, in the case of a series motor, as the speed increases, the increased counter emf *decreases* the armature current, and this decreased armature current in turn decreases the field excitation. Therefore, as the motor accelerates, the weakening field makes it necessary that the armature rotate still faster to generate sufficient counter emf to limit the armature current. At light loads, the motor literally races: the armature races to develop enough counter emf to limit the armature current, and, as the armature current is limited, the field is weakened, thus tending to decrease the counter emf, so that the motor has to run still faster. Large dc motors will usually race to destruction if not loaded, but fractional horsepower dc series motors, as well as universal motors, generally are designed to withstand these high speeds.

However, for personal safety, care should be taken to ensure that the motor is connected to a load to limit the speed when full voltage is applied.

APPLICATION CHARACTERISTICS

In the two articles that follow, the discussion is centered around conventional dc motors that operate from a conventional power source. Following this discussion is a section devoted to special-purpose motors and a section covering permanent-magnet motors.

12-7 Ratings

(1) Voltage. Standard voltage ratings for commercial fractional horsepower motors are 115 and 230 V. However, standard motors will operate successfully at any voltage within 10 percent of the rated voltage, although the operating characteristics will be affected somewhat. Subfractional horsepower and military-service motors usually have lower voltage ratings.

(2) Horsepower and speed ratings. Fractional horsepower dc motors are built in the same horsepower and speed ratings as 60-Hz single-phase induction motors, which are listed in Table 19-1. Smaller horsepower ratings are also built, but there are no universally accepted standards for these ratings.

Actual full-load operating speeds of individual motors, built in frame 42 and larger frames, may vary as much as plus or minus 7½ percent because of manufacturing variations. For motors built in smaller frames, the variation may go up to 10 percent.

12-8 Operating Characteristics

(1) Change in speed due to load. Direct-current motors are not nearly as constant in speed as split-phase and capacitor-start motors. Usually, however, the reduction in speed for shunt-wound motors is not greater than 25 percent and for compound-wound motors, is not greater than that shown in Table 12-1, taken directly from National Electrical Manufacturers Association (NEMA) standards[2]. *Speed regulation* is defined as the change in speed from no load to full load and is expressed as a percentage of full-load speed.

(2) Change in speed due to heating. As a shunt motor warms up, the no-load speed steadily increases because the resistance of the shunt field increases, decreasing the field current and flux. At full load, the speed may either increase or decrease with an increase in temperature because the increase in armature resistance with heating tends to cause

TABLE 12-1 Speed Regulation of Compound-Wound DC Motors Due to Load (Maximum Values)

Horsepower	Speed, rpm	Speed regulation, %
1/20–1/8	1725	30
1/20–1/8	1140	35
1/6–1/3	1725	25
1/6–1/3	1140	30
1/2–3/4	1725	22
1/2	1140	25

Source: From *NEMA Motor and Generator Standards,* NEMA Publ. no. MG1-1978, National Electrical Manufacturers Association, Washington, 1978.

the full-load speed to decrease. NEMA standards provide that this change in full-load speed as the motor warms up from ambient to normal operating temperature, expressed as a percentage of the latter speed, shall not exceed the values shown in Table 12-2.

12-9 Standard Terminal Markings. Standard terminal markings, as developed by NEMA, are given for shunt, compound or stabilized shunt, and series motors in Figs. 12-3, 12-4, and 12-5, respectively. These figures all show compensating-field as well as commutating-field windings since they are intended to cover both fractional horsepower and integral horsepower motors. Fractional horsepower dc motors (except some universal motors) do not generally use compensating windings, and they may or may not use commutating-field (interpole) windings.

TABLE 12-2 Speed Regulation of Shunt-Wound DC Motors Due to Heating (Maximum Values)

	Speed regulation, %			
	Insulation system class			
Enclosure type	A	B	F	H
Open	10	15	20	25
Totally enclosed	15	20	25	30

Source: From *NEMA Motor and Generator Standards,* NEMA Publ. no. MG1-1978, National Electrical Manufacturers Association, Washington, 1978.

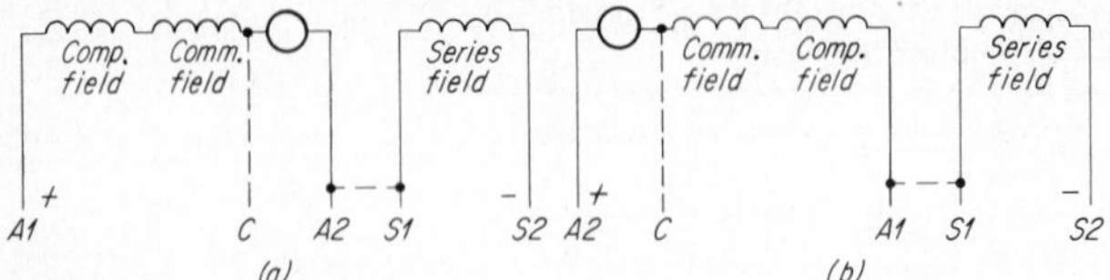

Fig. 12-5 Terminal markings and connections for dc series motors. (*a*) Counterclockwise rotation; (*b*) clockwise rotation. *(From NEMA Motor and Generator Standards, NEMA Publ. no. MG1-1978, National Electrical Manufacturers Association, Washington, 1978.)*

Guiding principles used in establishing the terminal markings may be helpful to the reader in understanding and using the diagrams. The markings, it will be noted, are combinations of capital letters and arabic numerals; generally, these are as follows:

Armature	A1, A2
Control signal lead, attached to commutating winding	C
Series field	S1, S2
Shunt field	F1, F2

When an armature lead passes through a commutating-field winding (interpole winding) before being brought out to the external circuit, the terminal marking is an "A," as shown in the three figures. However, when an armature lead passes through a series field and all internal connections are permanently made, the lead that is brought out is marked with an "S." (An example would be Fig. 12-5*a:* if the A2 and S1 leads were connected together permanently inside the motor, the S2 lead would still be tagged "S2.")

Direction of rotation is related to the numerals on the lead tags. When the windings are so connected to the line circuit that the current flows from a 1 to a 2 in *both* the armature and field circuits, the motor will rotate counterclockwise, facing the end opposite the shaft extension. When the windings are so connected that current flow is from 1 to 2 in one circuit, and from 2 to 1 in the other, the direction of rotation will be clockwise. This fact can be confirmed by studying the three figures.

SPECIAL-PURPOSE MOTORS

It is, of course, impossible to cover all kinds of special-purpose dc motors, but a few will be discussed.

12-10 Split-Series Motors. A split-series motor is represented schematically in Fig. 12-1*d*, which shows the field split into two sections. Sometimes, there are only two field coils, one on each pole; sometimes, two field coils are wound on each pole, and half the winding on each pole is then excited for either direction of rotation. The latter arrangement gives somewhat better balanced flux conditions but involves a few extra complications. Split-series motors are usually designed for plugging service. Simplicity of their control makes them ideal for such service.

12-11 Military-Service Motors. Very small dc motors are used for a very wide variety of applications on military and civilian aircraft. Such motors are characterized by light weight and are built to meet severe environmental requirements; generally they are for circuits of 28 V or less.

Two typical motors are illustrated in Fig. 12-6. They are nominally of 1⅜-in diameter and are usually split-series wound. They may be fitted with a brake capable of stopping the motor in three to five revolutions from a speed of 10,000 rpm, with a planetary gear reducer or with both. In lieu of a brake, a governor may be supplied. The smaller motor

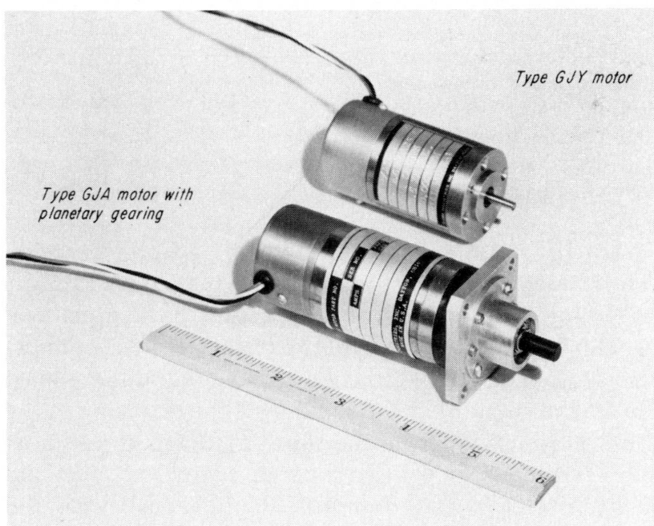

Fig. 12-6 Miniature split-series dc motor built to military specifications. The GJY motor is rated 0.01 hp at 10,000 rpm and weighs 6.2 oz; it is arranged for face mounting. The GJA motor is rated 0.02 hp at 10,000 rpm and is shown with a planetary reduction available in ratios from 18.78 to 21.808; it is arranged for flange mounting. *(Globe Industries Division of TRW, Inc.)*

shown (Globe type GJY) has a basic maximum length of 2.5 in, a continuous-duty rating of 1/100 hp at 10,000 rpm, a locked-rotor torque of 450 percent, an efficiency of 49 percent, and a no-load speed of 25,000 rpm. The larger motor shown (Globe type GJA) has a basic maximum length of 3 in, a continuous-duty rating of 1/60 hp at 10,000 rpm, a locked-rotor torque of 325 percent, a full-load efficiency of 50 percent, and a no-load speed of 22,000 rpm.

12-12 Electrically Reversible (Reversing) Motors. Where series characteristics can be used, the split-series motor is ideal because it has only three leads and can be controlled by a simple single-pole double-throw switch as shown in Fig. 12-1*d*. However, there are reversing applications in which series characteristics are neither permissible nor desirable; typical of such applications are hoists, machine tools, and electrode feeders for electric furnaces and arc welders. Figure 12-7 shows

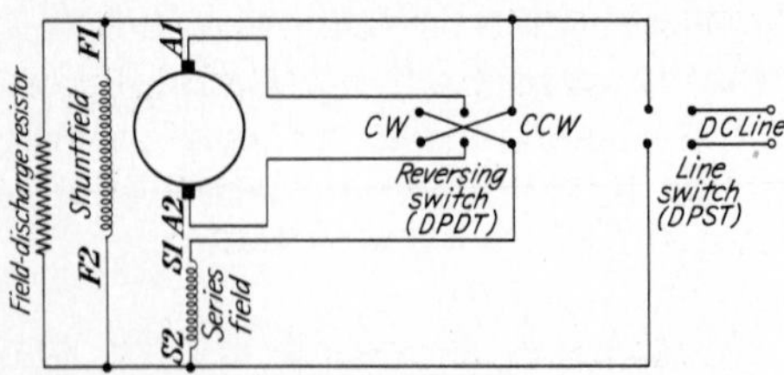

Fig. 12-7 Schematic diagram of connections for an electrically reversible (plug-reversing) compound-wound dc motor. If the line switch is not operated simultaneously with the reversing switch, the field-discharge resistor can be omitted.

the schematic connections for a compound motor arranged for this service. (Commutating-pole windings, if used, would be in the armature-winding circuit as shown in Fig. 12-4). Note that the armature connections are reversed while the shunt field is left connected to the line. It is preferable to leave the shunt field excited during the changing of the reversing switch to avoid interrupting the shunt-field circuit (see next paragraph). The line switch is shown as a separate switch, not mechanically coupled to, and preferably not operated simultaneously with, the reversing switch.

Control is simplified if the line switch is made single-pole and combined with the reversing switch; in this case only a three-pole double-throw switch is required. This arrangement has the added advantage that the motor is always deenergized in the open position of the three-pole switch. These advantages have made the use of this arrangement attractive and common. However, this arrangement has the serious disadvantage that the shunt field is opened at each reversal, causing a very high voltage surge across the field and a severe spark under the brushes, which may cause flashover between adjacent brushes. Voltage surges of several thousand volts have been measured on fractional horsepower motors, necessitating extra-heavy insulation of the field windings. This difficulty can be minimized by connecting a field-discharge resistor across the field as shown; a 25-W resistor often serves the purpose. In addition, special insulation of the field windings is required.

Many standard fractional horsepower dc motors can be used for this service, but it is safest to check with the manufacturer to determine (1) how many reversals the motor can stand per minute, if any, (2) what are the recommendations on the use of a field-discharge resistor, and (3) whether special insulation of field windings is required. Upper ratings of 1725- and 3450-rpm motors are more apt to be unsuitable for this service than motors rated lower in speed or horsepower or both. Another factor to be considered is the heavy inrush of current drawn by the motor at the instant of closing the reversing switch; this may be from one and one-half to two times the locked-rotor current.

12-13 Two-Speed Motors. An ingenious two-speed dc motor, formerly manufactured by Marathon, is shown in the schematic wiring and line connection diagram of Fig. 12-8. The motor is designed to operate as a cumulative compound motor on the low-speed connection. For high-speed operation, the motor is connected as a shunt motor, but the series field is connected in series with the shunt field in such a manner that it opposes the shunt field, thereby reducing the effective number of turns in the shunt field. As a result, the shunt field is weaker on the high- than on the low-speed connection; moreover, on the low-speed connection, the speed is reduced still more by the cumulative series field.

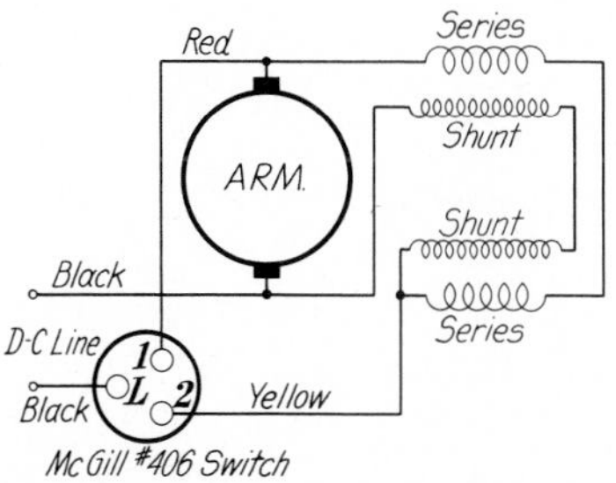

Fig. 12-8 Wiring and line connection diagram for a two-speed dc motor. *(Marathon Electric Manufacturing Corporation.)*

12-14 Traction Motors. Fig. 12-9 illustrates a dc traction motor designed primarily for use in low-voltage traction drives for battery-powered vehicles such as golf carts, industrial trucks, and floor-care machines. They are built in horsepower ratings from ½ to 4 hp; voltage ratings are 12, 24, and 36 V for up to 1½ hp. Field windings are series or compound. Performance characteristics of one such motor are given in Fig. 12-10.

Traction motors are also suitable for operating hydraulic pumps for power-steering and lifts and for blowers in heating and air-conditioning systems. The fields of such motors are series, shunt, compound, and permanent-magnet; the voltage ratings are up to 96 V.

Fig. 12-9 Cutaway view of a ball-bearing rigid-mounted dc series traction motor. *a*, Terminal studs for connecting to power supply; *b*, brush system, which provides easy brush replacement; *c*, commutator; *d*, wound armature with rectangular cross-section magnet wire; *e*, series field coil with square cross-section magnet wire; *f*, field pole fastened to magnetic steel frame. *(General Electric Company.)*

PERMANENT-MAGNET MOTORS

In the last three decades, much progress has been made in the development of permanent-magnet dc motors and these motors have attracted widespread interest and use.

12-15 Permanent-Magnet Motor Defined. A permanent-magnet dc motor is a dc motor in which the field excitation is supplied by a per-

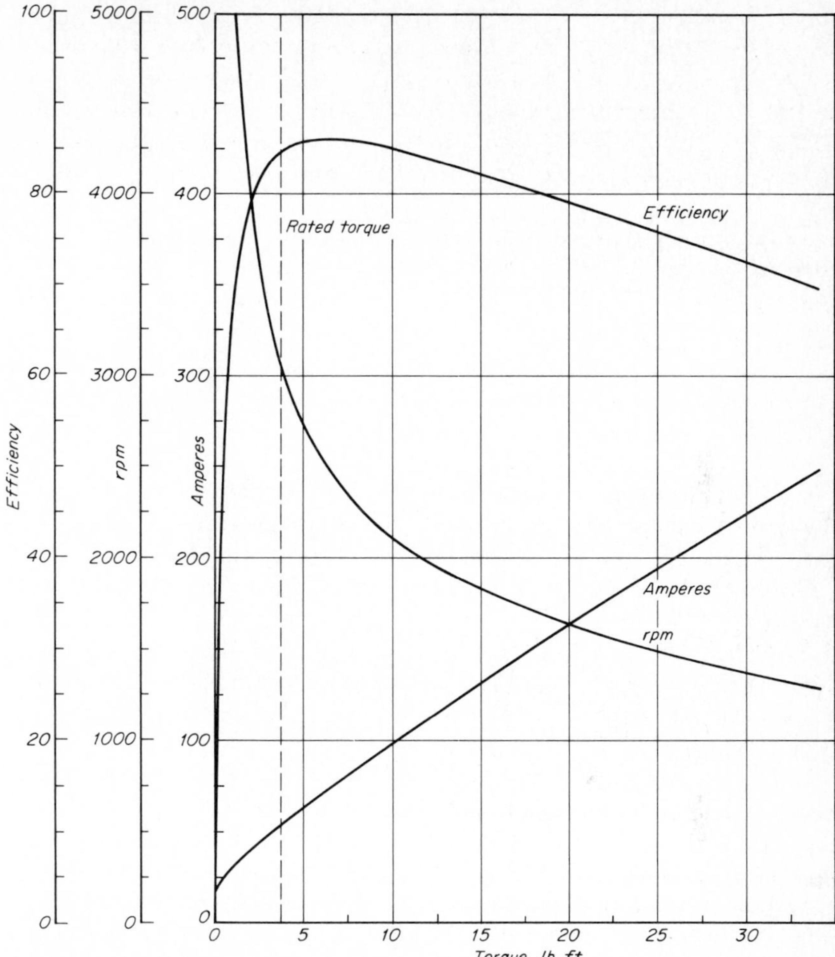

Fig. 12-10 Typical performance characteristics of a 2-hp 36-V 3300-rpm series-wound dc traction motor. *(General Electric Company.)*

manent magnet or magnets provided in the magnetic circuit of the motor.

12-16 Construction of a Permanent-Magnet Motor. One form of construction of a permanent-magnet motor is illustrated in Fig. 12-11. The permanent-magnet poles, which are barium ferrite ceramic magnets, are bonded by an epoxy adhesive to the outer field frame, which is fabricated from steel tubing. The steel frame is a part of the magnetic circuit, serving as the yoke; it also serves as the structural frame of the motor.

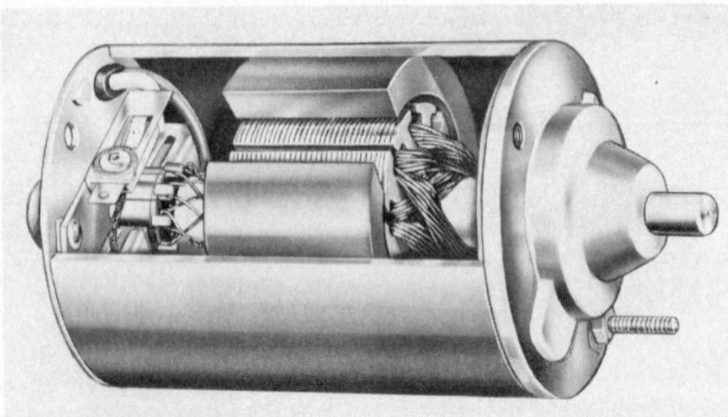

Fig. 12-11 Cutaway view of a permanent-magnet dc motor. Power ratings range from 1 to 40 mhp. Motor outside diameter is 2 in. *(American Bosch Electrical Products.)*

Another form of construction is quite similar to the one just described, except that the permanent-magnet material is cast in the form of a continuous ring, instead of in two pieces as shown.

The motor of Fig. 12-11 shows the magnets' having the same axial length as the armature. In some motors, the magnets are made axially longer than the armature, in order to deliver more flux to the latter. In some of these motors, the alnico poles are provided with soft-iron pole pieces to carry the flux axially from the overhang to a point directly over the armature.

12-17 Ratings, Performance, and Torque Characteristics. Permanent-magnet dc motors are generally offered in horsepower ratings from about 1 mhp up to 3 hp. Usually they are high-speed motors, with rated speeds from 2000 to 20,000 rpm. Mostly they are built for low-voltage operation, such as 6, 12, or 24 V. A performance curve for the motor of Fig. 12-11 is given in Fig. 12-12. Note that the speed-torque curve is a straight line, which is generally characteristic of this type of motor. This type of speed-torque curve is, of course, ideal for a servomotor. Note, too, that the current increases linearly with output torque.

Efficiencies in excess of 60 percent are not uncommon for the permanent-magnet motor because it has no field-excitation losses, and it has less loss and heating and greater efficiency than a comparable dc motor with a wound field.

12-18 Timing Motors. Small dc timing motors are offered with or without governors, depending upon the accuracy of speed required by the application.

(1) Basic timing motor. An ungoverned, basic timing motor is illus-

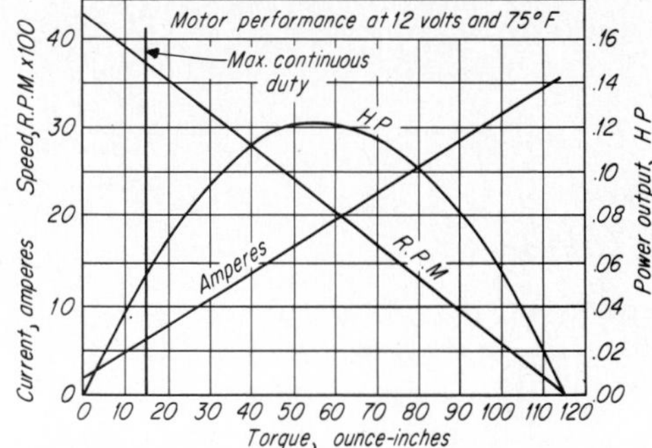

Fig. 12-12 Performance curves for a small permanent-magnet dc motor of the construction shown in Fig. 12-11. *(American Bosch Electrical Products.)*

trated in Fig. 12-13. The construction of this motor is quite unlike that of any motor previously discussed in this chapter, as can be seen by reference to the figure. The armature consists of a hollow aluminum (or inconel) cage, on which the coils are wound directly. An alnico field magnet is mounted on the bearing structure, *inside the hollow revolving armature.* The magnetic circuit is completed by the motor housing itself. Thus, there are no armature laminations, with their projecting teeth as in conventional motors, to cause cogging. When the motor is running, eddy currents are generated in the non-ferrous-metal cage. The effect of these eddy currents is to impose a dynamic load on the motor, substantially reducing its speed, but making the speed less sensitive to changes due to variations in applied load, because the dynamic loading due to eddy currents varies as the cube of the operating speed. In addition, the eddy-current loading of the rotor cage also affords an appreciable degree of temperature compensation, because the tendency for the motor speed to increase due to reduced copper resistance in the windings at reduced temperatures is offset by an increased eddy-current drag, since the resistance of the cup is also decreased by the reduced temperature. The rotor cage may be of aluminum or inconel, depending upon the characteristics desired.

Such motors are applied where the required speed regulation is of the order of plus or minus 10 percent at rated voltage; in addition, changes in the impressed voltage will cause an additional variation in speed which is proportional to the change in impressed voltage.

(2) Governed timing motor. A governed timing motor is illustrated in Fig. 12-14. The motor itself is the motor shown in Fig. 12-13. The governor, mounted outside the motor, has a balance wheel and hairspring which beat 15 times per second. The armature rotates at 2700 rpm and

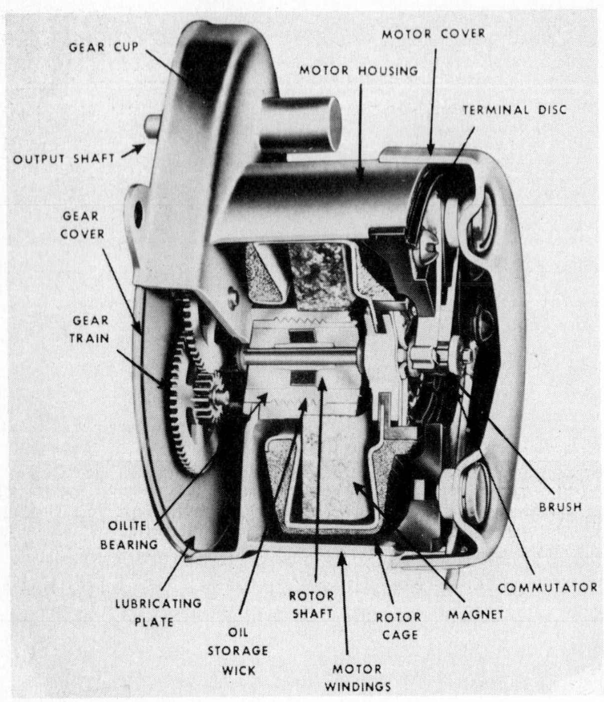

Fig. 12-13 A dc timing motor. Note unconventional construction. Output shaft speeds are 3000 rpm to ½ rph. Voltage ratings are 4.5 to 45 V. *(Haydon Switch and Instrument.)*

Fig. 12-14 A chronometrically governed dc timing motor; governor holds speeds to within 0.1 percent. *(Haydon Switch and Instrument.)*

is so designed that it would normally operate faster than this for all normal loads. A speed-regulating resistor, when in series with the armature circuit, is of such value as to make light-load speeds less than 2700 rpm. Contacts are provided so that this resistor can be cut in and out of the circuit to control the speed. In operation, these contacts are made by the escapement, initiating a pulse of current; the contacts are broken by a cam on the motor shaft. Thus the duration of the pulse of current depends upon the acutal armature position with respect to the balance wheel. This corrective pulse occurs at every beat of the escapement, or 15 times per second, or once for every third revolution of the armature.

Speed regulation, with a voltage change up to plus or minus 20 percent, is plus or minus 0.1 percent at 65°F. For a temperature range of −65°F to +165°F, speed regulation is only plus or minus 0.3 percent.

12-19 Brushless DC Motors. Solid-state electronic circuitry is used in lieu of commutators in some small dc motors. Such motors are truly brushless dc motors. They are not to be confused with so-called brushless dc motors which are actually ac motors operating on ac current obtained from an inverter or oscillator; one such motor is discussed in Art. 9-16. In most brushless dc motors the solid-state switches are transistors. For an explanation of the way that transistors work, the reader is referred to an elementary book on electronics.[3,4]

Most brushless motors have a permanent-magnet rotor and a polyphase stator; the most common stator winding is three-phase, connected delta or wye, with each phase displaced by 120°, although two-, four-, and six-phase windings are also used. The circuitry is such that either half-wave or full-wave rectification may be used. Half-wave systems use one transistor per phase; full-wave systems use two transistors per phase. The full-wave system is preferred because conduction is always continuous in one leg when the other is being commutated, thus giving less torque pulsation. Full-wave systems are more efficient than half-wave systems, since the losses are one-half as great. Wye connections are more efficient than delta; however, the delta connection generates less electromagnetic interference.

(1) Electro-craft motor. Figure 12-15 shows the schematic configuration of a three-phase full-wave transistor-commutated dc motor. A half-wave controller would have only one set of transistors, such as Q1, Q2, and Q3. Figure 12-16 is a representation of the torque function and logic sequence of the motor. When Q1 is energized between 0 and 60°, Q5 is also conducting and current is flowing from A to B. In the next sequence (60 to 120°) Q6 is energized and the current will flow from A to C; at the same time the current in leg B decreases to zero by conduction through D2.

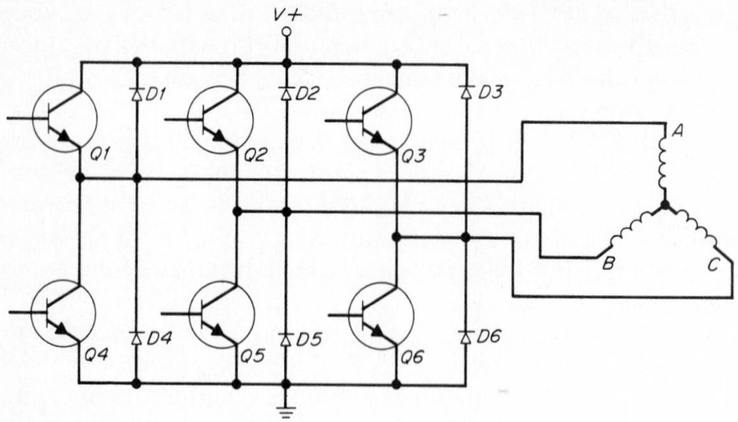

Fig. 12-15 Schematic configuration of a three-phase full-wave transistor-commutated dc motor. *(Electro-Craft Corporation.)*

The logic sequence in Fig. 12-16 is controlled by the angular position of the permanent-magnet rotor. Two methods used to control the logic sequence are by Hall-effect sensors* and by electro-optical sensors. Hall-effect sensors are mounted in the stator structure or on a terminal board, where they sense the polarity and magnitude of the permanent-magnet field in the air gap. The electro-optical switch, such as a light-emitting diode (LED) and a phototransistor, is controlled by a shutter mounted on the shaft, which generates precise angular encoding signals.

(2) Brailsford motor. Figure 12-17 shows the schematic arrangement of a transistor-commutated dc motor. The rotor consists of a permanent magnet which induces alternating voltages in the stator coils, causing a triggering of the transistors. Thus, the transistors become conductive at a frequency determined by the rotational speed of the rotor. However, if an alternating voltage is applied to the two extreme bottom terminals shown in the sketch, the rotor will synchronize with this voltage. A 3- to 5-W motor can be controlled with as little as 5 mW. Conversely, when the motor is running at its own speed, an ac voltage appears at these terminals; this voltage can be used as a speed-measuring indicator or to synchronize some other device with the motor.

In Fig. 12-17 the motor is shown in its starting position. Stationary parts are identified by numbers and rotating parts by letters. The rotor

*The Hall effect may be described briefly as follows. When current is passed through a thin slab of semiconductor material while a magnetic field is passing through the slab, a Hall-effect voltage is generated across the slab, perpendicular to the direction of current flow.

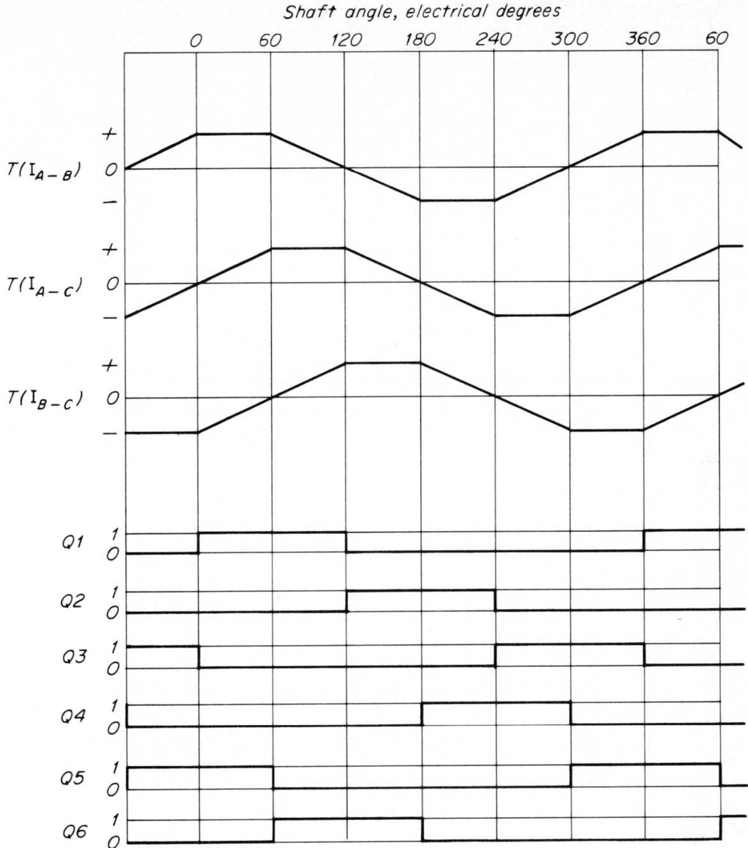

Fig. 12-16 Representation of the logic sequence and torque function of a brushless dc motor controlled by the circuit shown in Fig. 12-15. 1 = Q on; 0 = Q off. Hall-effect sensors, which are embedded in the stator, control current flow in the three-phase stator. Between 0 and 60° Q1 and Q5 are energized and current flows from A to B; between 60 and 120° Q1 and Q6 are energized and current flows from A to C. Progressive rotation results in current flow from B to C; B to A, C to A, and C to B during one revolution. *(Electro-Craft Corporation.)*

assembly includes a four-pole cylindrical magnet together with a starter switch. The starter switch consists of two brass members A and B, individually pivoted at points D and C, respectively. A spring, not shown, maintains tension between members A and B, keeping them in the positions shown, when the rotor is at rest. Four stationary flexible contact springs 1, 2, 3, and 4 are attached to the motor frame. When the rotor is at rest its magnetic field, as indicated by the letters N-S-N-S, will align with the extended portions of the stator poles in one of four posi-

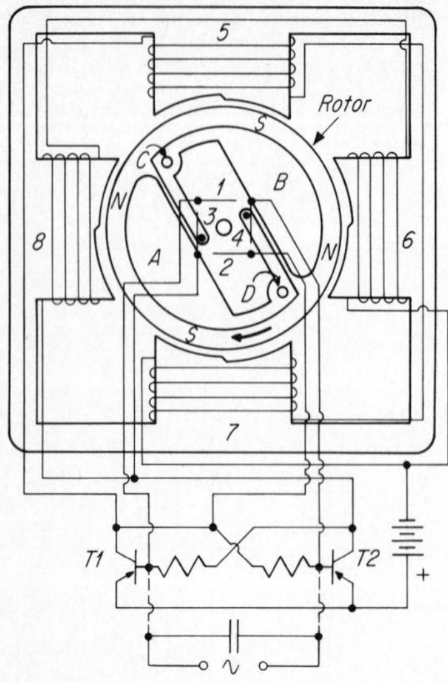

Fig. 12-17 Schematic diagram of a transistor-commutated small dc motor with permanent-magnet rotor. Contact pins engage springs only during starting. *(Brailsford and Company.)*

tions. In the figure, the south poles are aligned with poles 5 and 7. When battery voltage is applied, a circuit is completed through springs 3 and 4 to shunt the base of T2 with its collector, making T2 conductive, so that stator coils 5 and 7 become energized, with poles 6 and 8 becoming north poles and poles 5 and 7 becoming south poles. The rotor poles are repelled and turn one-quarter turn clockwise. In this position, contact is made through springs 1 and 2, which shunt the base of T1 with its collector, causing stator coils 6 and 8 to be energized, thus causing the rotor to turn another quarter turn clockwise. When the rotor has accelerated sufficiently, centrifugal force, acting on members A and B, causes them to pivot outwardly, withdrawing the contact pins from engagement with the springs, and the motor continues to run under transistor commutation. (Note that the pin contacting spring 4 is attached to member A, and the other pin to B, so that when the weights A and B move outward, the pins move inward.) Note that heavy-duty contacts are not required, since the switch serves only to trigger the transistor circuit. When the motor is running, the voltages induced in the stator coils trigger the transistors at the correct times, so that the contacts are no longer needed.

These motors are made for very low power consumption, ranging from 20 mW to just a few watts. For such low power inputs, the manufacturer reports exceptional efficiencies, ranging from 35 to 76 percent.

Brushless motors can have high torque-to-inertia ratios. Commutation is performed external to the rotating parts, and thus only the motor bearings are subject to mechanical failure. The controller can be located remotely from the motor and away from harmful environments; the motor and control have a lower level of radio frequency emission than do conventional commutator-type motors and controls. The speed-torque curve has the same characteristics as a conventional dc permanent-magnet motor.

SERVICE PROBLEMS

The following articles were written to apply generally to dc motors having fractional horsepower ratings and either conventional or electronic power supplies. To a lesser extent the remarks also apply to subfractional horsepower motors.

12-20 Armature Windings. Armature windings are discussed in Chap. 11, as well as in a previous work by Veinott.[1]

12-21 How To Change the Speed without Rewinding. The speed of a dc motor can be increased without rewinding

1. By inserting resistance in the field circuit (shunt or compound motors only).

2. By increasing the air gap. The air gap can be increased by grinding down the outside diameter of the armature, by grinding out the stator bore, or, if the poles are separate from the yoke, by grinding off a part of the poles, either from the face or from the body next to the yoke.

3. By putting shims of *nonmagnetic* material between the poles and the yoke if the poles are separate from the yoke.

The speed of a dc motor can be *decreased* without rewinding

1. By putting shims of magnetic material between the poles and the yoke if the poles are separate from the yoke. Shims may be made from any convenient shim stock of iron or steel (brass, copper, or lead should not be used). A shim that is approximately ⅛ in larger than the body of the pole at the yoke can be cut out and a hole put into this shim for the pole bolt.

2. By inserting resistance in the armature circuit. This procedure increases the speed regulation and makes the speed vary more with changes in load. Insertion of armature resistance does not affect the no-load speed appreciably.

12-22 How To Reverse a DC Motor. Electrically reversible motors are discussed in Art. 12-12. To connect a series, shunt, or compound motor for a definite direction of rotation, or to change the direction of rotation, refer to the diagrams of Fig. 12-3, 12-4, 12-5. Permanent-magnet motors are reversed simply by changing the polarity of the voltage applied to their terminals.

12-23 Finding the Neutral Position. A satisfactory way of finding the neutral position in a dc motor is by means of the *kick neutral method.* Connect a 15-V voltmeter across the brushes, and make and break the field circuit; if the brushes are off neutral, a deflection of the voltmeter needle will be observed. Shift the brushes until no deflection can be observed when the field circuit is made or broken. For more accurate results, it may be necessary to repeat this process, using a voltmeter having a full-scale reading of 1.5 V. Wedge-shaped brushes will contribute further to the accuracy of the setting.

If the motor is to operate in either direction of rotation or if it has a commutating pole, the brushes should be left in the neutral position. If the motor operates only in one direction, and if it has no commutating pole, the brushes may be shifted approximately 10° *against* the direction of rotation of the armature.

12-24 Brushes. The brushes usually selected for dc motors are those found best suited for the particular designs after exhaustive tests have been made of motors operating under conditions duplicating as nearly as possible those expected in actual service. For replacement, the service mechanic should obtain duplicates of the original brushes, securing the same make and grade, if possible. If the original type of brush cannot be obtained, one having as nearly as possible the same characteristics should be selected.

References

1. Veinott, Cyril G.: *Fractional Horsepower Electric Motors,* 2d ed., McGraw-Hill Book Company, New York, 1948.
2. *NEMA Motor and Generator Standards,* NEMA Publ. no. MG1-1978, National Electrical Manufacturers Association, 2101 L Street N.W., Washington, DC, 20037, 1978.
3. Jacobowitz, Henry: *Electronics Made Simple,* Garden City Books, New York, 1958.
4. Hibberd, Robert G.: *Solid State Electronics, A Basic Course for Engineers and Technicians,* McGraw-Hill Book Company, New York, 1968.

Bibliography

"A Brush Guide for Fractional h.p. Motors," *Appliance Eng.*, December 1982, p. 68.

DC Motors Speed Controls Servo Systems: An Engineering Handbook, Electro-Craft Corporation, 1600 Second Street South, Hopkins, MN 55343, 1980.

Gerlach, F. H.: "Better Commutators for Small Motors," *Electr. Manuf.*, December 1944, p. 124.

Ireland, James R.: "Ceramic PM Motor Design," *Electro-Technology,* February 1966, pp. 48–51.
Ireland, James R.: "Ceramic Permanent-Magnet Motors," McGraw-Hill Book Company, New York, 1968.
Kubler, E. F.: "The Armature Current Form Factor of a D-C Motor Connected to a Controlled Rectifier," *AIEE Trans.*, 1959, vol. 79, pt. III-B, pp. 764–770.
Pfaff, R. W.: "Characteristics of Phase-Controlled Bridge Rectifiers with D-C Shunt Motor Load," *AIEE Trans.*, 1958, vol. 77, pt. II, pp. 47–53.
Puchstein, A. F.: *The Design of Small Direct-Current Motors,* John Wiley & Sons, Inc., New York, 1961.
Rashidi, Abdul S.: "Better Motors with Rare-Earth Magnets," *Mach. Des.*, July 8, 1976, pp. 70–73.
Say, M. G., and E. O. Taylor: *Direct Current Machines,* Halstead Press Book, John Wiley & Sons, New York, 1980.
"Small Brushless D-C Motor Holds a Set Speed within 0.5%," *Prod. Eng.*, Sept. 11, 1967, p. 102.
"Small Motors of New Design Get The Most from Batteries," *Prod. Eng.*, Aug. 29, 1966, p. 57.
"Solid-State D-C Drives," *Electro-Technology,* March 1967, pp. 97–101.

CHAPTER

13

VARIABLE-SPEED DRIVE SYSTEMS

Fractional horsepower variable-speed drive systems have grown enormously in popularity over the last few decades, primarily because of the development of electronic devices for power conversion and control functions and because of the need to operate certain appliances over a wide speed range. Variable-speed systems have become popular for use in many common appliances, such as hand tools, fans, and mixers for home and laboratory use and for machining of various materials. Both dc and ac drive motors are employed, and both will be discussed in this chapter.

SYSTEMS USING A DC DRIVE MOTOR

Although there are many ways of obtaining an adjustable-speed dc motor which operates from an ac power supply, we shall limit our discussion here to those systems that use a dc shunt motor to drive the load; ac power is drawn from the mains and converted and controlled to make the dc drive motor perform in the desired manner. All such systems use essentially a Ward Leonard system, or some derivative or variation of it; hence we shall start our discussion by examining the basic Ward Leonard system itself.

13-1 The Ward Leonard System. A basic Ward Leonard system is represented schematically in Fig. 13-1. The drive motor, which drives the load, is a dc shunt motor. The power conversion unit consists of an ac motor, a dc generator, and a dc exciter. The armature of the drive motor is paralleled with the armature of the dc generator, and the field of the drive motor is excited directly by the constant-voltage dc exciter that supplies the field of both the drive motor and the generator. Thus, the system provides means for independently adjusting either the armature voltage or the field excitation of the drive motor.

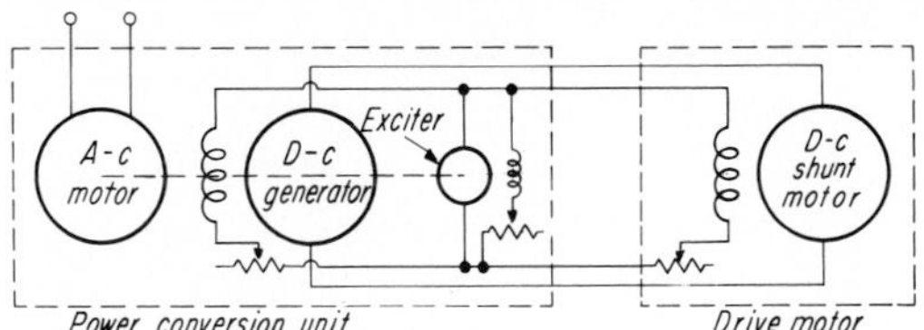

Fig. 13-1 Schematic representation of a Ward Leonard variable-speed drive system.

(1) Basic principles involved. *Base speed* of the motor is defined as the speed of the motor with full rated armature voltage and full rated field excitation applied. Two kinds of speed adjustment are available: variation of armature voltage and variation of field current. Reducing the armature voltage *reduces* the speed. Reducing the field excitation *increases* the speed.

(2) Armature-voltage-control speed range. Speed can be reduced to a very low value by reducing the armature voltage. Over this speed range, from base speed downward, the torque output can be essentially constant, for this represents constant armature current, constant losses in the armature, and constant heating. (If the cooling falls off too much at reduced speed, the output torque may have to be decreased slightly.) Thus, the horsepower output available varies directly as the speed, while the torque output is essentially constant over the speed range. This range gives the widest speed variation and is the one most commonly used.

(3) Field-control speed range. Reducing the excitation current of the drive motor reduces its field flux ϕ. As we can see from Eq. (12-5), the speed is inversely proportional to the flux ϕ. Also, from Eq. (12-1) we can see that, for a constant armature current, the torque output is proportional to the flux. Hence, for constant armature current and constant armature heating, in the field-control speed range the available torque output is inversely proportional to the speed, but the available horsepower output is essentially constant. The field-control range is not always made available in certain packaged drives, and when it is, it is often referred to as the extended speed range.

(4) Regulation, IR drop, and IR-drop compensation. Regulation is ordinarily defined as the change in speed from no load to full load, expressed as a percentage of full-load speed. In an adjustable-speed drive system, this drop in speed is more often expressed as a percentage of the base speed. As a motor is loaded, it has to draw more current to develop the necessary torque. This increase in armature current tends to affect the motor speed in at least three ways: by armature reaction, which tends to weaken the field and thereby to increase the speed [Art. 12-3(4)]; if the machine has a series field, by strengthening the field, thereby reducing the speed; by the *IR* drop, which reduces the counter emf, thereby reducing the speed. The effect of armature *IR* drop, at different speed settings, is illustrated in Fig. 13-2 in which the effect has been exaggerated for purposes of illustration. Note that the *IR* drop, for the same armature current, is the same at all speeds, but the relative importance of this drop increases at the lower speed settings. Consequently, to obtain good speed regulation at all speed settings, *IR*-drop compensation is incorporated into many speed-regulating systems (not shown in the illustration). Briefly stated, *IR*-drop compensation increases the impressed armature voltage by an amount just sufficient to compensate for the *IR* drop in the armature circuit, thus giving flatter speed regulation.

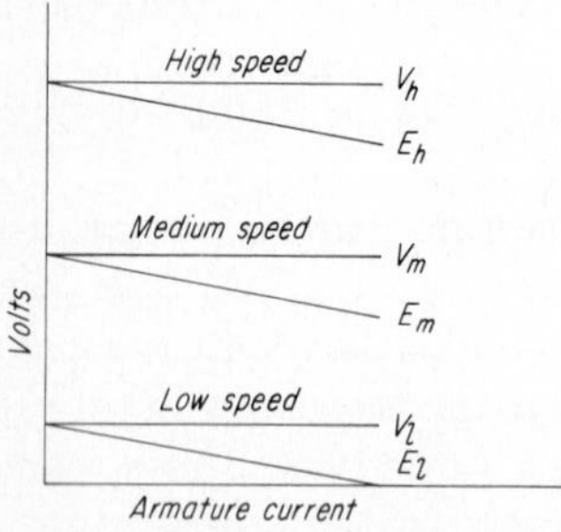

Fig. 13-2 Effect of armature *IR* drop on the counter emf at different speed settings. V = applied armature volts; E = counter emf = $I_a R_a$. *(Ward Leonard System.)*

(5) Reverse power flow. In the Ward Leonard system of Fig. 13-1, power can flow in either direction. Suppose that the drive motor is driving a heavy flywheel at base speed; now, let the excitation on the generator be reduced to, say, about half voltage. The drive motor suddenly becomes a generator; the generator in the power-conversion unit becomes a motor which carries the ac motor up above synchronous speed, making it an induction generator and pumping power back into the line. Or, this can occur if the load on the drive motor is an overhauling one, such as lowering a load on a hoist, for example.

(6) Constant-torque drives. If the field of the drive motor is held constant and a constant value of current is passed through its armature, the motor will develop a constant torque, independent of speed, within its range. By incorporating a current-sensitive regulator in the armature circuit to regulate the generator field, the Ward Leonard system of Fig. 13-1 can be made to produce a constant-torque output. Further, by changing the setting of the current regulator, different values of constant torque can be obtained.

(7) Feedback control. In Fig. 13-1 the speed of the drive motor depends upon the inherent regulation of the system. Suppose, however, that attached to the drive motor shaft were a tachometer which delivered a voltage proportional to the speed; now, suppose that this voltage signal from the tachometer were compared with some constant reference voltage such as a battery. If this difference in voltages, usually called the "error signal," were amplified and used in such a manner that it raised the armature voltage when the speed was low and lowered the armature voltage when the speed was too high, then the speed of the motor could be made independent of the inherent regulation of the machines; instead, it would depend upon how well this error signal corrected the speed error that caused it. Such a system is known as a closed-loop feedback control system.

(8) Electronic power-conversion units for Ward Leonard variable-speed drive systems. The rotating-machine power-conversion unit of Fig. 13-1 has, at different times, been replaced in whole or in part by electronic power-conversion units, making use of devices such as thyratron tubes, magnetic amplifiers, autotransformers, and rectifiers, both controlled and uncontrolled. How some of these are used forms the basis of the discussion that follows. But first rectifiers are considered.

13-2 Simple Rectifiers.

A *diode* is a simple two-terminal semiconductor device that has the property of being conducting in one direction only. Most popular today because of their small size for their rating and their low cost are silicon diodes. There are many ways to use them.

(1) Half-wave rectifier. The simplest way to rectify is to connect a rectifying element in series between the ac line and the dc load, as shown in Fig. 13-3. Since the diode will pass current in only one direction, the

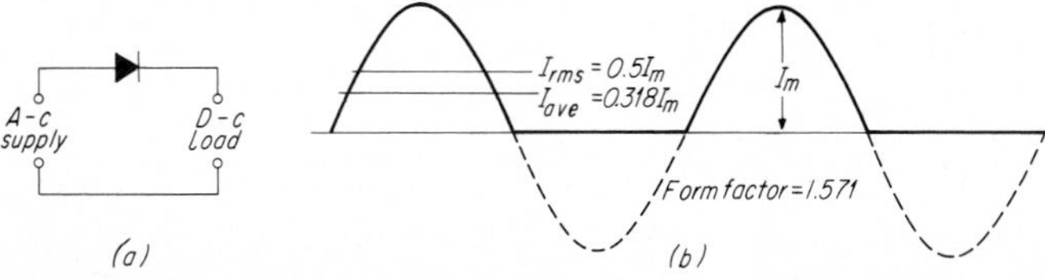

Fig. 13-3 A simple half-wave rectifier (uncontrolled). (*a*) Schematic circuit; (*b*) output wave form.

output wave form takes the shape shown in Fig. 13-3*b*; the negative halves of the sine waves are simply suppressed, leaving the wave form shown by the solid line.

(2) Full-wave rectifier. To obtain full-wave rectification, two diodes and a center tap in the ac power supply, as shown in Fig. 13-4, are used. The schematic circuit is given in Fig. 13-4*a* and the resultant wave form in Fig. 13-4*b*. This arrangement simply flops over the negative halves of the sine wave, eliminating the long "off" period. Note that this arrangement requires a center-tapped ac source, or an autotransformer with center tap has to be provided.

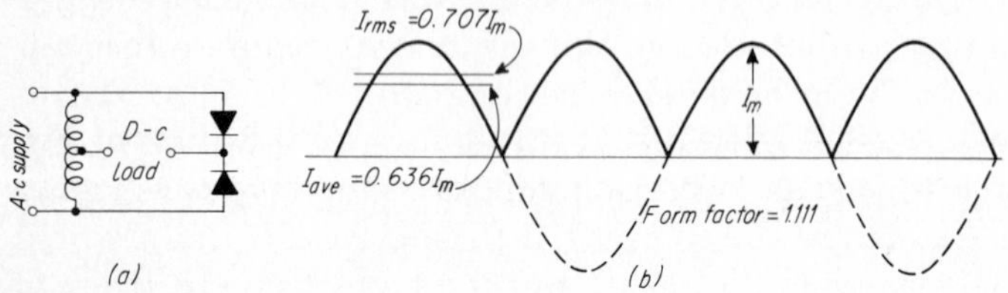

Fig. 13-4 A single-phase full-wave rectifier (uncontrolled). (*a*) Schematic circuit; (*b*) output wave form.

(3) Bridge-type rectifier. The autotransformer with its center tap can be eliminated by using four rectifying elements, connected in a bridge arrangement as shown in Fig. 13-5. Such a bridge connection gives a full-wave output, as shown in Fig. 13-4*b*.

13-3 Single-Phase Controlled Rectifiers. Controlled rectifiers use triodes, that is, three-electrode devices. Formerly it was a *thyratron* (a three-electrode gaseous tube) that was used; this has been replaced by a thyristor. Both these devices can carry current in only one direction, like the diodes described previously. However, the thyratron has an important distinguishing characteristic; so long as the grid is maintained at a sufficiently high negative potential, current cannot flow; once the negative grid potential has been reduced below its critical value, the tube becomes conducting in the forward direction. Once the tube has started to conduct, the grid has no further control over the plate cur-

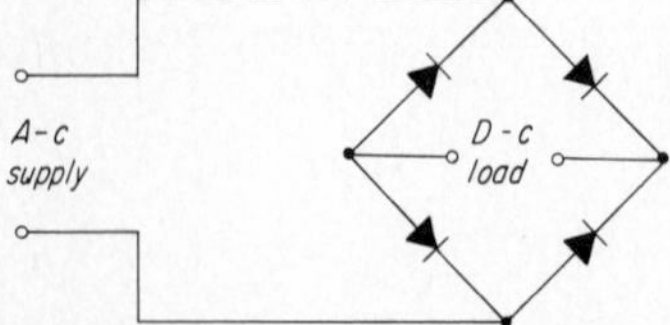

Fig. 13-5 A single-phase bridge rectifier giving full-wave rectification.

rent. By controlling the grid voltage, the initial firing of the tube can be controlled; conduction continues until the current ceases to flow. A silicon-controlled rectifier (SCR) or thyristor is a semiconductor which normally blocks current flow in both directions, but it can be triggered or fired into forward conduction by feeding a small current pulse into a third electrode, called the gate.

(1) A half-wave controlled rectifier. The schematic circuit of a half-wave firing-angle-controlled rectifier is shown in Fig. 13-6*a*. This is similar to the half-wave circuit in Fig. 13-3*a* except that the rectifying element is a thyristor or controlled rectifier, to which a firing circuit has been added. This firing circuit is arranged so as to delay the actual firing of the thyristor until a certain time angle after the zero point of the voltage wave has elapsed. This time angle is known as the firing angle. The resultant wave form of the output will be of the general form of Fig. 13-6*b*, wherein the firing angle is 90°. The heavy black line shows the output wave, and the dotted line the sine-wave input.

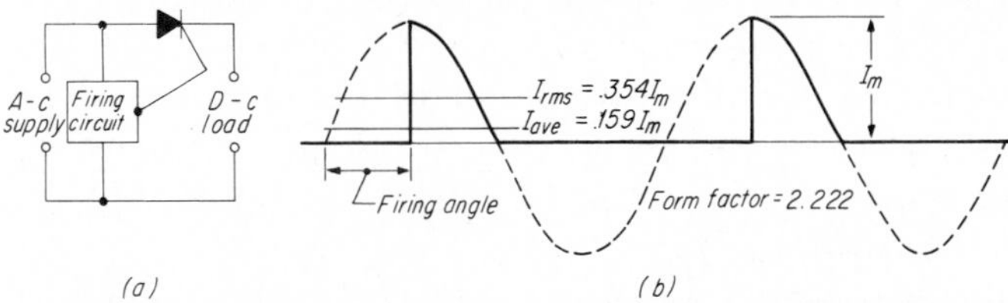

Fig. 13-6 A single-phase half-wave controlled rectifier that uses a thyristor (SCR) and firing angle of 90°. (*a*) Schematic circuit; (*b*) output wave form.

(2) Half-wave controlled rectifier, with back rectifier. When dc motors are operated from a rectified power source, it is common practice to add a back rectifier, as shown in Fig. 13-7. The back rectifier is bridged across the armature or the field of the motor in such a way that current can continue to flow even after the main rectifier has ceased to supply current. The inductance of the armature tries to make the current continue flowing, and the back rectifier provides a path for this current to continue flowing. The back rectifier protects the main rectifier from

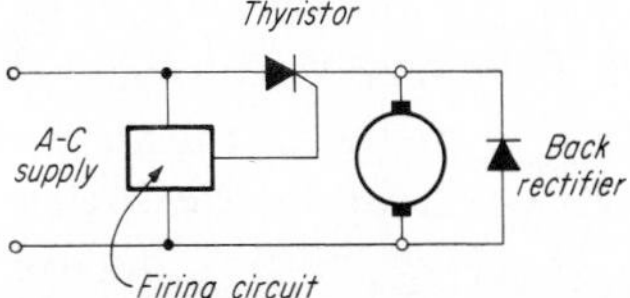

Fig. 13-7 A single-phase half-wave controlled rectifier, with back rectifier.

dangerous back voltages and provides a more uniform flow of armature current.

(3) Full-wave controlled rectifier. It is obvious that controlled rectifiers could be used in the circuit of Fig. 13-4 or 13-5, with appropriate firing circuits. Such an arrangement would give an output wave form similar to that of Fig. 13-6*b*, except that there would be twice as many pulses.

13-4 Form Factor. The form factor of a current wave is defined as the ratio of the root-mean-square (rms) value of the current to the average value of the current. The average and rms values, together with their ratios, were calculated for the output wave form of Fig. 13-6*b* for a number of different firing angles. They were also calculated for full-wave controlled rectifiers, which give twice as many pulses, each of the same shape, as those of Fig. 13-6*b*. The results are summarized in Table 13-1. The intermediate calculations were carried out to four places, but the results are rounded off to three places, and so the form factors given are not always the same as would be obtained by dividing the rms values in the table by the average values.

It should be pointed out that the values given in the table are the results of mathematical calculations on waves of the shapes shown in Fig. 13-6. In practice, the circuit constants may modify the actual form factors from the table values.

Measurement of form factor. The form factor of a rectifier-supplied load can be determined experimentally. The current is measured with a thermal or an electrodynamic ammeter, such as that used for measuring alternating current, and it is also measured with a permanent-magnet moving-coil (D'Arsonval) instrument. The form factor is simply the ratio of these two current readings.

13-5 Operation of DC Motors on Rectified Alternating Current. DC motors are likely to perform quite differently on rectified power than on conventional dc power, such as that supplied by a dc generator or by a battery.

(1) Effect on I^2R losses. Since it is average current that produces torque and rms current that produces heating, it follows that I^2R losses are increased by use of rectified power by an amount equal to $(I_{rms}/I_{av})^2$, that is, by an *amount equal to the square of the form factor*. This can be considerable; for example, from Table 13-1, the form factor for a full-wave unfiltered rectifier is 1.111, and this will increase the I^2R losses by a factor of $1.111^2 = 1.23$. For a half-wave rectifier, the form factor is 1.571, and the increase in losses is by the factor $1.571^2 = 2.47$. With controlled rectifiers, the increase in losses can be even greater, as can be seen by reference to the table. Proper filtering can reduce, but not eliminate, these extra I^2R losses, which can be considerable.

TABLE 13-1 Currents and Form Factors for Single-Phase Controlled Rectifiers

Firing angle	Full wave			Half wave		
	I_{av}	I_{rms}*	Form factor	I_{av}	I_{rms}*	Form factor
0	0.637	0.707	1.111	0.318	0.500	1.571
30	0.594	0.697	1.173	0.298	0.493	1.659
60	0.478	0.634	1.328	0.239	0.449	1.879
90	0.318	0.500	1.571	0.159	0.354	2.222
120	0.159	0.313	1.964	0.080	0.221	2.778
150	0.043	0.120	2.815	0.021	0.085	3.982

*Currents are expressed as a fraction of the peak value.

(2) Effect on commutation. Coils undergoing commutation are inductively coupled with the main field. If the latter is subject to a pulsating voltage, so that a pulsating flux is produced, a voltage will be induced in the coils undergoing commutation. Hence, operation on rectified ac power produces poorer commutation and more sparking at the brushes.

(3) Other effects. The extra I^2R losses produce extra heating. The pulsations in field flux generate additional eddy-current losses in the iron. Ripple currents may create bearing currents which, though feeble in magnitude, have been known to damage the bearings.

Before a dc motor is used with a solid-state power source, it is best to check to see whether the motor was intended for use on such a power supply. Good filtering can reduce all the effects discussed previously.

13-6 DC Motors Intended for Use on Adjustable-Voltage Electronic Power Supplies. Such motors are usually shunt-wound, with or without a stabilizing series-field winding. Nameplates for such motors may carry, in addition to the usual information, rated armature voltage, rated field voltage, armature rated-load amperes at rated speed, and rated form factor. The voltage and current ratings are based on *average values.* The standard National Electrical Manufacturers Association (NEMA)[1] ratings for fractional horsepower dc motors for use on single- and three-phase rectified power are the following:

Horsepower ratings. 1/20, 1/15, 1/12, 1/8, 1/6, 1/4, 1/3, 1/2, 3/4, 1 hp

Speed ratings. 3450, 2500, 1725, 1140 rpm

Armature voltages. 75, 90, 150, 180, 240 V

Field voltages. 50, 100, 150, 200, 240 V

Form factors. 2, 1.85, 1.7, 1.5, 1.4, 1.3, 1.2, 1.1

Momentary overload capacity. 150 percent rated current for 1 min

When such motors are tested, they should be tested on a power supply that will provide the rated voltage and the rated form factor at the rated load.

13-7 Variable-Speed Systems Using Electronic Power Conversion. There are a number of variable-speed drive systems available which use some form of electronic power conversion unit. Most of these are modeled after the Ward Leonard system, with various modifications. The ones discussed next may be considered representative.

(1) Servo-Tek system. One Servo-Tek drive is represented schematically in Fig. 13-8. Power for the drive-motor armature is obtained from the ac source through a bridge-type rectifier; by adjusting the autotransformer, the dc voltage impressed on the drive-motor armature is adjusted. The shunt field is supplied directly from the source through another bridge-type rectifier, so that its voltage remains constant, unaffected by adjustments of the armature voltage. This system is built in the full range of the fractional-horsepower ratings

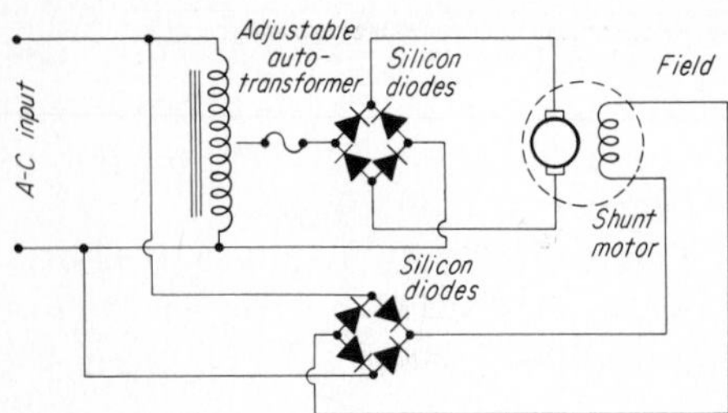

Fig. 13-8 A Ward Leonard variable-speed drive system using an autotransformer and two bridge-type simple diode rectifiers for the power-conversion unit. *(Servo-Tek Products Company.)*

(2) Bodine motor-control system. Figure 13-9 shows a solid-state controller for a fractional horsepower dc shunt motor. It is built in ratings from 1/50 to 1/4 hp and a speed range of 60 to 2200 rpm; the extended-range option [Art. 13-1(3)] provides speeds to 3200 rpm. Some units are designed for adjustable torque. Typical speed-torque curves are shown in Fig. 13-10. Dynamic braking is also provided.

(3) Electro-Craft Motomatic. Under the trademark Motomatic, Electro-Craft Corporation offers a closed-loop speed-control system which automatically adjusts the input to the motor so that an essentially constant speed is maintained even when the load on the motor changes or the line voltage fluctuates. Motomatic consists of a permanent-magnet motor generator with a single armature core containing both motor and generator windings; it is controlled by a transistorized high-gain feedback amplifier. From no-load to full-load, the variation in speed is less than 1 percent of set speed. Electro-Craft's E-350 system has a speed ratio of 1500:1 (7500:5 rpm) and is rated 1/10 hp at 7500 rpm. Other ratings are available with speed ratios of 1000:1 and 500:1, and up to

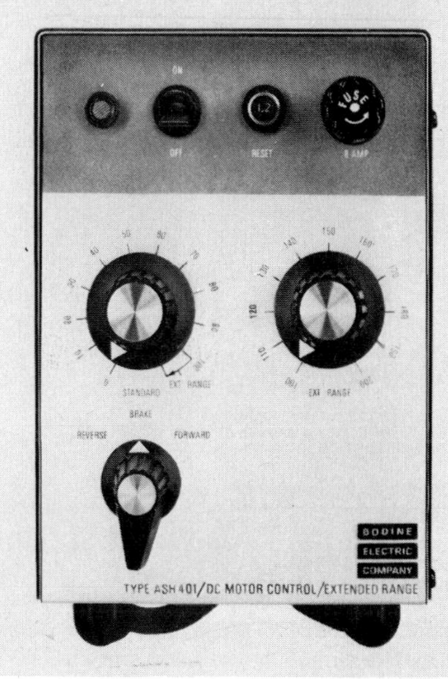

Fig. 13-9 A solid-state electronic controller for a fractional horsepower variable-speed drive system. *(Bodine Electric Company.)*

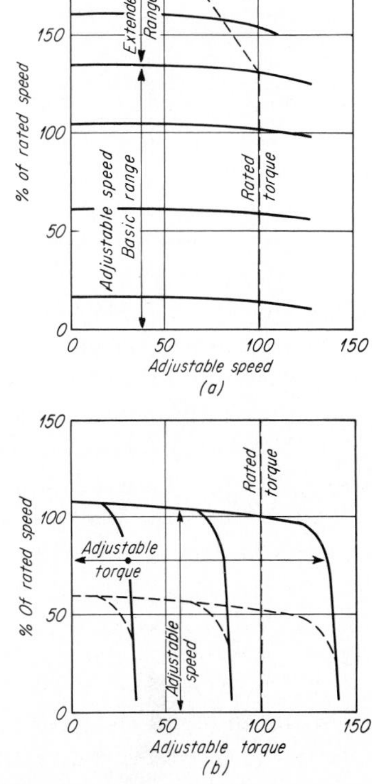

Fig. 13-10 Speed-torque curves for (*a*) adjustable-speed and (*b*) adjustable-torque variable-speed dc drive systems. *(Bodine Electric Company.)*

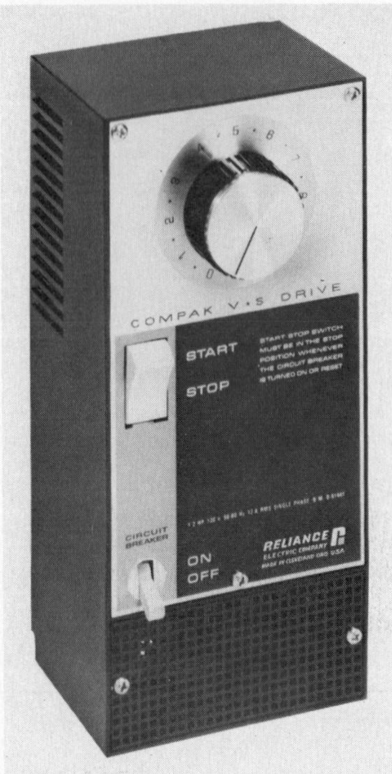

Fig. 13-11 A Compak V-S adjustable-speed controller for dc fractional horsepower motors. *(Reliance Electric Company.)*

½ hp. Integral gearheads are available on some models; dynamic braking and reversibility are optional.

(4) Reliance Compak V-S drive. Figure 13-11 shows a controller and electronic power supply, all self-contained in the unit shown. The motor field is supplied through a thyristor and back rectifier. Speed control is effected by controlling the firing angle of the thyristor. Units rated ⅙, ¼, and ½ hp use half-wave rectification. During each half cycle when the armature current is shut off, a signal is picked up from the counter electromotive force (emf) of the armature and used as a feedback signal to hold the speed constant. Regulation is 5 percent from no load to full load over a 20:1 speed range. With reduced load or intermittent duty, a speed range of as much as 100:1 can be achieved. Standard base speed is 1725 rmp. Units above ½ hp use full-wave rectification.

SYSTEMS USING AN AC DRIVE MOTOR

As discussed in Chap. 7 the operating speed of an induction motor can be varied by changing the voltage impressed on the windings, for this changes the slip at any given load. Another older method of voltage control was by use of an autotransformer. Today, a more popular means of voltage control, and hence speed control, of an ac motor is by the use of a solid-state device, such as a triac. Another means of effecting speed control is by changing the speed of the rotating field itself; this is done by changing the supply frequency by means of a solid-state inverter. Both methods are discussed in this chapter.

13-8 Triac. The triac, an acronym for *tri*ode *ac* switch, was developed to replace two back-to-back silicon controlled rectifiers (SCR), but with a common gate electrode. Whereas a single SCR gives half-wave output, a triac gives full-wave output. The symbol for an SCR is shown in Fig. 13-12*a*, and the symbol for a triac is shown in Fig. 13-12*b*.

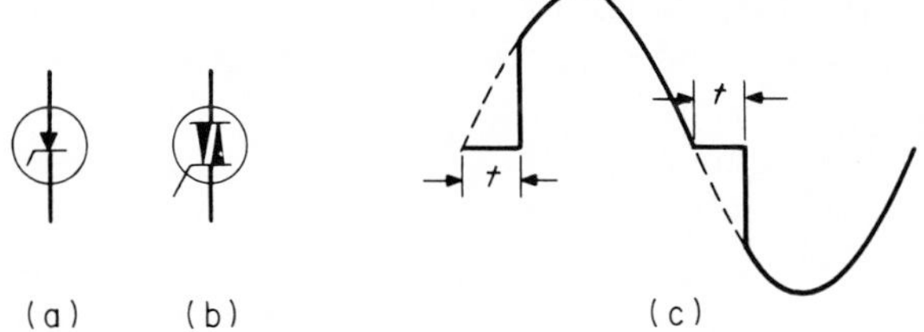

Fig. 13-12 (*a*) Symbol for a silicon-controlled rectifier (SCR); (*b*) symbol for a triac; (*c*) voltage wave form showing effect of time delay *t* in triggering a triac.

When a signal of sufficient amplitude is applied to the gate of a triac, the triac will trigger and convert from a nonconducting state. Once conducting, it will continue to conduct until the current is zero in magnitude, at which time the triac again becomes nonconducting. Figure 13-12*c* shows the effect on a sine wave of voltage due to a time delay in the triggering of a triac. Note that there are two triggerings per cycle, one in the positive half-cycle and one in the negative half-cycle. By gate-circuit design the time delay is made variable, so that the rms voltage is made adjustable over a wide range of values. As is the case with other methods of voltage control, the triac method of speed control is limited to use with shaded-pole, capacitor-run, and universal types of ac motors. In some early triac circuit designs the sharp wave front caused noise in motors, predominantly at low speed settings; however, by proper matching of control to motor this problem has been all but eliminated. Often the triac circuit does not fire at the same time in the positive and negative directions, resulting in a dc offset voltage impressed on the motor, which causes a braking effect and additional heating of the windings. Triac control permits a continuously variable range of speed adjustment. It is widely used for permanent-split and shaded-pole ceiling fan motors, as well as for universal motors used in hand drills. When a triac is used with a permanent-split or shaded-pole motor, it is subject to the general limitations discussed in Art. 7-11.

13-9 Inverter. An inverter is a device that converts dc power into ac power. There are two basic types of inverters, pulse width modulation (PWM) and six-step. Single-phase power is taken from the mains in both types. In the PWM system the alternating current is fed into a full-wave rectifier and filter network whose output is fixed dc voltage and is then fed into a PWM inverter. In one six-step system alternating current is fed into an SCR bridge and filter network whose output is variable direct current and then into a six-step inverter; another six-step system has the alternating current fed into a rectifier and filter network whose output is fixed direct current and then into a chopper and filter whose output is variable direct current and then into the six-step inverter.

A PWM inverter takes the fixed dc voltage, chops it into variable-width pulses, reverses the direction, and repeats the pulses, completing one cycle. Figure 13-13 illustrates a PWM inverter output-voltage wave form. A six-step inverter chops the variable direct current into square waves, reverses the direction, and repeats the square wave to complete one cycle. By varying the frequency of reversals in either system, the frequency of the ac output is varied. Figure 13-14 shows a schematic diagram of a simple three-phase inverter-motor circuit using six SCRs; Fig. 13-15 illustrates the three-phase six-step voltage wave

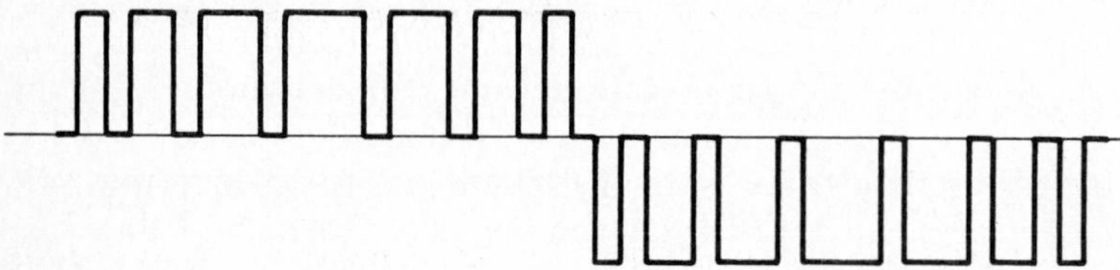

Fig. 13-13 Square-shaped wave form of voltage with variable pulse width.

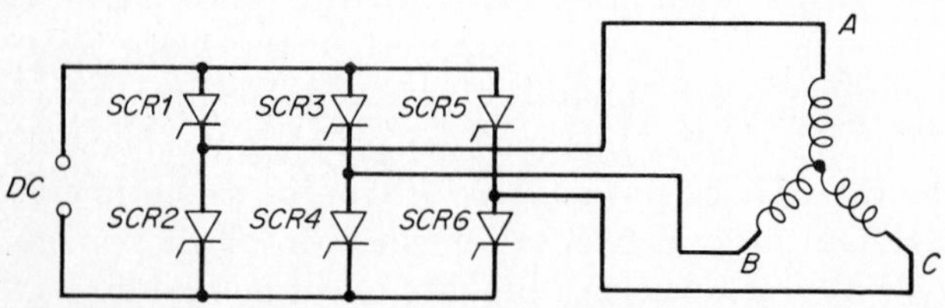

Fig. 13-14 Schematic diagram of a simple inverter motor circuit using six SCRs.

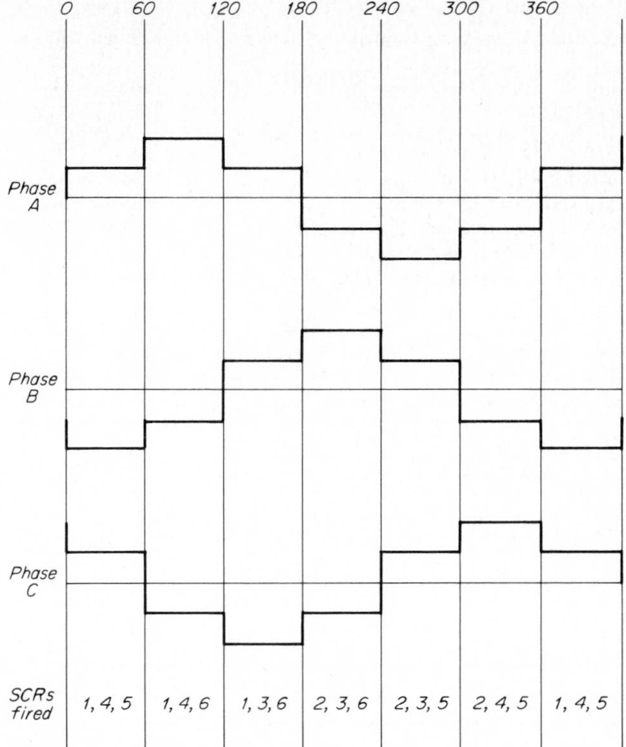

Fig. 13-15 Wave form and firing sequence of voltage impressed on the three-phase motor shown in Fig. 13-14 from the output of a six-step inverter.

form impressed on the motor. In both systems, the current wave form approximates a sine wave.

The normal operating frequency is from 6 to 120 Hz. The inverter output, when first switched on, starts at a low frequency, in the order of the percentage slip of the motor at full load and at 60 Hz, to prevent damage to the inverter circuitry; then the frequency ramps to the set value. Below 60 Hz the inverter output voltage varies with the frequency to maintain a fixed-volts-per-hertz ratio; the motor is then a constant-torque motor below 60 Hz. Above 60 Hz the inverter output voltage is constant; the motor is rated constant horsepower in this range of frequencies. Some inverters are rated more than 120 Hz, even up to 400 Hz.

Reference

1. *NEMA Motor and Generator Standards*, NEMA Publ. no. MG1-1978, National Electrical Manufacturers Association, 2101 L Street N.W., Washington, DC 20037, 1978.

Bibliography

Adem, A. A.: "Seven Solid-State Circuits for Controlling Universal-Motor Speed," *Mach. Des.*, Jan. 5, 1967, pp. 118–123.

Arnold, Frank Jr.: "Putting Together Your Own High-Performance DC Motor," *Drives & Controls Int.*, January–February 1983.

Bedford, B. D. and R. G. Hoft: *Principles of Inverter Circuits,* John Wiley & Sons, Inc., New York, 1964.

Erdman, D. M., H. B. Harms, and J. L. Oldenkamp: "Electronically Commutated DC Motors for the Appliance Industry," *Proceedings of Industry Applications Society,* Institute of Electrical and Electronics Engineer, 345 East 47th Street, New York NY 10017, October 1984, pp. 1339–1345.

Kosow, Irving L.: *Control of Electric Machines,* Prentice-Hall, Inc., Englewood Cliffs, N.J., 1973.

Kreuter, Rodney A.: "Controlling DC Power with Pulse-Width Modulation," *Pop. Electron.* June 1979, pp.62–63.

Scholey, Douglas: "Induction Motors for Variable Power Supplies," *IEEE Trans. Ind. Appl.*, vol. IA-18, no. 4, July–August 1982, pp. 368–372.

Venturini, M., and M. Maytum: "Permanent Magnet DC Motor Control," *Powerconversion Int.*, July–August 1983, pp. 55–57.

CHAPTER 14

POLYPHASE INDUCTION MOTORS

In integral horsepower sizes, induction motors are almost all polyphase. In fractional horsepower sizes, the reverse is true, for most of them are single-phase. The principal reason is simply that, in most installations in which fractional horsepower motors are used, polyphase alternating current is not available. If polyphase power does happen to be available, single-phase motors can always be used, but the converse is not true. (There are on the market static devices designed to operate polyphase motors on single-phase circuits but, in spite of claims to the contrary, polyphase motors operated from single-phase lines with such devices generally give a performance markedly inferior to that of comparable polyphase or single-phase motors, and at a higher overall cost.)

In general, in fractional horsepower sizes, single-phase induction motors sell at lower prices than polyphase induction motors, principally because of their larger activity; price is another reason that single-phase motors are often used even in some cases in which polyphase motors could be used. On the other hand, in many military applications, polyphase motors are used in small sizes in order to gain minimum weight and maximum reliability.

14-1 Essential Parts of a Polyphase Induction Motor. Essential parts of a polyphase induction motor are as follows:

1. A primary winding is usually placed on the stator, but sometimes it is placed on the rotor. The primary winding is the one connected directly to the source of power, regardless of whether it is on the stator or on the rotor. A two-phase motor has two electrically distinct windings, displaced by 90 electrical degrees. In the three-phase induction motor, there are three windings, or phases, displaced 120 electrical degrees from each other. These three phases may be connected in the circuit by any one of the common connections discussed in Art. 1-14 and illustrated in Fig. 1-7.

2. A secondary winding is usually placed on the rotor; and it (*a*) may be of squirrel-cage construction or (*b*) may have phase windings like the primary winding (such windings are rare in fractional horsepower sizes).

3. No switches, short-circuiters, or other mechanisms are required, except that collector rings and brushes are required for wound-rotor induction motors.

14-2 Principles and Characteristics of the Polyphase Induction Motor. The principle of operation of a polyphase induction motor is discussed in Chap. 2, in which it is shown that polyphase windings, excited by polyphase currents, set up a rotating magnetic field which causes the rotor to follow it and thereby act as a motor. The speed of this rotating

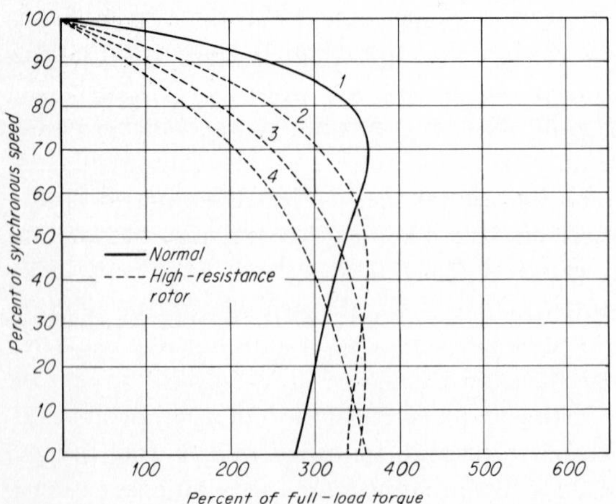

Fig. 14-1 Typical speed-torque curves of a fractional horsepower polyphase induction motor. Dashed curves represent high-resistance rotors.

field is known as the synchronous speed. The motor always runs more slowly than synchronous speed, and the difference between the actual operating speed and synchronous speed is known as the *slip*. The slip is usually expressed as a percentage of the synchronous speed.

A typical speed-torque curve for a fractional horsepower polyphase induction motor is shown in Fig. 14-1 by the solid-line curve. When higher resistance rotors are used, the speed-torque curves take on shapes such as curves 2, 3, or 4. Breakdown torques are usually about 40 percent higher than the figures for capacitor-start motors in Table 19-1. Polyphase induction motors, in fractional horsepower sizes, are provided with 40 percent more breakdown torque than single-phase motors, primarily in order to keep full-load slip within reasonable limits and to provide more locked-rotor torque. Locked-rotor torque is about 75 percent of breakdown torque. Thermal protection of polyphase motors is discussed in Art. 17-21.

WINDINGS AND CONNECTIONS

Many small polyphase motors today are machine-wound in the factory for reasons of cost. Such windings generally resemble in appearance single-phase concentric windings such as those discussed in Chap. 3, except that there are three windings, displaced 120 electrical degrees apart, instead of two windings 90° apart. Even skein windings are sometimes used on small polyphase motors. In the repair shop, where winding machines are not usually available, the conventional lap windings are easier to use. For that reason, the discussion that follows is confined almost entirely to such lap windings. If a machine-wound motor has to be rewound in the repair shop, an equivalent lap winding can usually be used to replace it.

14-3 Lap Winding for Polyphase Motors. A lap winding consists of a number of identical coils. For fractional horsepower motors, the coils are wound from insulated magnet wire and are later manually inserted in the slots and connected. If wound on a collapsible mold, as in Fig. 14-2, they are shaped like a racetrack; wound on other molds, they may be diamond-shaped. Usually there are as many stator coils as there are slots, or, in other words, two coil sides per slot, making what is known as a two-layer winding. All the coils usually are identical in number of turns and size of wire, and the phase coils are generally better insulated than the other coils. (A *phase coil* is a coil adjacent to a coil of a different phase.) The typical way in which such a winding progresses is shown in Fig. 3-1. The coils are connected in groups; e.g., in Fig. 3-1 there are

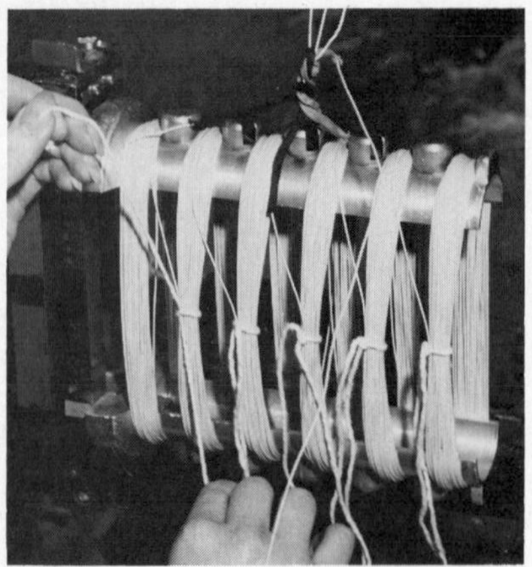

Fig. 14-2 Coils for a polyphase motor wound on a collapsible mold of light weight which permits fast winding. Six coils, enough for two groups in this case, are wound from a continuous strand of wire. Start and finish leads of each group are marked by dark- and light-colored sleeves threaded onto the wire before winding. *(Westinghouse Electric Corporation.)*

four coils per group. A *group* is simply a given number of individual coils connected in series, with only two leads brought out, and the stator winding is then connected from these groups. So far as the winding itself is concerned, the lap winding used in a polyphase motor is very similar to the lap winding used in a dc armature. However, in the dc winding all the coils are connected in series, and a closed winding is formed, whereas in polyphase motors the coils are connected in a number of groups.

In the mush-wound coils used in fractional horsepower motors, it is a fairly common practice to wind all the coils of a single group from a continuous strand of wire (see Fig. 14-2). If there are the same number of coils in all groups, this is a very desirable practice because it eliminates a number of connections and simplifies the final connections of a wound stator. If, however, the number of coils is not the same in all groups (see Art. 14-9), it may be less confusing to bring out the connections from each individual coil and to connect them in groups after the stator is wound.

14-4 Number of Groups and Number of Coils per Group. In the ordinary winding (as distinguished from a consequent-pole winding), there is one group per phase per pole. Therefore,

$$\text{Number of groups} = \text{number of poles} \times \text{number of phases}$$

$$\text{Number of coils per group} = \frac{\text{total number of coils}}{\text{number of groups}}$$

If the number of coils per group works out to be a whole number, each group has this number of coils, but if a fractional number is obtained, an unequal grouping of the coils must be used, as explained in Art. 14-9 and shown in Tables 14-1 and 14-2. Windings that have an integral number of coils per group are often referred to as integral-slot windings. Similarly, windings with a fractional number of coils per group are often called fractional-slot windings. Consequent-pole windings have half as many groups as the corresponding conventional windings.

14-5 "Throw-Up" and "Throw-Down" Windings. In the usual two-layer lap winding, every coil has one side in the bottom of a slot and the other side in the top of some other slot. This means that when the first coils are inserted, only the bottom coil side can be placed in the slot, and the top coil side must be left inside the stator bore, but not in a slot, until the winding is nearly completed. In the case of two-pole induction motors, where the throw is large, there would normally be a large number of these coil sides in the bore. As the bore of a two-pole motor is relatively small, the large number of coil sides interferes with the winding process so that it is a fairly common practice to wind the coils *throw down.* In the throw-down winding, both coil sides of all the first coils are inserted in the slots, so that the first coils wound have both sides in the bottoms of the slots. This point is illustrated in Fig. 14-3*b*. As a result, the last coils wound have both coil sides in the tops of the slots. This throw-down winding is compared with the conventional, or throw-up, winding, which is shown in Fig. 14-3*a*. The throw-down winding

TABLE 14-1 Balanced Three-Phase Lap Windings (Maximum number of parallel circuits and number of coils per group for equal and unequal groupings)

	2 poles		4 poles		6 poles		8 poles		10 poles		12 poles	
Number of slots or coils	Circuits	Coils per group	Circuits	Coils per group	Circuits	Coils per group	Circuits	Coils per group	Circuits	Coils per group	Circuits	Coils per group
6	2	1										
9	1	1½										
12	2	2	4	1								
15	1	2½	1	1¼								
18	2	3	2	1½	6	1						
21	1	3½	1	1¾	0							
24	2	4	4	2	0		8	1				
27	1	4½	1	2¼	3	1½	1	1⅛				
30	2	5	2	2½	0		2	1¼	10	1		
33	1	5½	1	2¾	0		1	1⅜	1	1⅒		
36	2	6	4	3	6	2	4	1½	2	1⅕	12	1
39	1	6½	1	3¼	0		1	1⅝	1	1³⁄₁₀	0	
42	2	7	2	3½	0		2	1¾	2	1⅖	0	
45	1	7½	1	3¾	3	2½	1	1⅞	5	1½	3	1¼
48	2	8	4	4	0		8	2	2	1⅗	0	

Note: 0 indicates no balanced winding is possible. If the coils per group is a whole number, all groups have the indicated number of coils. If the coils per group is not a whole number, refer to Table 14-2 for the grouping.

TABLE 14-2 Balanced Three-Phase Lap Windings with Unequal Coil Grouping

Average number of coils per group	Group Number																													
	1	2	3	4	5	6	7	8	9	10	11	12	13	14	15	16	17	18	19	20	21	22	23	24	25	26	27	28	29	30
$1\frac{1}{10}$	2	1	1	1	1	1	1	1	1	1	2	1	1	1	1	1	1	1	1	1	2	1	1	1	1	1	1	1	1	1
$1\frac{1}{8}$	2	1	1	1	1	1	1	1	2	1	1	1	1	1	1	1	2	1	1	1	1	1	1	1						
$1\frac{1}{5}$	2	1	1	1	1	2	1	1	1	1	2	1	1	1	1															
$1\frac{1}{4}$	2	1	1	1	2	1	1	1	2	1	1	1																		
$1\frac{3}{10}$	2	1	1	2	1	1	2	1	1	1	2	1	1	2	1	1	2	1	1	1	2	1	1	2	1	1	2	1	1	1
$1\frac{3}{8}$	2	1	2	1	1	2	1	1	2	1	2	1	1	2	1	1	2	1	2	1	1	2	1	1						
$1\frac{2}{5}$	2	1	2	1	1	2	1	2	1	1	2	1	2	1	1															
$1\frac{1}{2}$	2	1	2	1	2	1																								
$1\frac{3}{5}$	2	2	1	2	1	2	2	1	2	1	2	2	1	2	1															
$1\frac{5}{8}$	2	2	1	2	2	1	2	1	2	2	1	2	2	1	2	1	2	2	1	2	2	1	2	1						
$1\frac{3}{4}$	2	2	2	1	2	2	2	1	2	2	2	1																		
$1\frac{7}{8}$	2	2	2	2	2	2	2	1	2	2	2	2	2	2	2	1	2	2	2	2	2	2	2	1						
$2\frac{1}{4}$	3	2	2	2	3	2	2	2	3	2	2	2																		
$2\frac{1}{2}$	3	2	3	2	3	2																								
$2\frac{3}{4}$	3	3	3	2	3	3	3	2	3	3	3	2																		
$3\frac{1}{4}$	4	3	3	3	4	3	3	3	4	3	3	3																		
$3\frac{1}{2}$	4	3	4	3	4	3																								
$3\frac{3}{4}$	4	4	4	3	4	4	4	3	4	4	4	3																		
$4\frac{1}{2}$	5	4	5	4	5	4																								
$5\frac{1}{2}$	6	5	6	5	6	5																								
$6\frac{1}{2}$	7	6	7	6	7	6																								
$7\frac{1}{2}$	8	7	8	7	8	7																								

Note: If a parallel connection is used, the short-throw or top-to-top connection must be used with the grouping given in the table.

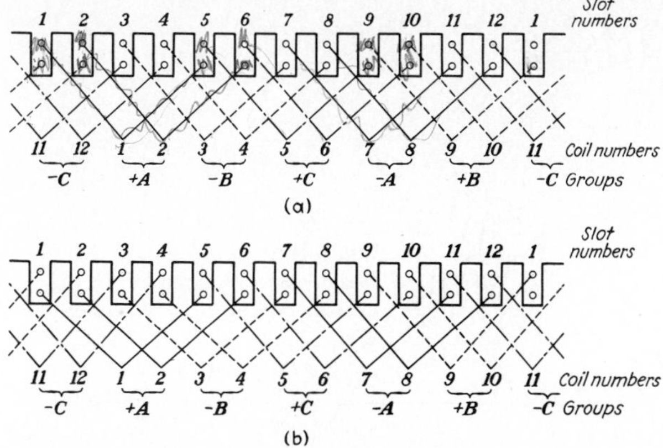

Fig. 14-3 Two types of polyphase windings. (*a*) "Throw-up" winding; (*b*) "throw-down" winding.

does cause a slight electrical unbalance but usually not enough to have serious consequences.

14-6 Short-Throw and Long-Throw Connections. Two types of connections between the groups are recognized:

1. Short throw
2. Long throw

In the short-throw connection, each group is connected to another group displaced 180 electrical degrees. In the long-throw connection, a group is connected to another group displaced 360 electrical degrees away. In single-phase windings, there are similarly two such connections, as discussed in Art. 4-12.

When parallel connections are used, it is desirable to use the long throw if possible, in order to balance the magnetic pull on the rotor. However, in the case of unequal grouping of coils (Arts. 14-9 and 14-10), it is usually necessary to use the short throw. If the stator is series-connected, it makes no difference whether the long or short throw is used; whichever is the more convenient connection may be employed.

14-7 Stator (or Rotor) Connection Diagrams: Single-Voltage and Three-Phase. Typical stator connection diagrams for windings used in fractional-horsepower polyphase motors are given in Figs. 14-4 and 14-5. These diagrams are reproduced, by permission, from Dudley's book *Connecting Induction Motors.*[1,*] Many other connection diagrams for

*For numbered references, see References at end of this chapter.

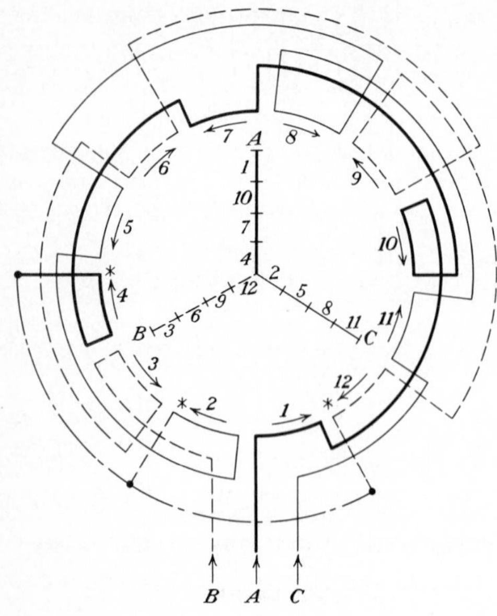

Fig. 14-4 Four-pole three-phase stator connection diagram, windings connected series Y, short throw. *(After A. M. Dudley, and S. F. Henderson, Connecting Induction Motors, 4th ed., McGraw-Hill Book Company, New York, 1960.)*

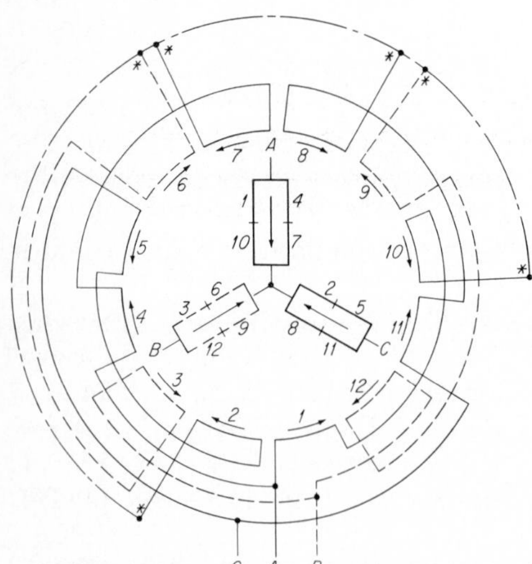

Fig. 14-5 Four pole three-phase stator connection diagram, windings connected parallel Y, short throw. *(After A. M. Dudley and S. F. Henderson, Connecting Induction Motors, 4th ed., McGraw-Hill Book Company, New York, 1960.)*

polyphase induction motors are given by Dudley[1], by Braymer and Roe,[2] and also by Veinott.[3]

14-8 How To Check a Diagram or Winding. There are so many possible methods of connecting polyphase induction motors that it is impossible to give them all in a book of this scope. It is, therefore, desirable that service personnel know how to check the correctness of a diagram, for they may often be called upon to make up their connection diagrams.

A method for checking the correctness of a three-phase stator connection diagram for a conventional winding is illustrated in Fig. 14-4. Current is assumed to flow *into* each of the three leads, as shown by the arrows. The winding is traced three times, once from each lead to the star point; as we trace through each group, an arrow is marked beside the group to indicate the direction in which the pencil traced when going through that group. Now, if the three-phase diagram is correct—as it is in the figure—no two adjacent arrows will point in the same direction. Unless this direction does so alternate with every group, the diagram is not correct.

Diagrams for consequent-pole windings can be checked by following the procedure outlined. However, in this case, all arrows must point in the *same* direction.

14-9 Windings with Unequal Coil Grouping. Table 14-1 show the possibilities for balanced three-phase lap windings for windings of from 2 to 12 poles, inclusive, for 6 to 48 slots or coils. In each case in which a balanced winding is possible, the number of coils per group has been figured and given, and in a large number of cases this number comes out a fractional number. For the latter cases, Table 14-2 is given to show how the coils may be grouped. To illustrate: assume that it is desired to put a four-pole winding into 15 stator slots (a common arrangement for synchros). The number of coils per group is 1¼, from Table 14-1. In Table 14-2 we find that with 1¼ coils per group, the grouping is 2-1-1-1-2-1-1-1-2-1-1-1. This means that the first group in phase *A* has 2 coils per group; then, progressing around the stator, the next three groups have 1 coil each; the next group, 2 coils; the next three groups, 1 coil each; and so on, progressing completely around the stator. If there are 1¼ coils per group, this grouping is the same whether the winding is 4 poles in 15 slots, 8 poles in 30 slots, or 12 poles in 45 slots. Other combinations are described in a previous edition of this book.[3]

In Table 14-1 under "Circuits" is given the maximum number of parallel circuits into which it is possible to divide a balanced winding for the combination shown. If a parallel winding is used, it is not necessary to connect into so many parallel paths as the maximum number indicates, but the number of parallel paths used may be any factor of the maximum number of possible parallels. For example, if 6 parallels are possible, 6, 3, or 2 parallel paths may be used. If 10 parallels are possible, 10, 5, or 2 parallels may be used.

In general, when the coil grouping is unequal, the short-throw connection must be used for parallel-connected windings.

An explanation of the way that the coil groupings given in Table 14-2 were determined appears in Art. 14-10 of a previous work[3] for the benefit of those who may have occasion to work out combinations that are not listed.

14-10 Irregular Lap Windings. In Table 14-1 it was shown by a 0 that balanced windings are not possible with certain numbers of slots, notably for 6- and 12-pole motors. In many cases, however, there is a possible winding arrangement which is nearly enough balanced to use. Three such arrangements are given in the following table:

Slots or coils	Poles	Grouping
24	6	2-1-1-1-1-2-1-2-1 and repeat
48	6	2-3-3-3-3-2-3-2-3 and repeat
48	12	2-1-1-1-1-2-1-2-1 and repeat

These winding arrangements are satisfactory for ordinary squirrel-cage motors However, if an irregular winding arrangement such as one of these is used for either the primary or secondary winding of a wound-rotor induction motor, a serious loss in torque, due to this irregular winding, may be encountered. This reduction may be as much as 25 percent, as reported by Hellmund and Veinott.[4] An analysis of the harmonic content of this winding has since been published.[5]

14-11 Dual-Voltage Motors. A stator connection diagram for a four-pole dual-voltage motor is given in Fig. 14-6. A schematic phasor diagram is included with the figure. The terminal markings of all of these diagrams are in conformance with the latest standards.[6]

It is to be noted here that there are certain combinations of slots and poles into which it is not possible to put a dual-voltage winding; in these arrangements only one circuit per phase is possible. These combinations are shown in Table 14-1. For example, a two-pole winding in 9, 15, 21, 27, etc., slots, or any other combination in which the maximum number of circuits is one, cannot be connected for dual voltage.

14-12 Effective Series Conductors per Phase. The total number of effective series conductors per phase in a polyphase lap winding is given by the expression

$$C_e = \frac{S_c \times 2 \times T}{m \times q} \times k_p \times k_d \tag{14-3}$$

where S_c = total number of stator coils

T = number of turns per coil

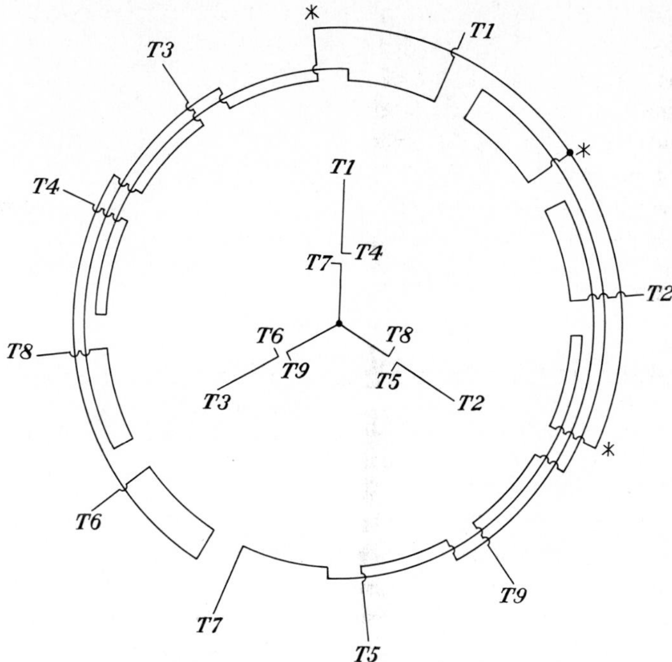

Fig. 14-6 Four-pole three-phase dual-voltage stator connection diagram, windings connected series and parallel star, short throw.

m = number of phases
q = number of parallel circuits
k_p = chord factor (pitch factor)
k_d = distribution factor (breadth factor)

Chord factor is the ratio of the voltage generated in a coil to the voltage that would be generated in a full-pitch coil of the same number of turns. Usually it can be found in Table 14-3, or it can be calculated from the expression

$$k_p = \sin\left(\frac{\text{slots throw} \times 90}{\text{slots per pole}}\right) \tag{14-4}$$

The preceding expression is valid even when the number of slots per pole is a fractional number, such as 6.75, for example.

Distribution factor takes into account the fact that the voltages generated in all coils of a group are not in time phase with each other because they are displaced in space from one another. It can be found for nearly all practical cases by reference to Table 14-4. How these factors were developed and how to compute them for more general cases are covered in another work.[7]

TABLE 14-3 Chord Factors for Polyphase Windings

Throw	Slots per pole												
	24	18	16	15	14	13	12	11	10	9	8	7	6
1–25	1.000	0.866											
1–24	0.998	0.906											
1–23	0.991	0.940	0.831										
1–22	0.981	0.966	0.882										
1–21	0.966	0.985	0.924	0.866									
1–20	0.947	0.996	0.956	0.914	0.847								
1–19	0.924	1.000	0.981	0.951	0.901	0.833							
1–18	0.897	0.996	0.995	0.978	0.944	0.884	0.793						
1–17	0.866	0.985	1.000	0.995	0.975	0.935	0.866	0.756					
1–16	0.832	0.966	0.995	1.000	0.994	0.971	0.924	0.841	0.707				
1–15	0.793	0.940	0.981	0.995	1.000	0.993	0.966	0.910	0.809				
1–14	0.752	0.906	0.956	0.978	0.994	1.000	0.991	0.960	0.891	0.766			
1–13	0.707	0.866	0.924	0.951	0.975	0.993	1.000	0.990	0.951	0.866	0.707		
1–12	0.659	0.819	0.882	0.914	0.944	0.971	0.991	1.000	0.988	0.940	0.831		
1–11	0.609	0.766	0.831	0.866	0.901	0.935	0.966	0.990	1.000	0.985	0.924	0.782	
1–10		0.707	0.773	0.809	0.847	0.884	0.924	0.960	0.988	1.000	0.981	0.901	
1–9		0.643	0.707	0.743	0.782	0.833	0.866	0.910	0.951	0.985	1.000	0.975	0.866
1–8			0.698	0.669	0.707	0.749	0.793	0.841	0.891	0.940	0.981	1.000	0.966
1–7			0.570	0.616	0.624	0.663	0.707	0.756	0.809	0.866	0.924	0.975	1.000
1–6				0.530	0.532	0.566	0.609	0.655	0.707	0.766	0.831	0.901	0.966
1–5									0.588	0.643	0.707	0.782	0.866
1–4											0.570	0.624	0.707

TABLE 14-4 Distribution Factors for Polyphase Windings

Slots per pole per phase, or number of coils per group	Two-phase	Three-phase Conventional	Three-phase Consequent-pole
1	1.000	1.000	1.000
2	0.924	0.966	0.866
3	0.911	0.960	0.844
4	0.906	0.958	0.837
5	0.904	0.957	0.833
6	0.903	0.956	0.831
Infinity	0.900	0.955	0.827
1½	0.911	0.960	0.844
All other fractional-slot windings	0.900	0.955	0.827

14-13 Machine-Wound Concentric Windings. For reasons of economy, many fractional horsepower polyphase induction motors are machine-wound in the factory. In general, such windings are limited to integral-slot windings, that is, to cases in which the number of slots per pole per phase is an integer. Usually they are wound as concentric windings. A wide variety of arrangements is possible, more than can be discussed here. When such a motor has to be rewound, it is usually practicable to rewind the motor by developing concentric coils, just as for a single-phase motor, using identical wire sizes and turns; this, of course, is the safest procedure. Often, however, it may be more economical and feasible to replace the original winding with a conventional lap winding. In some cases, it is possible to find a lap winding that is *exactly equivalent* (in the sense that these words are used in Art. 3-1); in other cases, this is not possible. In the latter case, the replacing lap winding should at least have the same number of effective series conductors per phase and the same wire size as the concentric winding it replaces. How to figure the distribution factor and number of effective conductors of a concentric winding is explained in Art. 4-41; the procedure described is equally applicable to the concentric windings used in polyphase motors. Figuring the effective conductors of a polyphase lap winding is explained in Art. 14-12.

Three examples of polyphase concentric windings are illustrated in Figs. 14-7 to 14-10.

(1) Chain winding. An excellent example of the appearance of a machine-wound four-pole polyphase stator is illustrated in Fig. 14-7. This winding is a chain winding of the type that has been used in large machines almost as long as induction motors have been built. This arrangement is shown in schematic form in Fig. 14-8. Note: There is only one coil side in each slot; there are only 6 groups, as in a consequent-pole winding, and not 12, as there are in a conventional four-pole three-phase winding. The winding itself is single-layer, and there

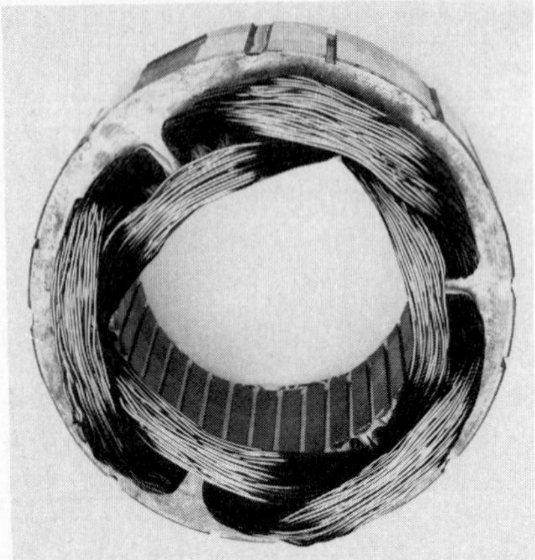

Fig. 14-7 A four-pole three-phase chain-wound stator. There is only one coil side per slot, and there are only six groups as in a consequent-pole winding. *(Emerson Motor Division.)*

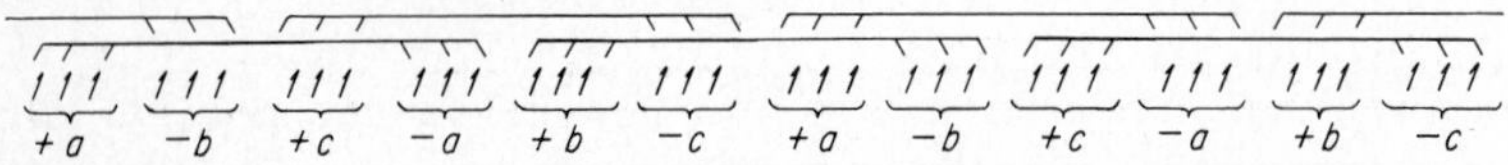

1 1 1 1 1 1 1 1 1 1 1 1 1 1 1 1 1 1 1 1 1 1 1 1 1 1 1 1 1 1 1 1 1 1 1 1
+a −b +c −a +b −c +a −b +c −a +b −c

Fig. 14-8 Chain winding for a four-pole 36-slot three-phase stator. Slots are uniformly filled. No slot contains coils connected to different phases.

are two layers or levels of coil-end connections. In practice, a machine with three needles, spaced 120° apart, first winds three coils 120 (mechanical) degrees apart. The next operation is to wind three more coils, making the coil-end connections just outside the ends of the three coils wound before.

The distribution factor of this arrangement is

$$k_d = \frac{\sin 70^\circ + \sin 90^\circ + \sin 110^\circ}{3}$$

$$= \frac{0.940 + 1.000 + 0.940}{3} = 0.960$$

This winding is exactly equivalent to a full-pitch two-layer lap winding which has half as many turns per coil as there are turns per slot in the chain winding. This arrangement is suitable for four-pole motors, but not for two- or six-pole motors.

(2) Conventional concentric winding. Figure 14-9 illustrates a general

+a −a +a −a
2 1 1 2 2 1 1 2 2 1 1 2 2 1 1 2
−b +b −b +b
1 2 2 1 1 2 2 1 1 2 2 1 1 2 2 1
−c +c −c +c
1 2 2 1 1 2 2 1 1 2 2 1 1 2 2 1

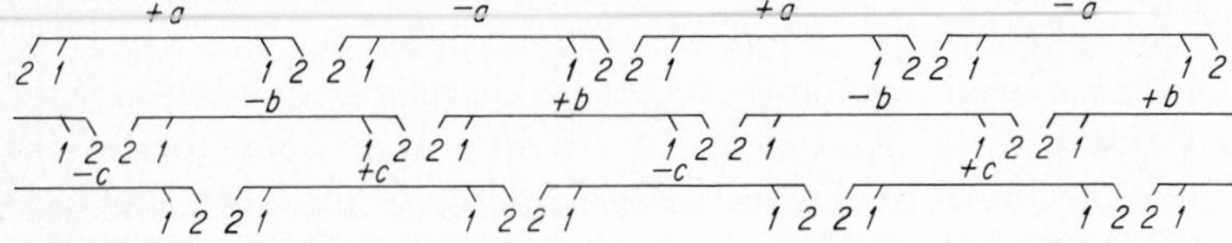

Fig. 14-9 Concentric winding for a four-pole 26-slot three-phase stator. Slots are uniformly filled. Every third slot contains coils connected to different phases.

concentric winding of conventional arrangement, that is, one which has one group per pole per phase—12 groups for the winding shown. Each phase has four groups, one per pole, which are wound, or inserted, one phase at a time, giving three layers or levels of coil-end connections. Slots are uniformly filled, but every third slot contains coils connected to different phases. The distribution factor of this arrangement is

$$k_d = \frac{\sin 60° + 2 \sin 80°}{3} = \frac{0.866 + 2 \times 0.985}{3} = 0.945$$

This winding is exactly equivalent to a two-layer lap winding of eight-ninths the pitch and a number of turns per coil equal to the number of turns corresponding to "1" in the figure. This type of arrangement is suitable for any number of poles.

(3) Full-pitch equivalent concentric winding. Figure 14-10 shows an arrangement for a two-pole motor which is exactly equivalent to a full-pitch two-layer lap winding. It has, like a conventional winding, six groups, or one per phase per pole. Slots are uniformly filled and no slot contains coils connected to different phases. This arrangement is applicable to any number of poles. The distribution factor of this arrangement is

$$k_d = \frac{\sin 67.5° + \sin 82.5°}{2} = \frac{0.924 + 0.991}{2} = 0.957$$

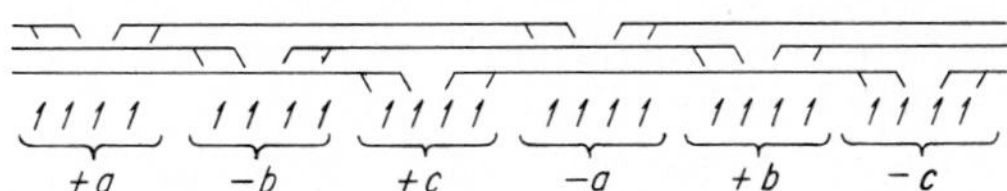

Fig. 14-10 Concentric winding for a two-pole 24-slot three-phase stator. Slots are uniformly filled. No slot contains coils connected to different phases.

THERMALLY PROTECTED MOTORS

A general description of thermal protective devices and how they function is given in Arts. 17-16 to 17-21. How they are used in three-phase motors is the subject of the following paragraphs.

14-14 Single-Voltage Motors. There are two methods of protection commonly used in three-phase fractional horsepower motors. One uses a three-phase protector, normally mounted in the end shield; the other employs multiple single-phase protectors laced to the end turns. Three-phase protectors are neutral-opening; that is, since the protector is connected at the common point, or wye, the disk itself forms the star point. This means they are suitable for wye-connected motors only and not for delta-connected motors. A typical schematic diagram is shown in Fig. 14-11. On-winding protectors may be used in either wye- or delta-connected motors and are connected on the line side of the windings. The most common practice is to use two single-phase on-winding protectors, each laced to and connected in a different phase. Typical schematic diagrams are shown in Fig. 14-12; in either method of protection three external leads are sufficient for connection to the power supply.

Three-phase motors with on-winding protectors are normally limited to those used in air-moving applications. A wye-connected motor with two on-winding protectors cannot have a neutral wire, since one phase would not be protected; a motor with a three-phase protector cannot have a neutral wire, since there is no terminal on the protector to which to connect the neutral.

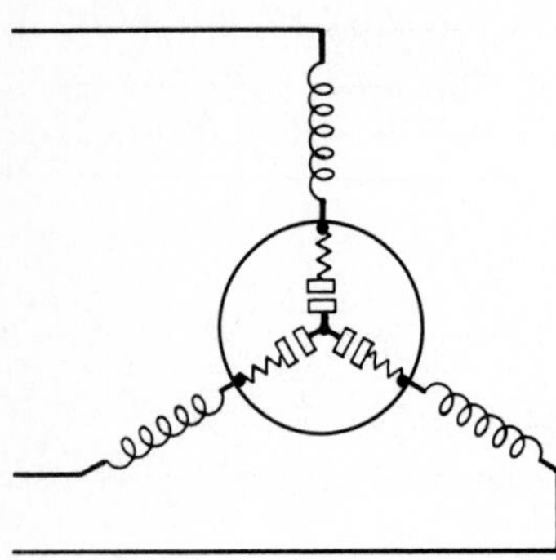

Fig. 14-11 Schematic wiring and line connection diagram for single-voltage wye-connected three-phase thermally protected motor with three-phase protector connected at the star point.

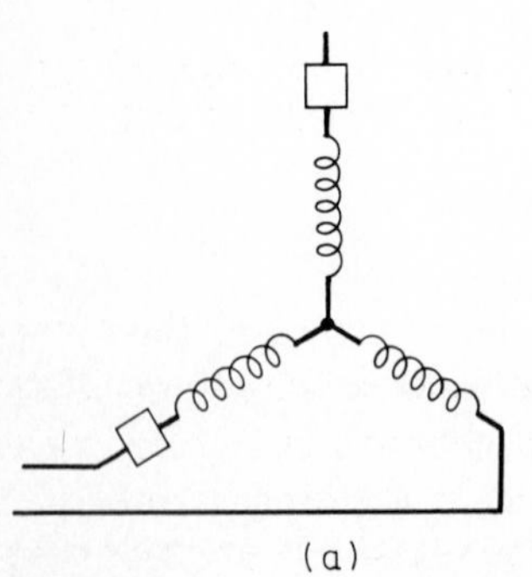

(a)

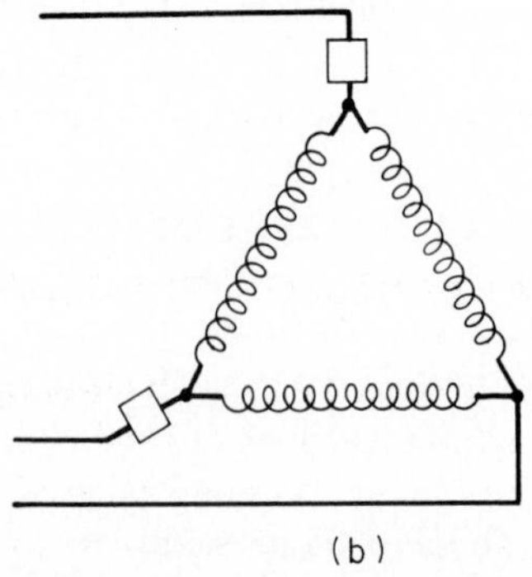

(b)

Fig. 14-12 Schematic wiring and line connection diagrams for single-voltage three-phase thermally protected motors with two on-winding protectors. (*a*) Wye-connected; (*b*) delta-connected.

14-15 Dual-Voltage Motors. Dual-voltage motors almost always use three-phase protectors; such protectors have six terminals, as opposed to three terminals for single-voltage motors. Typical schematic diagrams are shown in Fig. 14-13. As described in Art. 5-15 for capacitor-start motors the winding is divided into two sections, each of which is presumed to carry the same current on either voltage connection. Thus, the auxiliary heater in each phase of the protector is connected in series with only one section of each phase of the winding, and the contacts are connected in series with each phase. When the device opens, the entire winding is disconnected from the line. Further, the auxiliary heater carries the same current on both voltage connections, but the disk carries full current on both. Thus, the heater effect of the disk has to be relatively low compared to the heater to obtain comparable protection on both voltage connections. A dual-voltage thermally protected motor requires 12 external leads for connection to the power supply.

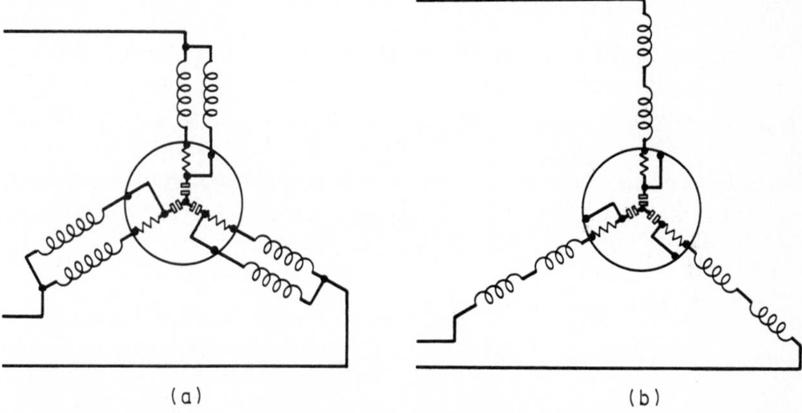

Fig. 14-13 Schematic wiring and line connection diagrams for a dual-voltage three-phase thermally protected motor. (*a*) Lower voltage; (*b*) higher voltage.

MISCELLANEOUS SERVICE PROBLEMS

14-16 Rewinding or Reconnecting Polyphase Motors. Many possibilities exist for rewinding or reconnecting a polyphase motor for changes in voltage, frequency, number of phases, number of poles, torques, and horsepower rating. Some relevant information is given in the following paragraphs, but readers interested in further study are referred to an earlier work by Veinott, in which some 13 pages are devoted to this subject.[3]

(1) Change in voltage. The number of effective conductors should be increased or decreased in direct proportion to the increase or decrease in applied circuit voltage. For small changes of 10 percent, or sometimes even 15 percent, rewinding is usually not necessary, for the same winding may generally be used with satisfactory results. A two-parallel winding may be connected in series for double line voltage. A three-phase wye-connected winding can be connected for 57 percent voltage by connecting it in delta.

(2) Change in frequency. Most 60-Hz fractional horsepower polyphase motors will operate satisfactorily on 50 Hz; in fact, the usual practice is to rate polyphase motors for 50/60 Hz. Strictly speaking, however, for the same horsepower, the effective conductors or turns per coil should be increased 10 percent when changing from 60 to 50 Hz.

(3) Change in number of phases. To change from a three-phase series star (wye) to a two-phase series connection, the number of turns per coil should be increased from 15 to 20 percent for the same line-to-line voltage.

14-17 Operation of a Three-Phase Motor on an Unbalanced Voltage Supply. Ideally, a three-phase motor will always be connected to a voltage-balanced three-phase system. However, since the three-phase system may also supply single-phase power with unbalanced loads, unbalanced phase voltages may appear at the terminals of the three-phase motor. The National Electrical Manufacturers Association (NEMA)[6] defines the percent voltage unbalance as

$$\text{Percent voltage unbalance} = 100 \times \frac{\text{maximum voltage deviation from average voltage}}{\text{average voltage}}$$

With voltages of 220, 215, and 210 V, the average is 215 V, the maximum deviation from average is 5 V, and the percent unbalance = 100 × 5/215 = 2.3 percent.

When a three-phase motor is operated on an unbalanced voltage supply, the effect is equivalent to the introduction of a negative sequence voltage, which produces a flux rotating against the rotation of the rotor in the air gap, which tends to produce high currents. The effect on motor performance is to decrease the breakdown and locked-rotor torques; reduce the full-load speed, greatly unbalance the currents at operating speeds, but only in the order of voltage unbalance at locked-rotor torque; and increase the temperature rise when operating at a particular load. Motor horsepower should be derated by the factor shown in Fig. 14-14 when operated on unbalanced voltages. Operation

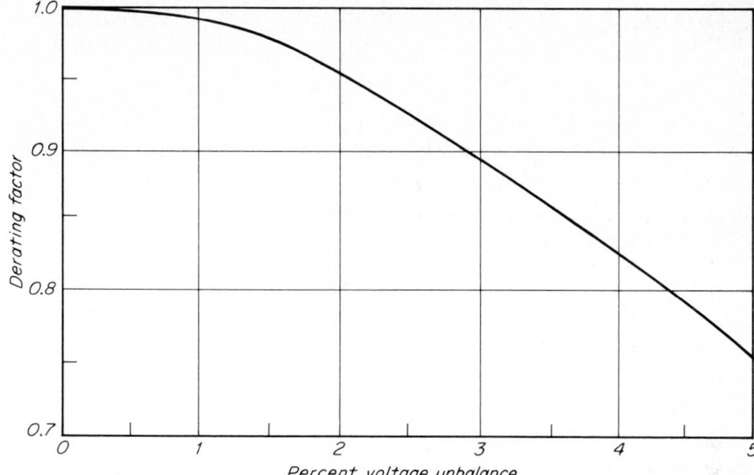

Fig. 14-14 Horsepower derating factor due to unbalanced voltage. *(From NEMA Motor and Generator Standards, NEMA Publ. no. MGI-1978, National Electrical Manufacturers Association, Washington, 1978.)*

of a motor above a 5 percent voltage unbalance condition is not recommended.

References

1. Dudley, A. M., and S. F. Henderson: *Connecting Induction Motors,* 4th ed., McGraw-Hill Book Company, New York, 1960.
2. Braymer, Daniel E., and A. C. Roe: *Repair Shop Diagrams and Connecting Tables for Induction Motors,* McGraw-Hill Book Company, New York, 1927.
3. Veinott, C. G.: *Fractional Horsepower Electric Motors,* 2d ed., McGraw-Hill Book Company, New York, 1948.
4. Hellmund, R. E., and C. G. Veinott: "Irregular Windings in Wound-Rotor Induction Motors," *Electr. Eng.*, February 1934, pp. 342–346.
5. Veinott, C. G.: "Spatial Harmonic Magnetomotive Forces in Irregular Windings and Special Connections of Polyphase Windings," *IEEE Trans. Power Appar. Syst.*, vol. 83, 1964, pp. 1246–1253.
6. *NEMA Motor and Generator Standards,* NEMA Publ. no. MG1–1978, National Electrical Manufacturers Association, 2101 N Street N.W. Washington DC 20037, 1978.
7. Veinott, C. G.: *Theory and Design of Small Induction Motors,* McGraw-Hill Book Company, New York, 1959. (Out of print, but available from University Microfilms, P. O. Box 1647, Ann Arbor, Mich. 48106.)

Bibliography

Braymer, Daniel E., and A. C. Roe: *Repair Shop Diagrams and Connecting Tables for Induction Motors,* McGraw-Hill Book Company, New York, 1927.

Braymer, Daniel E., and A. C. Roe: *Rewinding and Connecting Alternating-Current Motors,* McGraw-Hill Book Company, New York, 1932.

Woll, R. F.: "Effect of Unbalanced Voltage on the Operation of Polyphase Induction Motors," *IEEE Trans. Ind. Appl.*, vol. IA-11, no. 1, January–February 1975, pp. 38–42.

CHAPTER

15

STEPPER MOTORS

Sensational strides in the field of automatic control have generated a great demand for a number of small rotating electric machines built in subfractional and fractional horsepower sizes. In general, such machines may be different applications of a familiar type, modifications or adaptations of a familiar type, or a novel and hitherto more or less unfamiliar new type. This chapter and the next are devoted to an examination and discussion of many such machines. Varieties and availabilities of a number of such machines are shown in Table 15-1. Stepper motors, a popular innovation, afford simple means for controlled remote positioning of many devices. They are used in clocks, elapsed-time indicators, pen drives for plotting curves, indexing control, drivers for watt-hour meters, and many other applications.

15-1 Stepper Motor Defined. A stepper motor is an electromagnetic device designed to convert a series of input power pulses into discrete angular movements, one for each power pulse. Depending upon the design of the motor and its intended use, the power pulses may consist of current pulses of the same polarity, or alternate pulses of opposite polarity. The sequential pulses may all be delivered to the same motor winding, or successively to different coils in the motor. Pulses do not have to be delivered at any fixed rate. The machine is capable of continuous rotation.

15-2 Types of Stepper Motors. Many stepper motors, especially those that are multipolar, are constructed basically like a synchronous inductor motor, such as described in Chap. 10. Such motors use a permanent magnet in the rotor; the stator may use a shading coil to fix the direction of rotation or it may actually use two coils, mutually displaced 90 electrical degrees. Some permanent-magnet stepper motors use a rotor consisting of a solid alnico cylinder that is magnetized for two or for four poles; such motors use a more or less conventional stator winding. Other permanent-magnet stepper motors use two stator coils, each enclosed in a cup with multiple pairs of poles and with the rotor magnetized with the same number of pairs of poles in one stator section. Some stepper motors use the variable reluctance principle. A two-wire stepping motor was developed over 30 years ago by Sigma Instruments, Inc., and offered under the trade name Cyclonome. This motor is essentially an inductor motor with permanent magnets in the stator structure. A detailed description of the Cyclonome motor is given in an earlier edition of this book.[1,*]

15-3 Stepper Motor Terminology. Common terminology used to describe the operation of stepper motors is an follows:

Step. The angular rotation produced by the output shaft each time the motor receives a step pulse.

Step angle. The rotation of the output shaft caused by each step, measured in degrees.

Steps per second. The number of steps that the motor takes in 1 s.

Steps per revolution. The total number of steps required for the output shaft to make one complete revolution.

Holding torque. The torque required to rotate the output shaft a full step with the motor energized and in a standstill condition.

Step response. The time, usually in milliseconds, for the motor to make one complete step.

Start-without-error rate. The stepping rate at which a motor with no external load inertia can start and stop without losing a step.

Slew rate. The maximum step rate at which a motor will run under a

*For numbered references, see References at end of this chapter.

TABLE 15-1 Rotating Machines for Control Systems

Manufacturer*	Stepper motors	Servo motors	Timing motors
Aeroflex Laboratories, Inc. Magnetic Products Div. 35 South Service Road Plainville, Long Island, NY 11803		X	
Airpax/North American Philips Controls Corp.	X	X	X
American Bosch Electrical Products		X	
Barber-Colman Company		X	
Bodine Electric Company	X	X	
Brevel Motors, Inc.	X		
Clifton Precision Div., Litton Systems, Inc. Marple at Broadway Clifton Heights, PA 19018	X	X	
Eastern Air Devices	X	X	
ECM Motor Company	X		X
Electro-Craft Corporation		X	
General Electric Company		X	X
General Time Controls, Inc. 135 South Main Street Thomaston, CT 06789			X
Globe Motor Div., TRW Inc.	X	X	
Haydon Switch and Instrument, Inc. P.O. Box 3329 Waterbury, CT 06705	X		X
Hoover Electric Company		X	
Hurst Mfg. Corp. Box 326 Princeton, IN 47670	X		X
IMC Magnetics Corporation		X	
Indiana General Motor Products		X	
Inertial Motor Corporation		X	
Japanese Products Corporation	X	X	X
Mallory Timers Co. Emhart Electrical/Electronic Group P.O. Box 986 Indianapolis, IN 46206			X
Motor Technology, Inc.		X	
MPC Products 7426 North Linder Avenue Skokie, IL 60077		X	
Muirhead, Inc. 1101 Bristol Road Mountainside, NJ 07092	X	X	
Northland, Div. Scott & Fetzer	X		
Oriental Motor Company, Ltd.	X		
Pacific Scientific, Motor and Control Division		X	
The Pittman Corporation		X	

TABLE 15-1 Rotating Machines for Control Systems *(Continued)*

Manufacturer*	Stepper motors	Servo motors	Timing motors
H. K. Porter Company, Inc.		X	
Portescap U.S. Box 99 Trooper Road Worcester, PA 19490	X	X	X
Sigma Instruments, Inc. 170 Pearl Street Braintree, MA 02184	X		
The Singer Company, Kearfott Div.	X	X	
The Superior Electric Co. 383 Middle Street Bristol, CT 06010	X		
UMC Electronics, Inc.		X	
Vernitech Div., Vernitron 300 Marcus Blvd. Deer Park, NY 11729	X	X	
Welco Industries, Inc.	X		

*See Table 1-1 for manufacturers' addresses not listed in this table.

rated load once the motor has started rotating. This is usually substantially greater than the start-without-error rate.

Ramping. A control method used to vary the pulse rate to accelerate from zero steps per second to the running rate or from any step rate to a different rate whether it is accelerating or decelerating.

SHADED-POLE STEPPER MOTORS

15-4 Unidirectional Motors. Figure 15-1 shows the basic control circuit for a stepper motor which uses a single center-tapped coil, plus a shading coil to fix direction of rotation. Such a motor, which is essentially a timing motor of the inductor type with a permanent-magnet rotor, has been offered by Haydon Switch and Instrument, Inc., Water-

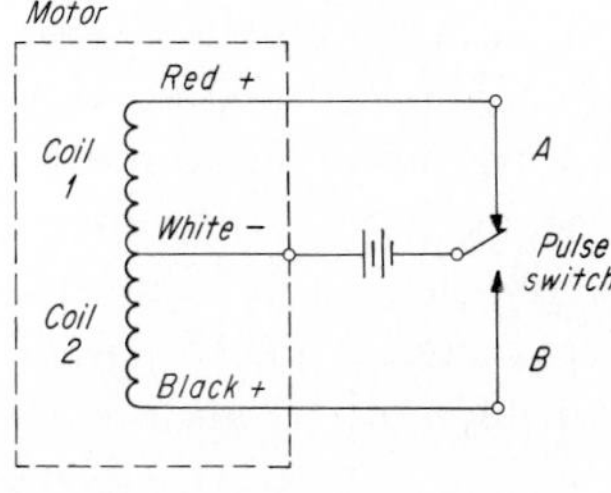

Fig. 15-1 Basic control circuit for a unidirectional shaded-pole stepper motor. *(Haydon Switch and Instrument, Inc.)*

bury, Connecticut. The negative side of the battery is connected to the center terminal, and a pulse switch connects the positive terminal alternately from one outside terminal to the other. The pulse switch is shown in the figure as mechanical, but it could be any form of switching circuit.

The motor has 24 poles, and each pulse advances the rotor one pole pitch, or 15 mechanical degrees, requiring 24 pulses per revolution. Pulses may be applied at any rate up to a maximum of 80 per second, which would give a maximum speed of 200 rpm; each pulse must have a minimum duration of 12.5 ms. A running torque of 0.2 oz-in is developed. Some holding torque is provided by the permanent magnets when the stator coil is not energized, somewhat more if it is. This motor is also supplied with integral gear reductions, giving various reductions up to 100:1.

For high-speed slewing or homing, the center tap can be disconnected, and an ac voltage of any frequency up to 40 Hz may be applied to the outer terminals of the coil.

Another mode of energizing such a motor is given by Kavanaugh and Perkins:[2] a steady dc signal is applied continuously to one-half of the coil; stepping is accomplished by applying a pulse across the other half of the winding, of the opposite polarity, and of double the strength of the steady pulse. One step is achieved when the pulse is applied and one when it is removed. One revolution is thus achieved for each 12 pulses.

15-5 Bidirectional Motors. Bidirectional shaded-pole stepper motors, which are essentially the same as the one described in the preceding article, are built with two motors in the same housing. One stator is used for one direction of rotation, and the other stator for the opposite direction of rotation.

PERMANENT-MAGNET STEPPER MOTORS WITH TWO-PHASE WINDINGS

Most permanent-magnet stepper motors use the popular synchronous-inductor construction, which was described in Chap. 10. These are characterized by a rotor employing a permanent magnet and two stator phases, effectively displaced 90 electrical degrees. Because of their construction, there may actually be only two physical coils, but the motors are so arranged that the two coils produce magnetomotive force (mmf) fields displaced 90 electrical degrees from each other. Motors of this type are readily bidirectional.

15-6 SLO-SYN Stepper Motors. SLO-SYN is a registered trademark of The Superior Electric Company. The construction of these motors is generally like that shown in Fig. 10-12, and their principles of operation, since they are a type of ac motor, are generally as described in Art. 10-19.

(1) AC operation. The motor may be operated from a two-phase supply, or from a single-phase source, using the circuit shown in Fig. 15-2. A phase-shifting network, consisting of a resistor and capacitor, is

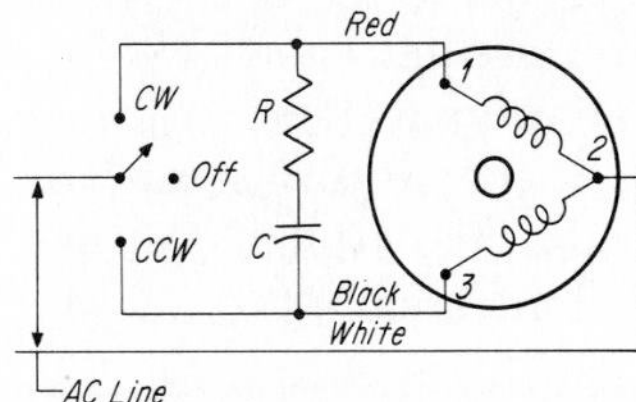

Fig. 15-2 Schematic connections for a three-wire SLO-SYN motor operated as an electrically reversible synchronous motor. *(The Superior Electric Company.)*

required when operating a SLO-SYN motor from a single-phase source. The motors have 250, 100, or 36 poles and hence operate at 28.8, 72, or 200 rpm, respectively, on a 60-Hz source. Like any other synchronous inductor motor, it can accelerate and synchronize only a limited amount of inertia.

(2) DC stepping: 200 steps per revolution. Figure 15-3 shows how two

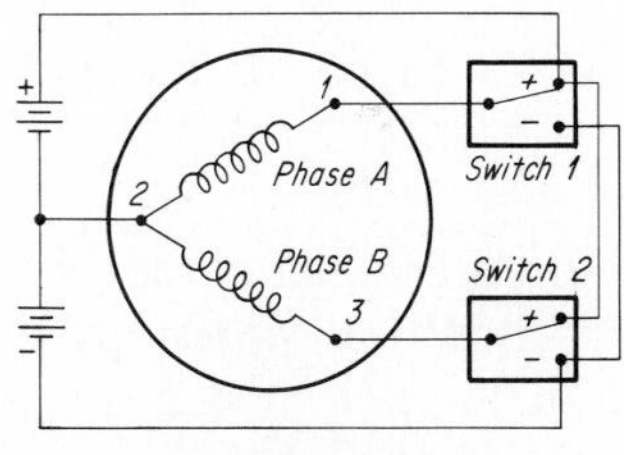

Switching sequence *

Step	Switch #1	Switch #2
1	+	+
2	−	+
3	−	−
4	+	−
1	+	+

* To reverse direction, read chart up from bottom.

Fig. 15-3 Schematic connections and switching sequence for a three-wire SLO-SYN motor operated as a stepper motor; there are 200 steps per revolution. *(The Superior Electric Company.)*

batteries and switches may be used to step the motor with dc pulses. Each step advances the rotor one-half of a pole pitch, or 90 electrical degrees. (For a more complete explanation, the reader should refer to Art. 2-6 and to Fig. 2-4. There, phases A and B correspond to phases A and B in Fig. 15-3, and steps 1 and 2 in Fig. 15-3 to times 2 and 4, respectively, in Fig. 2-4, which shows clearly how the magnetic field advances 90 electrical degrees with this change.) In a 100-pole motor each step advances the rotor 1.8 mechanical degrees. It should be noted that the motor circuits are inductive, and it may be necessary to use supplemental means to prevent destructive arcing at the switches. The motor can stop, start, reverse, or step at varying speeds without losing any steps. At low stepping rates, the available torque will be approximately the rated torque of the motor, but the available torque drops off as the stepping rate is increased, as illustrated in Fig. 15-4.

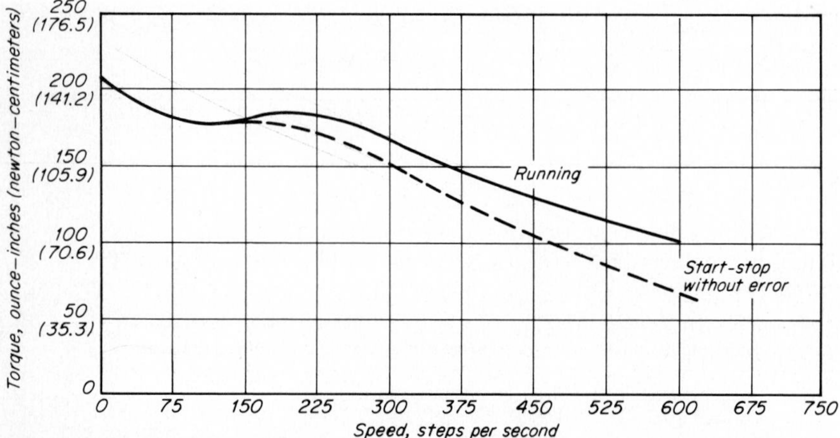

Fig. 15-4 Effect of stepping rate on available torque of a typical 1.8° SLO-SYN stepper motor. Typical torque versus speed characteristics. *(The Superior Electric Company.)*

(3) DC stepping: 100 steps per revolution. In the preceding paragraph, one phase was reversed for each step, thus requiring two steps to advance the magnetic field—and rotor—one pole pitch; hence, since the motor has 100 poles, 200 steps were required to advance the rotor one complete revolution. Proper direction of rotation is assured, since the field only advances 90 electrical degrees per step. If the direction of current in a single winding is reversed, the magnetic field moves 180 electrical degrees, or one pole pitch, but the *direction* of travel would be uncertain, unless auxiliary means were used to predetermine it. Shaded-pole stepping motors, discussed in Arts. 15-4 and 15-5, reverse a single coil, and the rotor advances a pole pitch for every pulse; the shading coil establishes the direction of rotation. Figure 15-5 shows an arrangement in which both phase windings are kept in series and each step supplies alternate polarity pulses, thus advancing the rotor one pole pitch per pulse, or 100 pulses per revolution. Direction of rotation is provided by shunting a capacitor across one of the two phases. For

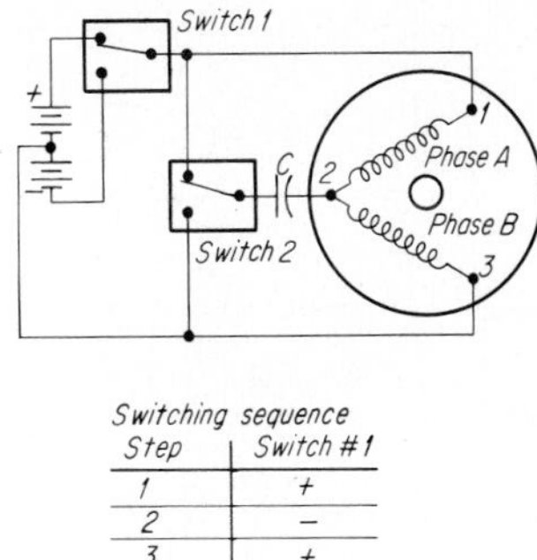

Switching sequence

Step	Switch #1
1	+
2	−
3	+
4	−

Fig. 15-5 Schematic connections and switching sequence for 100 steps per revolution. *(The Superior Electric Company.)*

example, with switch 2 in the position shown, at the start of the pulse, B receives the initial current through the capacitor until the latter becomes charged, when the currents in A and B become equal; however, the initial pulse through B provides a directional starting torque. Switch 2 is used only to predetermine direction of rotation; if only a single direction of rotation is needed, switch 2 can be omitted entirely. Note that the pulsing circuit for this arrangement is much simpler than that of Fig. 15-3. An alternate method, requiring 100 steps per revolution, is to connect the motor for ac operation, as in Fig. 15-2 (resistor is unnecessary), and supply pulses of alternate polarity where the ac line is shown. In this case, the no-load positions at the end of each pulse are not the same for clockwise rotation as for counterclockwise rotation; this is because, after the initial transient dies out, only one phase carries current, and the phase that carries current is different for the two different directions of rotation.

(4) Bifilar motors. If the two phase windings are both center-tapped, the current can be sent through alternate halves of a phase to obtain alternate magnetic polarities in the same fashion as illustrated in Fig. 15-1. The connections and switching sequence of such a motor are illustrated in Fig. 15-6. The motor shaft advances 200 steps per revolution, or 1.8° per step, when a four-step input sequence is used and 400 steps per revolution, or 0.9° per step, when an eight-step input sequence is used. The four-step and eight-step input sequences are shown in Fig. 15-6*a* and *b*, respectively. Since current is maintained on the motor windings when the motor is not being stepped, a high hold-

Four-step input sequence* (full-step mode)				
Step	SW1	SW2	SW 3	SW 4
1	On	Off	On	Off
2	On	Off	Off	On
3	Off	On	Off	On
4	Off	On	On	Off
1	On	Off	On	Off

*Provides clockwise rotation as viewed from nameplate end of motor. To reverse direction of motor rotation, perform switching steps in the following order: 1, 4, 3, 2, 1.

(a)

Eight-step input sequence* (half-step mode)				
Step	SW 1	SW 2	SW 3	SW 4
1	On	Off	On	Off
2	On	Off	Off	Off
3	On	Off	Off	On
4	Off	Off	Off	On
5	Off	On	Off	On
6	Off	On	Off	Off
7	Off	On	On	Off
8	Off	Off	On	Off
1	On	Off	On	Off

*Provides clockwise rotation as viewed from nameplate end of motor. To reverse direction of motor rotation, perform switching steps in the following order: 1, 8, 7, 6, 5, 4, 3, 2, 1.

(b)

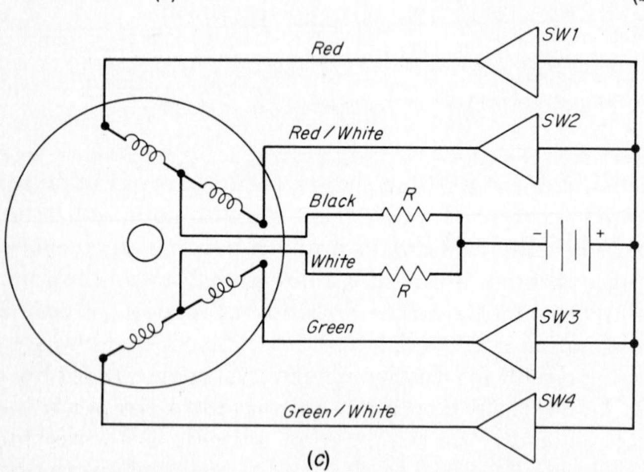

Fig. 15-6 Schematic connections and input for four-step and eight-step sequences for 200 and 400 steps per revolution SLO-SYN stepper motor. (*a*) Four-step input sequence; (*b*) eight-step input sequence; (*c*) SLO-SYN stepper motor with six leads. *(The Superior Electric Company.)*

ing torque results. Figure 15-6*c* shows a motor with six leads; however, it can be connected as a five-lead motor by connecting the black and white leads together and changing the resistor to one-half the resistance value and twice the wattage rating of that required for the six-lead connection.

(5) Transient voltage suppression. Since motor circuits are inductive, with dc stepping, it may be necessary to protect the stepping switches from the inductive surges. The most common method for suppressing transient voltages uses shunting diodes as illustrated in Fig. 15-7. (See

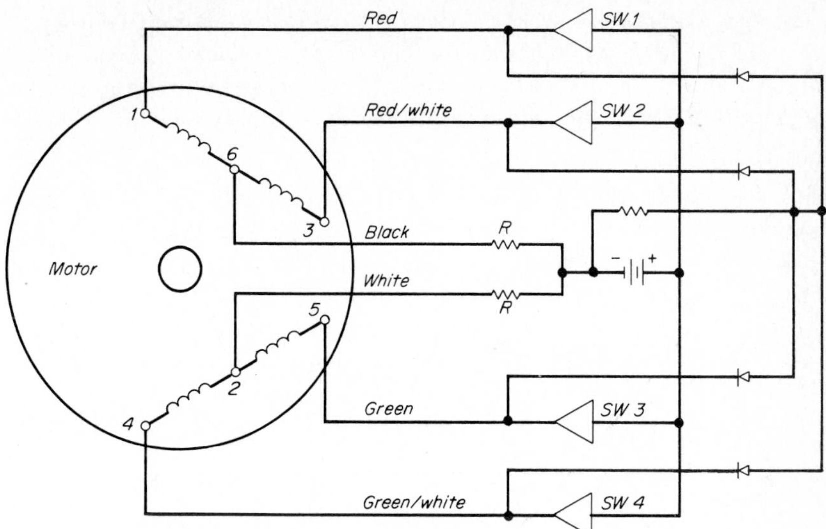

Fig. 15-7 Protection of stepping switches from inductive surges. *(The Superior Electric Company.)*

Art. 13-2 for the definition of the term diode.) Capacitors, with values between 0.1 and 1.0 mfd can also be used in place of the diodes.

(6) Performance characteristics and sizes. SLO-SYN motors for ac motor applications are built in various sizes with rated torque ranging from 25 to 1800 oz-in. Weights vary from 1¼ to 45 lb. Standard voltage ratings are 120 and 240 V on 50- and 60-Hz sources. At low stepping rates (30 steps per second or less), the stepping torque may be equal to or greater than rated torque, but the holding torque is substantially greater. Planetary gear assemblies are also available, with stepdown ratios ranging from 4:1 to 2708:1. SLO-SYN motors for dc operation are offered in a range of frame sizes from 2.2- to 6.6-in diameter and with standard step angles of 0.72, 1.8, and 5°. They can be driven at rates up to 20,000 steps per second. The minimum holding torque varies from 53 to 1800 oz-in, and the typical time for a single step for a 1.8° motor is from 2 to 70 ms.

(7) Resonance. When a stepping motor is operated at its natural frequency, typically 90 to 160 steps per second, an increase in the audio and vibration level of the motor may occur. The frequency at which the resonance occurs varies widely, depending on the characteristics of the load. In some cases, the motor can oscillate or lose steps; however, an increase in the inertial loading will usually allow operation at this frequency.

(8) Power supplies (translators). Prepackaged power supply translators are available for SLO-SYN motors. The translators provide sequencing and switching logic for bidirectional control of the stepping motors at rates up to 10,000 full steps or 20,000 half steps per second.

15-7 Haydon Stepper Motors. Haydon Switch and Instrument, Inc., makes a line of stepper motors with an integral gear-train speed reducer under the trade name Big Inch. The single-phase unipolar series 31300 is a two-wire unidirectional stepper motor which converts electric impulses into discrete angular steps of the output shaft without control logic. During each pulse the rotor turns 360° (180° with power applied and 180° with power removed.) The step rate is from 0 to 80 pulses per second, with a 12-ms minimum pulse width. Available output step angles range from 0.36 to 120° in either clockwise or counterclockwise rotation. The series 33700 motor is a two-phase bidirectional four-wire stepper which uses the switching sequence shown in Fig. 15-3. The stator is oriented to produce a 90° rotor step angle; the operating range is from 0 to 400 pulses per second, with the output shaft step angle between 0.09 and 30°. The series 31600 motor is bifilar-wound, four-phase, and bidirectional. The schematic connections and four-step and eight-step sequencing are similar to those shown in Fig. 15-6. The rotor steps 90° with each pulse; the operating range is from 0 to 320 pulses per second, with the output shaft step angle between 0.09 and 30°.

15-8 Airpax Permanent-Magnet Stepper Motors. Airpax/North American Philips Controls Corporation makes two-phase stepper motors with two stator cups, each formed around a stator coil. Stator-pole pairs, which are mechanically displaced by one-half a pole pitch, become alternately energized north and south magnetic poles. Between the two stator-coil pairs the displacement is one-quarter of a pole pitch. The permanent-magnet rotor has the same number of pole pairs as one stator-coil section. The rotor moves one-quarter of a pole pitch per winding polarity change. A two-phase motor with 12 pole pairs per stator-coil section would thus move 48 steps per revolution or 7.5° per step. Figure 15-8 shows a cutaway view of a permanent-magnet stepper motor; Fig. 15-9 shows a layout of the same stepper motor with a given stator polarity and rotor position, corresponding to step 1 in Fig. 15-10. Continuing the sequence causes the rotor to rotate in a clockwise direction; reversing the sequence reverses the direction of rotation.

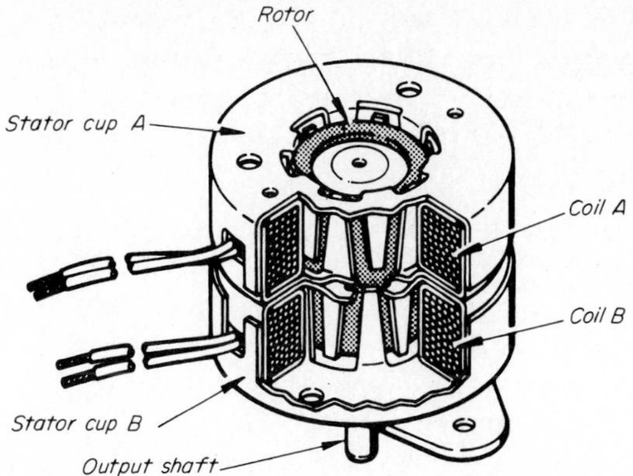

Fig. 15-8 Cutaway view of a two-phase stepper motor with two stator cups. The permanent-magnet rotor has the same number of pairs of poles as one stator coil section. *(Airpax/North American Philips Corporation.)*

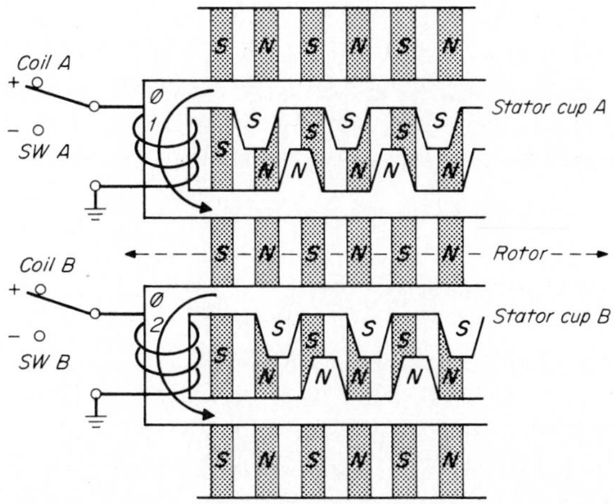

Fig. 15-9 Layout of the stepper motor in Fig. 15-8, showing a given stator polarity and rotor position. *(Airpax/North American Philips Corporation.)*

Standard step angles for Airpax stepper motors are 7.5, 15, and 18°; however, step angles from 0.72 to 90° are available. The no-load or constant-load accuracy of each step is 6.5 percent and is noncumulative; therefore, a 7.5° stepper motor will position to within 0.5°, whether the rotational movement is 7.5° (1 step) or 7500° (1000 steps.)

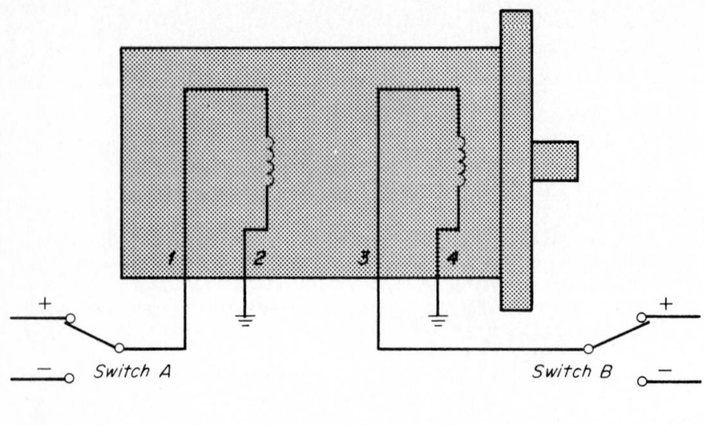

Clockwise rotation ↓ / Counterclockwise rotation ↑

Step	Coil A	Coil B
1	+	+
2	+	−
3	−	−
4	−	+
1	+	+

Fig. 15-10 Schematic connections and four-step switching sequence for the stepper motor shown in Fig. 15-8. *(Airpax/North American Philips Corporatioin.)*

VARIABLE-RELUCTANCE STEPPER MOTORS

15-9 Principle of the Variable-Reluctance Stepper Motor. A variable-reluctance stepper motor is represented schematically in Fig. 15-11. There are three stator coils, or phases, located 120° apart, that is, equally spaced around the circumference. The rotor is of soft iron, with

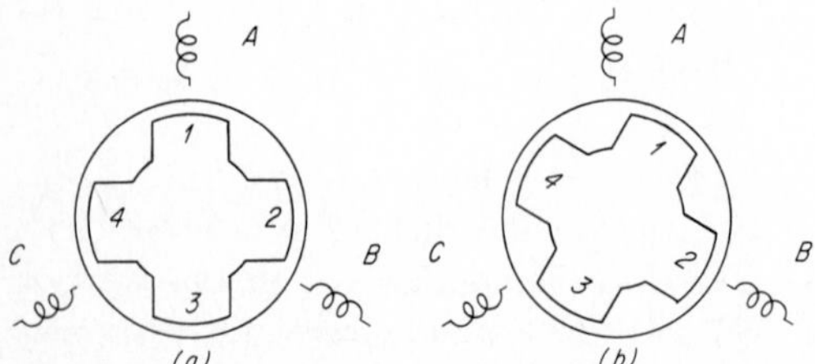

Fig. 15-11 Principle of the variable-reluctance stepper motor. (*a*) Phase A energized; (*b*) phase B energized.

four salient poles, as shown. When phase A is excited, rotor teeth 1 and 3 line up with the axis of phase A, as shown in Fig. 15-11*a*. If, now, coil B is excited and the excitation removed from coil A, the rotor will move 30° clockwise to assume the position shown in Fig. 15-11*b*. While coil B exerts a pull on both 2 and 3, the attraction is much greater for 2, which is but 30° away, than for 3, which is 60° away; thus, the direction of rotation is assured. If, now, coil C is energized while the current is taken away from B, the coil will move another 30° clockwise. Hence, if we supply current pulses to phases A, B, C, A, etc., the rotor will step 30° with each current pulse, requiring 12 pulses to complete a revolution.

It should be obvious that the variable-reluctance motor just described has no holding torque at all unless current is flowing through at least one of the coils; this characteristic is in contrast to that of a permanent-magnet stepper motor which does have a definite holding torque with no excitation, because of the holding power of the permanent magnet. To assure no loss or gain in position, coil B must be energized before the excitation is removed from coil A, and coil C must likewise be energized before the excitation is removed from B. If coil A is deenergized before power is applied to B, torsion in the load coupling could turn the rotor counterclockwise just enough (16° would suffice) so that when current was supplied to coil B it would attract poles 3 and 1, thus turning counterclockwise. It should be clear that if the pulses are applied to phases A, C, B, A, etc., the rotor will run counterclockwise.

It can also be shown that if the rotor in Fig. 15-8 had had eight equally spaced teeth, the rotor would move 15° for each pulse. For a pulsing sequence of A, B, C, A, etc., rotation would be counterclockwise or clockwise for a pulsing sequence of A, C, B, A, etc. Thus, with eight rotor teeth and three stator coils, 24 steps would be required per revolution.

15-10 Kearfott Variable-Reluctance Stepper Motors. Kearfott Division, The Singer Company, makes variable-reluctance stepper motors in sizes 8, 11, 15, 18, 20, 23, and 35, varying from ¾- to 3½-in diameter. These have either three-phase or four-phase stator windings; the rotors are of soft iron, with a multiple number of rotor teeth and slots. Standard rotation is 15° per step, or 24 steps per revolution. Maximum stepping rate (at no load) varies from 200 to 1000 pulses per second, giving a maximum rpm of 500 to 2500. Holding torque is from 0.5 to 90 oz-in. Sizes 11 and 20 are also made with a 10° step angle, with a maximum stepping rate from 550 to 900 pulses per second. Holding torque is from 1.1 to 37 oz-in.

Their size 23 small-angle stepper motors rotate at 200, 168, or 72 steps per revolution, or 1.8, 2.14, and 5° per step, respectively. Maximum stepping rates of 400, 350, and 260 steps per second, respectively, give 120, 125, and 217 rpm.

The Kearfoot Division also supplies a logic drive circuit for its variable-reluctance stepper motors. All of the necessary components are mounted on one card and connected together by means of a printed circuit.

References

1. Veinott, C. G.: *Fractional- and Subfractional-Horsepower Electric Motors,* 3d ed., McGraw-Hill Book Company, New York, 1970.
2. Kavanaugh, Richard J., and Robert W. Perkins: "Timing and Stepper Motors," *Mach. Des.,* Dec. 16, 1965, pp. 102–105.

Bibliography

Baty, Gordon: "Automating Production Operations with Stepping Motors," *Automation,* February 1968, pp. 74–77.

Dailey, Jack R.: "Computer Simulation of Stepping-Motor Performance," *Electro-Technology,* March 1967, p. 60.

Fredriksen, T. R.: "Closed-Loop Stepping Motor," *Mach. Des.,* Sept. 16, 1965, p. 202.

Henry-Baudot, J.: *Les Machines électriques en automatique appliquée,* Dunod, Paris, 1967.

Slingland, Ed: "Small Steps Turn into Big Inprovements," *Powerconversion Int.,* October 1983, pp. 20–25.

CHAPTER

16

SERVOMOTORS

Servomotors, sometimes called control motors, are motors which are specially designed and built primarily for use in feedback control systems; all have a high speed of response. They may be used for self-balancing recorders, remote-positioning devices, tracking and guidance systems, process controllers, or a host of other applications of a similar nature. They are made for dc as well as for ac operation; often rather special and unconventional constructions are used. Frequently, other devices are built integral with them. Such devices may include tachometers of many types and descriptions, damping devices of one kind or another, brakes, and, of course, reduction gears. Manufacturers of servomotors are listed in Table 15-1. At least two different types of ac servomotors are readily recognizable: standard and damped.

STANDARD LOW-INERTIA AC SERVOMOTORS

16-1 AC Servomotor Defined. An ac servomotor is an induction motor with two primary windings mutually displaced in magnetic position from one another by 90 electrical degrees and a low-inertia high-resistance rotor giving a speed-torque curve substantially linear in shape from no-load speed to standstill. It is designed for operation with a constant voltage applied to the *fixed phase* and an adjustable voltage impressed on the other phase, called the *control phase.* It may be operated from either a two-phase or a single-phase source of power. For operation on single-phase circuits, a capacitor is normally connected in series with the fixed phase.

16-2 Essential Parts of a Servomotor. Essential parts of a standard ac servomotor are virtually the same as those of a permanent-split capacitor motor (see Art 7-2), except that a capacitor is not needed when the motor is operated from a two-phase supply. There are two primary windings, called the fixed phase and the control phase, respectively. These windings are displaced from one another in space by 90 electrical degrees. The squirrel-cage rotor is usually small in diameter, to keep the mechanical inertia as low as possible, and of high resistance in order to obtain the ideal of a linear speed-torque curve as closely as possible. The fixed phase is usually a winding with two leads; the control phase may be a two-lead winding, a center-tapped three-lead winding, or a four-lead winding in two sections, like a dual-voltage winding. When used on single-phase circuits, a capacitor is connected in series with the fixed phase. An outline drawing of a typical motor is shown in Fig. 16-1. To the basic motor itself may be attached one or more of the following as needed by the particular application: reduction gears, tachometer, damper, and holding brake.

16-3 Principles of Operation. A servomotor may be operated as either a two-phase or a permanent-split capacitor motor; in either case, the voltage is held constant on the fixed phase and is varied on the control

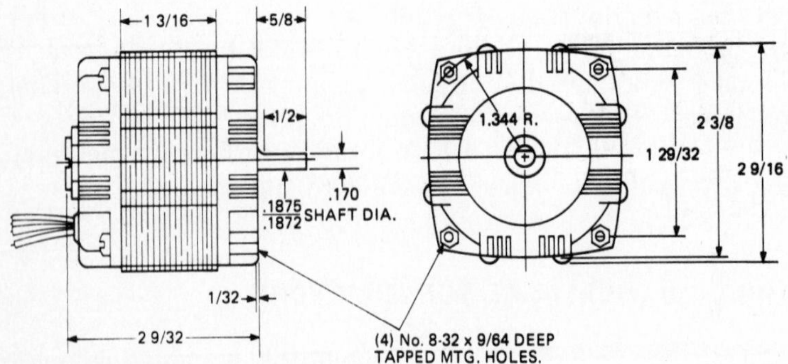

Fig. 16-1 Outline drawing of a typical ac servomotor (model KL direct-drive servomotor). *(Hurst Manufacturing Corporation.)*

phase. When the voltage impressed on the control phase is zero, while the fixed phase is excited, the control-phase winding should be short-circuited in order to keep the motor from starting and operating as a single-phase motor, in case the motor is given a sudden displacement by external means.

(1) As a two-phase motor. Principles of operation of a polyphase induction motor were discussed quite fully in Chap. 2. Typical speed-torque curves were presented in Fig. 14-1; this figure also shows the effect of high rotor resistance on the shape of the speed-torque curve. Servomotors are usually designed to have even more rotor resistance than shown by curve 4 of Fig. 14-1, so that the breakdown torque occurs at a slip well in excess of unity (standstill conditions) and the speed-torque curve from no-load to standstill approaches the ideal of a straight line more closely. A practical example of such a speed-torque curve is given in Fig. 16-2.

Now, if the voltage impressed on both phases of a two-phase motor is varied, the torque, at any slip, varies as the square of the voltage. If, however, the voltage impressed on only *one* phase (the control phase) is varied, while the voltage impressed on the other phase (the fixed phase) is held constant, the torque, at any slip, varies as the *first power* of the voltage. In general, this is the way an ac servomotor is used: voltage is held constant on one phase while it is varied on the other. The effect of varying the control-phase voltage is illustrated graphically in Fig. 16-2.

(2) As a single-phase motor. When two-phase power is not available, a suitable capacitor is connected in series with the fixed phase as shown in Fig. 16-2, and the combination is connected to a constant voltage. Output is controlled by applying a variable voltage to the control phase. At any slip, torque varies approximately as the first power of the voltage applied to the control phase, as illustrated in Fig. 16-2.

16-4 Certain Performance Characteristics Explained. There are certain performance characteristics, beyond the usual ones for conventional motors, that are of particular interest to users of servomotors. A few of these are explained next, though the list is by no means complete. Proper application of servomotors is a science in itself and is beyond the scope of this book; however, many references on this subject are given in the Bibliography and References at the end of this chapter.

(1) No-load speed. No-load speed is usually given in rpm and is denoted by N_0.

(2) Stall torque. Locked-rotor torque is usually given in ounce-inches. If it is given in gram-centimeters, divide by 72 to obtain torque

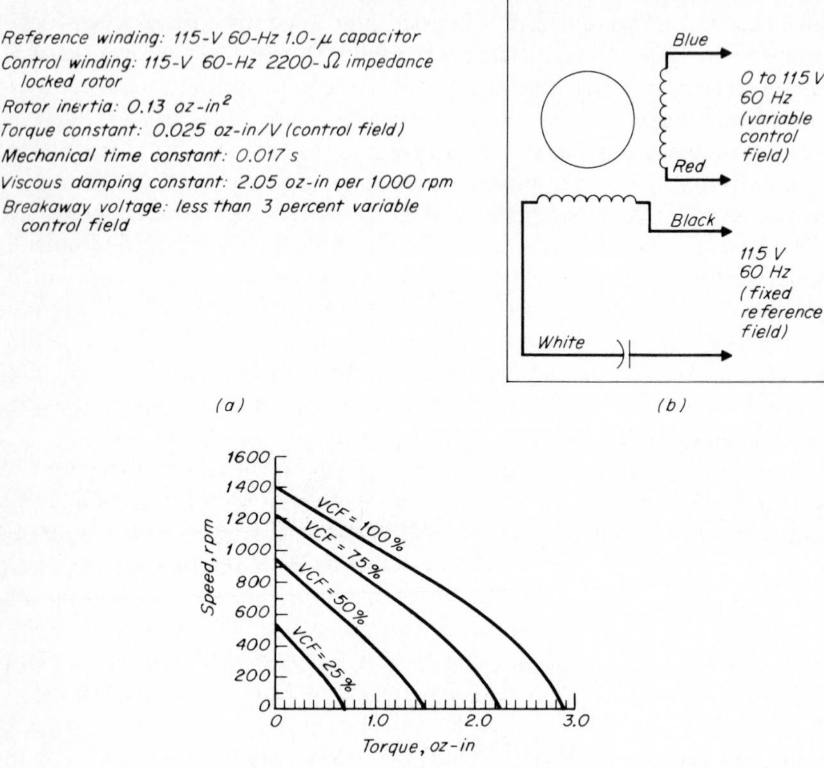

Fig. 16-2 Performance data, schematic diagram of speed-torque curves of an ac servomotor. (*a*) Selected typical data for servo operation; (*b*) typical hook-up for Model KL motor; (*c*) typical speed vs. torque at various control field voltages (VCF) as a percentage of rated voltage for Model KL servomotor. *(Hurst Manufacturing Corporation.)*

in ounce-inches. (See Table A-3 for conversion factors for other units of torque.)

(3) Rotor inertia. This is the rotational moment of inertia about the shaft and is usually given in g-cm^2 for ac motors and in g-cm-s^2 for dc motors. If it is given in oz-in^2, multiply by 182.9 to obtain the inertia in g-cm^2.

(4) Theoretical acceleration. This is the acceleration at standstill, expressed in radians per second per second, or simply rad/s^2. It is calculated as follows:

$$\text{Acceleration} = \frac{\text{stall torque}}{\text{inertia}} \times 70{,}620 \qquad (16\text{-}1)$$

where stall torque is given in ounce-inches and inertia in g-cm^2.

(5) Mechanical time constant. The mechanical time constant is the time for the rotor to accelerate from standstill up to 63.2 percent of no-load speed with full voltage suddenly applied to the control phase, and with no external load. It is usually expressed in seconds or milliseconds. In calculating this constant, it is assumed that the torque decreases linearly from zero speed to no-load speed, causing the speed of the motor to rise exponentially with time. This mechanical time constant is calculated as follows:

$$\text{Mech. time const.} = \frac{\text{inertia} \times \text{no-load speed} \times 1.485}{\text{stall torque} \times 1{,}000{,}000} \qquad (16\text{-}2)$$

where inertia is given in g-cm^2 and stall torque in ounce-inches.

(6) Reversing time constant. This is the time required to go from no-load speed in one direction to 63.2 percent of no-load speed in the opposite direction, when the control-phase voltage is suddenly reversed in time phase by 180°, and there is no external load or inertia. In general, it is approximately 1.7 multiplied by the mechanical time constant.

16-5 Performance Characteristics of Standard Servomotors. There are probably more servomotors made to military specifications (MIL) for 400-Hz operation than for 60-Hz. Performance characteristics of a number of such motors are tabulated in Table 16-1.

Many servomotors are equipped with built-in reduction gearing. National Electrical Manufacturers Association (NEMA)[1,*] standards show gear ratios of from 6:1 to 1800:1; some manufacturers go to even higher ratios. NEMA motors are nearly square in shape, with a bolt-circle diameter of 2$\frac{11}{16}$ in. Mea[2] shows servomotors ranging in sizes up to as large as 7 in in diameter, with locked-rotor torques up to as high as 1700 oz-in.

16-6 Wiring Diagrams and Terminal Markings. Terminal markings and lead connections for a five-lead servomotor with a center-tapped control phase operated from a single-phase supply are shown in Fig. 16-3. In practice, a way is needed for adjusting the voltage applied to the control phase; when this voltage is zero, the phase should be short-circuited to keep the motor from starting and running as a single-phase motor in case the motor is given a sudden displacement. The autotransformer, represented schematically in Fig. 16-3, serves both purposes.

*For numbered references, see References at end of this chapter.

TABLE 16-1 Characteristics of 400- and 60-Hz Standard Servomotors[°]

Size	Nominal OD, in	Frequency, Hz	Voltage			No-load speed, rpm	Locked-rotor				Rotor inertia, g-cm^2	Theoretical acceleration, rad/s^2	Time constant, s	Weight, oz
			Fixed phase	Control phase	Minimum starting		Torque, oz-in	Watts per phase	Impedance, Ω Fixed	Impedance, Ω Control				
5	½	400	26	26	0.6	9,500	0.10	2.1	335	335	0.175	40,500	0.020	0.68
8	¾	400	26	26	0.7	6,900+	0.18	2.1	254	254	0.46	29,300	0.026	0.95
	¾	400	26	26	0.7	10,000+	0.205	2.1	200	200	0.18	86,100	0.013	1.5
	¾	400	26	33	0.9	6,600+	0.316	3.6	116	187	0.18	124,000	0.005	1.5
10	15/16	400	26	26	0.9	6,500+	0.30	3.1	157	157	0.46	46,000	0.015	1.45
	15/16	400	26	26	1.5	3,500+	0.34	3.75	148	148	0.46	52,000	0.009	1.45
	15/16	400	115	115	3.0	10,000+	0.28	2.8	3,500	3,500	0.46	44,000	0.024	1.7
11	1 1/16	400	115	115	3.0	6,200+	0.60	3.9	2,175	2,175	1.07	39,450	0.017	4.5
	1 1/16	60	115	36	1.0	3,000+	0.60	3.9	3,108	308	1.07	39,450	0.0085	4.5
13	1¼	400	115	36	1.0	10,800	2.1	17.0	72.3	71.4	3.0	57,000	0.021	4.4
15	1 7/16	400	115	115	3.0	5,000	1.45	6.8	1,030	1,030	3.3	32,600	0.017	7.3
	1 7/16	60	115	36	1.1	3,000+	2.25	6.8	1,610	158	3.3	51,400	0.007	7.3
18	1¾	400	115	115	3.0	5,000	2.25	9.9	640	640	4.0	39,700	0.012	12.2
	1¾	400	115	33	1.0	9,500	8.0	60.0	179	14.7	3.3	190,000	0.006	12.0
20	2	400	200	115	1.6	10,000	8.6	65.0	467	153	4.0	152,000	0.007	17.2
23	2¼	400	115	115	3.0	10,000	6.5	40.0	275	275	12.0	36,000	0.003	29.0
	2¼	400	115	115	3.0	10,000	9.0	50.0	228	228	12.0	53,000	0.002	29.0
	2¼	60	115	36	2.0	3,200	22.0	65.0	213	21.8	31.0	57,000	0.007	43.0

[°] From The Singer Company, Kearfott Division.

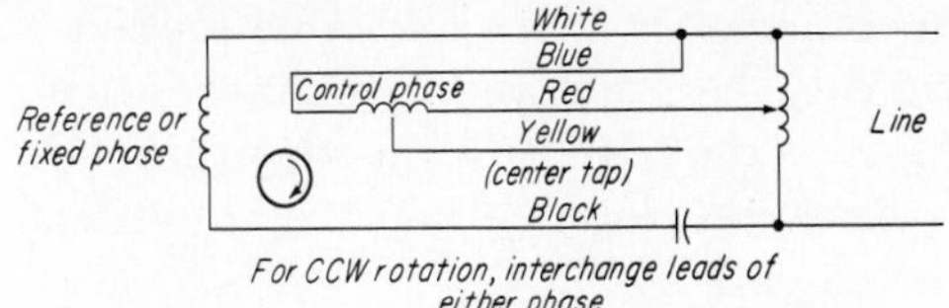

Fig. 16-3 Schematic wiring and line connection diagram for a single-phase low inertia servomotor with center-tapped control phase and five terminal leads. *(From NEMA Motor and Generator Standards, NEMA Publ. no. MG1-1978, National Electrical Manufacturers Association, Washington, 1978.)*

For two-phase operation, the connections are the same as just discussed except that the capacitor is omitted while the fixed phase is connected to one of the supply phases and the control phase to the other through voltage-adjusting means.

MIL S-22432a and SAE ARP 497 identify the leads as follows:

Phase	Leads
Fixed	Yellow, white
Control:	
2-lead windings	Black, red
3-lead windings	Black, black/red, red
4-lead windings	Black, green, red/black, red

16-7 Drag-Cup Servomotors. Another form of "standard servomotor" uses a drag-cup construction, as illustrated in Fig. 16-4. The stator is of conventional construction, like that of any of the motors so far discussed in this chapter. However, the rotor element consists of a cup of a nonmagnetic conducting material such as copper or one of its alloys. The magnetic circuit is completed through stationary laminated disks, held centrally within the stator bore, as shown in the figure. One could describe the drag-cup rotor as a special form of squirrel-cage rotor in

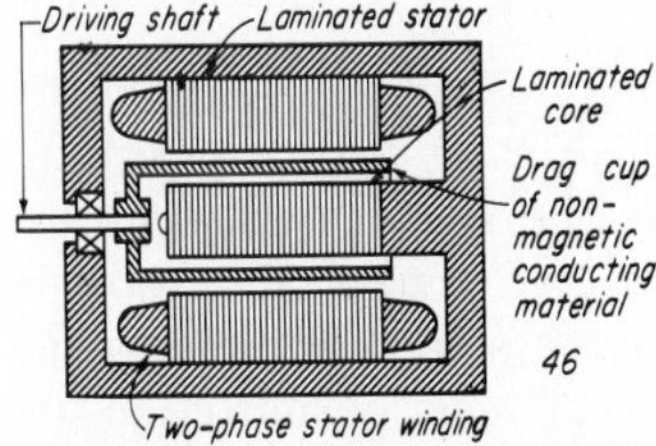

Fig. 16-4 Cross-sectional representation of a drag-cup ac servomotor. *(Product Engineering.)*

which the rotor teeth are removed, the rotor core or yoke section is held stationary, and the squirrel-cage bars and end rings are replaced by a cylindrical cup. Initially, this was the popular form of construction for servomotors, until high-performance motors of more conventional rotor construction were later developed.

Performance characteristics of three typical designs of drag-cup motors are given in Table 16-2. Drag-cup motors do have the advantage of being free from all cogging effects and do have a low starting voltage.

DAMPED AC SERVOMOTORS

Servomotors have a certain amount of inherent damping which is sufficient for many applications but not others, so that additional damping has to be supplied. The additional damping may be in the form of a viscous damper, an inertial damper, or a tachometer, most of whose damping effect is due to the tachometer voltage fed into the control system. Before explaining specific types of dampers, we shall first discuss what is meant by damping and why it is needed.

16-8 Damping and Why It Is Needed. Servomotors are used in feedback control systems. Essentially, any feedback control system is designed to regulate some desired output, in accordance with some reference quantity which may be fixed or varying in magnitude. The servomotors are used to actuate directly or indirectly something that influences the output being regulated by the system. Of major importance for the successful operation of a control system, of which the servomotor is a most vital link, are the following:

1. The motor must respond rapidly to changes in control-phase voltage.

2. The motor must be stable, that is, it must not oscillate or overshoot.

Rapid response is obtained by having a high ratio of torque to inertia in order to give an initial rapid acceleration [Art. 16-4(4)]. Overshooting is minimized and stability achieved by the use of a damping or retarding torque that increases with rotational speed. That is, the damping torque exerts no retarding torque at standstill, and so the motor can get away to a fast start, but the motor is not permitted to run away or to overshoot because of the damping torque, which gets progressively higher as the speed increases. How such damping torque can be developed, either by the motor itself or by external means, is the subject of Arts. 16-9 to 16-12.

TABLE 16-2 Characteristics of Drag-Cup Servomotors[°]

			Voltage				Locked-rotor							
									Impedance, Ω					
Size	Nominal OD, in	Frequency, Hz	Fixed phase	Control phase	Minimum starting	No-load speed, rpm	Torque, oz-in	Watts per phase	Fixed	Control	Rotor inertia, g-cm²	Theoretical acceleration, rad/s²	Time constant, s	Weight, oz
10†	0.937	400	10	20		Variable		0.25	370	1144	0.12			3.0
18	1¾	60	115	115	5	2875	0.41	3.65	2740	2770	3.7	7800	0.037	7.6
24	1⅞	400	55	115	2	4750	0.45	2.85	1360	1835	2.5	12,700	0.044	11.8

°From Rotating Components Div., Aeroflex Laboratories Inc.

†Special design, powered by the output of two size 8 synchros motors.

16-9 Internal Damping of a Servomotor. The way that internal damping of a servomotor is effected can perhaps best be understood by reference to Fig. 16-5. Suppose the motor developed the same torque at all speeds, as shown by curve *a*. In this case, when the motor is given a signal to start, it starts to accelerate at a high rate which, if maintained,

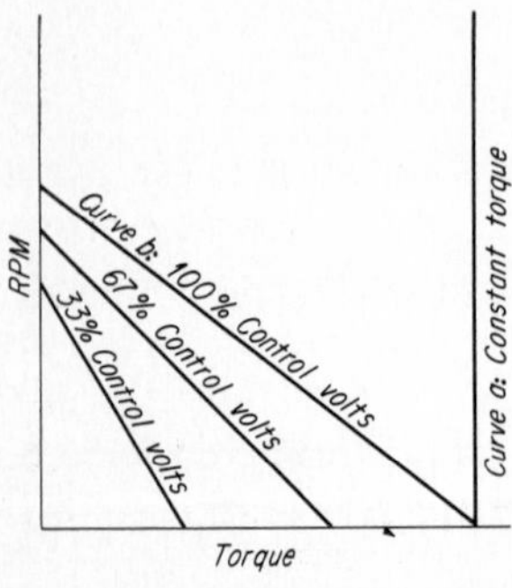

Fig. 16-5 Speed-torque curves of an "ideal" servomotor showing the effect of the control-phase voltage on the internal damping.

would quickly carry the motor and its connected load up to destructively high speeds; the system would have to depend upon reducing the control-phase voltage to zero quickly enough to prevent the motor from running away. However, the speed torque of the actual motor is more like the straight line *b* in Fig. 16-5. Note that *b* is exactly the same curve that would be obtained by subtracting from *a*, at every speed, a torque proportional to the speed of the motor. With a speed torque such as shown in *b*, the motor *starts* just as fast as it would without damping, but as it comes up to speed, the net torque decreases, as if reduced by a viscous damper (viscous damping is explained in Art. 16-10), so that the acceleration decreases also and the speed rises more slowly; when a speed at which the damping torque equals the motor torque is reached, the acceleration vanishes and the motor runs at a constant speed. Thus the shape of the speed-torque curve *b* makes the motor act as if it were damped; this damping is known as *internal* or *inherent damping*.

(1) Internal damping constant. The internal damping coefficient (or viscous-friction coefficient) F of a servomotor, assuming a linear speed-torque curve, is given by

$$F = \frac{\text{stall torque} \times 6.74 \times 10^5}{\text{no-load rpm}} \qquad (16\text{-}3)$$

where F is measured in dyne-centimeter-seconds per radian and stall torque in ounce-inches.

(2) Effect of control-phase voltage on internal damping. As the control-phase voltage is reduced, the locked-rotor torque decreases faster than the no-load speed, as shown in Figs. 16-2 and 16-5, so that the internal damping coefficient decreases with decreasing control-phase voltage. This means that the system is inherently less stable at reduced control-phase voltages.

(3) Speed of response. Davis[3] says that the maximum frequency to which a servomotor can respond is given by

$$\text{Frequency} = \frac{1}{2\pi \times \text{time constant}} \tag{16-4}$$

where frequency is expressed in hertz and the time constant in seconds. This means that a motor with a time constant of 0.01 s could respond to signal frequencies up to $1/(0.01 \times 2\pi) = 15.9$ cycles per second.

16-10 Viscous-Damped Servomotors. Viscous-damped servomotors are servomotors to which a viscous damper has been added. The damper is usually a drag-cup generator operating in a constant field set up by a permanent magnet. In such a generator, the voltage induced in the drag cup (and hence the drag-cup current) is directly proportional to the speed. Thus, the retarding torque developed by the generator is directly proportional to the speed, and the energy absorbed by the damper is proportional to the square of the speed. This kind of damping is called viscous damping because it is like the damping effect of a viscous liquid, which exerts a force on a body being moved through it which is proportional to the velocity of the moving body.

Although the viscous damper adds an important measure of stability to the system, it does absorb energy, so that the no-load speed and the speed and efficiency at all output torques are reduced.

16-11 Inertial-Damped Servomotors. Inertial-damped (also called inertially damped) servomotors are now used in many high-performance servo control systems. A sectional view of one is given in Fig. 16-6. Stator and rotor are the same as for a standard servomotor, but the motor has a double shaft extension. A high-inertia permanent-magnet flywheel is mounted on its own bearings on the shaft; it is coupled to the shaft through the drag cup, which is mounted rigidly on the shaft and positioned in the field of the permanent magnet on the flywheel. To understand the action, consider the rotor and flywheel at rest, with the fixed phase excited. Now, let power be applied suddenly at the control phase; the rotor starts instantly but the flywheel, because of its inertia, remains motionless momentarily. Since the drag cup is moving through a magnetic field, a drag torque, proportional to speed, tends to restrain the rotor and to accelerate the flywheel. In other words, at this instant, the rotor senses an external viscous damper. This same damping torque is, meanwhile, accelerating the flywheel; as it comes up to speed, the relative velocity of the rotor, with respect to the flywheel,

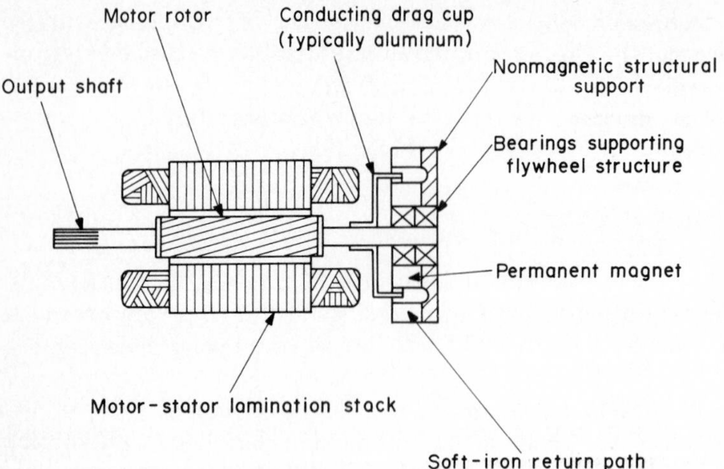

Fig. 16-6 A cross-sectional representation of an inertial-damped ac servomotor showing essential parts. *(Vernitron Corporation.)*

falls off, as does the damping torque as a result. When the motor is operating at any constant speed for a period of time, the flywheel will run at nearly the same speed, and there will be no damping. But for every change in speed of the servomotor, in either direction from normal, a damping torque will be developed until the flywheel is brought to the new speed; the direction of this damping torque will always be in the direction that opposes the *change* in rotor speed, and the magnitude of this torque will be proportional to the relative speed of the rotor with respect to the flywheel. An inertial-damped servomotor is illustrated in Fig. 16-7. An analysis of the performance of such a motor is given by Diamond.[4]

16-12 Tachometer Damping. Tachometers, of one type or another, are sometimes built integral with servomotors. The load of the tachometer

Fig. 16-7 A 60-Hz inertial-damped size 11 servomotor. Unit has a no-load rpm of 3000 and a stall torque of 0.60 oz-in, and it requires 3.5 W input. *(The Singer Company, Kearfott Division.)*

itself acts to some extent as a viscous damper. In addition, the output of the tachometer, which is proportional to the speed of the servomotor, is generally fed back into the feedback control system in such a way as to achieve a damping effect upon the system as a whole.

DC SERVOMOTORS

Some interesting innovations in dc motor construction have been developed in order to make them more suitable for servo applications.

16-13 Printed-Circuit Motors. In France in the late fifties, a new and innovative form of construction for dc motors was developed: the printed motor. An exploded view of a printed motor is shown in Fig. 16-8. The field structure is an eight-pole permanent magnet, as shown. The magnetic circuit is completed through the flux-return plate, which also provides the additional function of supporting the brushes. The armature is disk-shaped and contains no iron, and armature conductors are produced on both sides of the disk by printed-circuit techniques; connections from side to side are made by plated-through holes. Brushes bear directly on the armature conductors, which thus serve as a commutator. Because there are no armature slots or teeth, there are no cogging effects due to variations in magnetic reluctance; torque irregularities due to a finite number of conductors are virtually eliminated by skewing the conductors. Since the armature conductors are not surrounded by iron, as in conventional construction, the armature inductance is very small indeed, giving a very small electrical time constant. Because there are no rotating magnetic parts, the inertia of the

Fig. 16-8 Exploded view of a printed-circuit motor showing from right to left the magnet structure, disklike armature, and flux-return plate. Brushes bear directly on the winding through the return-plate structure. *(Printed Motors, Incorporated.)*

armature is very small, giving a low mechanical time constant. The exposed construction of the armature conductors gives them excellent heat-dissipation characteristics.

Printed motors are used in one type of application in which intermittent motion is required, as in paper-tape readers. They are also used as positioning servomotors, in which the lack of any preferred armature position and smoothness of torque are advantages. Printed motors are used as velocity servos, in which speed variations have to be controlled, either within a single revolution or over several revolutions. Some are designed to be operated in ambient temperatures up to 750°F. Also, because of their flat shape, low-cost versions are used to open and close windows in automobile doors.

16-14 Motors with Surface-Wound Armatures. Another interesting and significant development in dc servomotors is the use of surface-wound slotless armatures, in which the armature coils are secured to the outer surface of round armature laminations. (It is a curious fact that this was one of the very earliest forms of armature construction and was abandoned years ago. However, it has been revived to obtain optimum performance for servomotors.) Motors of this construction use a permanent-magnet stator, though shunt fields are often found in larger servomotors. The rotor is very long and of small diameter to keep down mechanical inertia and centrifugal stresses on the conductors. Coils are affixed to the outer surface of the armature laminations, as illustrated in Fig. 16-9, using resins and glass banding tape. This construction gives a smooth torque that is free of cogging variations, low armature inductance, low armature inertia, and excellent heat-transfer characteristics. Motors of this construction are built by Yaskawa Electric Company of Japan and marketed under the trade name Minertia motors. The General Electric Company also builds dc servomotors with this form of construction, which are called Hyper Servo Motors.[5]

16-15 Moving-Coil Motors. Still another construction to reduce mechanical inertia and to improve speed of response is exemplified in

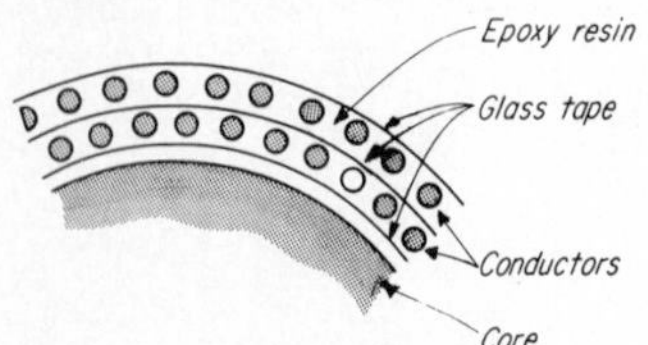

Fig. 16-9 Schematic sectional view of a surface-wound armature. *(Yaskawa Electric Manufacturing Company.)*

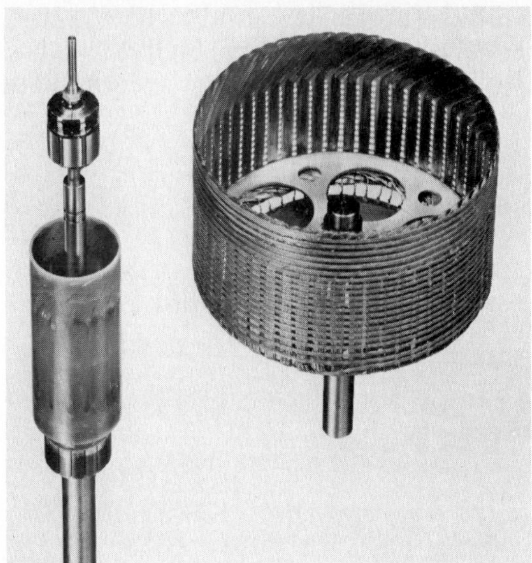

Fig. 16-10 Two iron-free moving-coil armatures for servomotors. At left, a slim resin-molded moving-coil winding and integrally coupled tachometer-generator armature. At right, an open "basket-type" construction designed for high-power dissipation when air-cooled. *(Electro-Craft Corporation.)*

the moving-coil motor, in which the only moving parts of the armature are the coils themselves. Two such armatures are illustrated in Fig. 16-10. The armature coils are "basket-wound" to form a cup-shaped iron-free armature. Excitation is supplied by a high-flux permanent-magnet stator. Low armature inductance, high acceleration rate, and smooth cogging-free torque are design features of the motor.

Moving Coil Motor, or MCM, is a registered trademark of Electro-Craft Corporation. Their model M-1438 has the following ratings, with and without air cooling:

	With air cooling	Uncooled
Continuous stall torque, oz-in	170	84
No-load speed, rpm	10,000	4,500
Electrical time constant, ms	0.16	0.16
Mechanical time constant, ms	0.69	0.69
No-load acceleration at maximum continuous torque, rad/s^2	215,000	120,000
No-load acceleration at maximum pulse torque, rad/s^2	950,000	950,000
Outside diameter, in	5.0	5.0
Axial length, in	4.75	4.75

The manufacturer claims a life expectancy of 1.5×10^9 stop-start cycles at 150 per second, or 3×10^8 revolutions at 1750 rpm with 60-oz-in output can be achieved before maintenance is required when it is operated under normal conditions.

References

1. *NEMA Motors and Generators Standards,* NEMA Pub no. MG1-1978, pt. 18, National Electrical Manufacturers Association, 2101 L Street N.W., Washington, D.C. 20037, 1978.
2. Mea, Anthony M.: "Servomotors," *Mach. Des.,* Dec. 16, 1965, pp. 94–97.
3. Davis, Sidney: "Rotating Components for Automatic Control," *Prod. Eng.,* November 1953, pp. 129–160.
4. Diamond, Al: "Inertially Damped Servo Motors, Performance Analysis," *Electro-Technology,* July 1965, pp. 28–32.
5. Blake, Bond M.: "Meet a Brand-New Servo Motor with Startling Performacne," *Control Eng.,* May 1966, p. 81.

Bibliography

Aherne, John M.: "The Direct-Drive D-C Servo," *Electro-Technology,* April 1963, p. 54.

Anderson, R. L., and J. C. Duté: "Precision Velocity Servo," *Electro-Technology,* March 1967, p. 56.

Bartlett, Peter M., and Harold Shankwitz: "The Specification of Brushless DC Motors as Servomotors," *Powerconversion Int.,* September 1983, pp. 20–25.

Henry-Baudot, J.: *Les Machines électriques en automatique appliquée,* Dunod, Paris, 1967.

Kelling, Leroy U. C.: "Servo Components for Numerical Control," *Mach. Des.,* Oct. 14, 1965, p. 234.

Lifschitz, Joshua: "Ripple in D-C Servo Motors," *Electro-Technology,* May 1965, p. 37.

Lowitt, P. M., and S. M. Shinners: "Optimum Design of High-Power Servos," *Electro-Technology,* January 1965, pp. 41–47.

Thaler, G. J., and W. A. Stein: "Transfer Function and Parameter Evaluation for D-C Servo Motors," *AIEE Trans.,* vol. 74, pt. II, 1955, p. 410.

Wood, Richard D.: "Optimum Compensation for High Performance Servomotors," *Powerconversion Int.,* July–August 1983, pp. 12–15.

CHAPTER

17

GENERAL CONSTRUCTIONAL FEATURES

Many constructional features of small electric motors are often common to more than one type of motor. Some of these features are discussed in this chapter.

BEARINGS

Volumes have been written about bearings, a very large subject in itself. A whole issue of *Machine Design* was devoted to this subject.[1,*] In the four articles that follow, we shall limit the discussion to those arrangements commonly used in fractional and subfractional horsepower motors.

17-1 Sleeve Bearings: Constructions. Bearings are one of the most vital parts of the motor and, except for the starting switch, are the only wearing parts in most types of fractional horsepower electric motors. Typical constructions are discussed next.

*For numbered references, see References at end of this chapter.

(1) Emerson Motor Division. A typical sleeve-bearing construction is illustrated in Fig. 17-1. The bearing itself is a short babbit-lined steel tube, which is formed from a flat strip by rolling. It is provided with a window to distribute the lubricant. The bearing is pressed into the end shield and machined to size. The bearing cavity in the end shield is

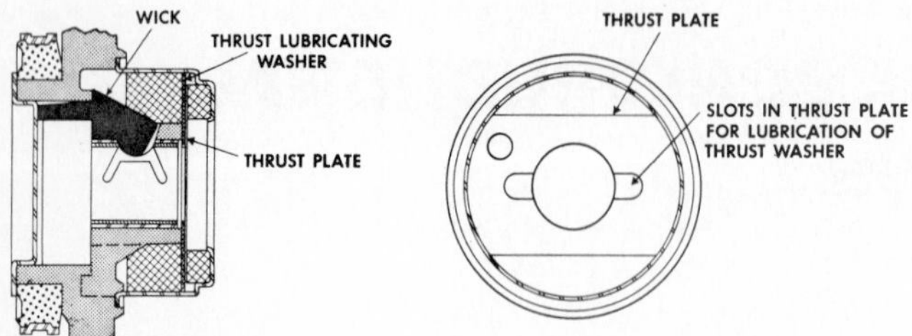

Fig. 17-1 Sleeve-bearing construction for all-angle operation. *(Emerson Motor Division.)*

loaded with die-cut felt, which presses against and holds the wick in the bearing window and against the shaft journal. Rotation of the shaft causes the journal to wipe oil from the wick and pump it through the grooves out to both ends of the bearing, where it is slung out to the felts and back up to the wick and journal, completing the cycle. The inner bearing cap has a formed lip to retain oil when the shaft is in a vertical position. On the shaft-extension end of the motor the outer bearing cap also has a formed lip; a neoprene washer, which is positioned between the bearing and cap, revolves with the shaft to sling oil out to the felt. The manufacturer refers to this construction as all-angle mount.

(2) General Electric: Form G. Figure 17-2*a* illustrates the sleeve bearing construction of Form G5 motors. The bearing itself is babbit-lined and is machined after being pressed into the end shield and provided with wide-angle oil grooves to distribute the lubricant evenly over the entire length of the bearing surface. The construction was designed to give greater axial and radial thrust capabilities over previous designs, with a 0.669-in diameter bearing, a spring-tempered thrust plate, and a molded thrust washer that is locked securely on the shaft. A redesigned felt feeder wick and injectable oil reservoir combine to give larger lubricant capacity than did previous designs. This construction was introduced in 1975 to replace the Form G, which is illustrated in Fig. 17-2*b*. Both are designed to permit continuous operation of the motor in any position; hence the maker refers to them as designed for all-angle operation.

(3) Permawick lubrication. In recent years a new wicking material, composed of a blend of cellulose fibers and oil, has been developed and marketed under the trade name of *Permawick*. Careful selection of fiber

(*a*)

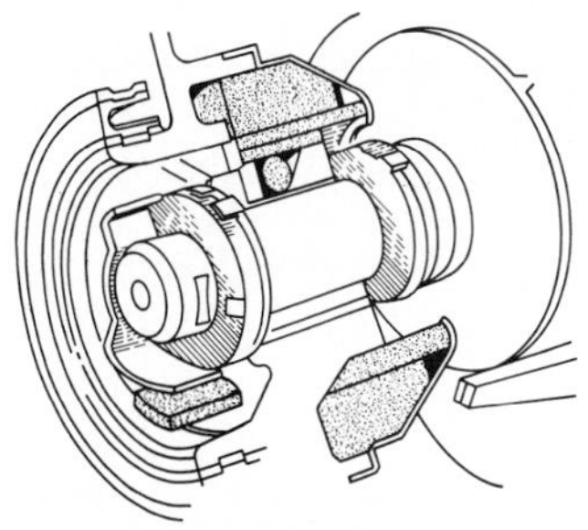

(*b*)

Fig. 17-2 Sleeve-bearing construction for Form G motors that are designed for all-angle operation. (*a*) After 1975; (*b*) before 1975. *(General Electric Company.)*

material and size, combined with appropriate oils, alters the wicking characteristics to suit the bearing conditions for which the motor is designed. The ability of Permawick to be extruded, without oil separation, directly into the end-shield cavity surrounding the bearing permits automation of assembly operations. It is now widely accepted and used by the fractional horsepower motor industry. Construction details, recommended by the Permawick Company, are given in Table 17-1 and illustrated in Fig. 17-3.

(4) Reliance Electric Company. A sectional view of a Reliance Electric Company sleeve bearing using Permawick is illustrated in Fig. 17-4. Oil in the reservoir can readily be replenished when necessary through the oil cup in the top of the bearing housing. It is designed for all-angle operation.

(5) Self-aligning bearings. Many small motors in the lower horsepower ranges use self-aligning bearings. One such construction is illustrated in Fig. 9-11. A porous bronze bearing, surrounded by an oil-soaked wick, has an outer surface approximately spherical in shape, permitting it to line up with the shaft. Misalignment is thus impossible.

Self-aligning bearings are becoming more popular in larger fractional horsepower sizes. The General Electric Company has introduced a line of laundry motors with porous self-aligning aluminum bearings. The Emerson Motor Division is using porous bronze bearings in its shaft-

TABLE 17-1 Construction Details for Permawick-Lubricated Sleeve Bearings (Recommendations of the Permawick Company)

Component	Wick-fed bearing design		
Load	Up to 25 lb/in^2	25–125 lb/in^2	125–250 lb/in^2
Bushings, type	Porous bushing	Steel-backed window bushing	Steel-backed window bushing
Material	Bronze or iron	Tin-babbit SAE-12	Lead-bronze, 25% lead SAE-794
Length-to-diameter-ratio	1.0–2.5 to give load not over 25 lb/in^2	0.9–1.2 regardless of load	0.9–1.2 regardless of load
Window size	None	6–8% of developed area*	10–12% of developed area*
Journal finish	A critical feature is the surface finish of the journal; ideally, it should be under 12 μ but never more than 16 μin		
Thrust plate	A spring-steel thrust plate with a minimum hardness of 55 Rockwell C which butts against the bushing is always recommended		
Thrust washer	Under 20-lb thrust, the materials that are recommended are (1)molybdenum-disulfide-impregnated nylon, (2) nylon, and (3) phenolic-impregnated linen with rubber on one side. Above 20-lb thrust, lead-bronze (25% lead) should be used.		

*Dependent upon the type of Permawick used.

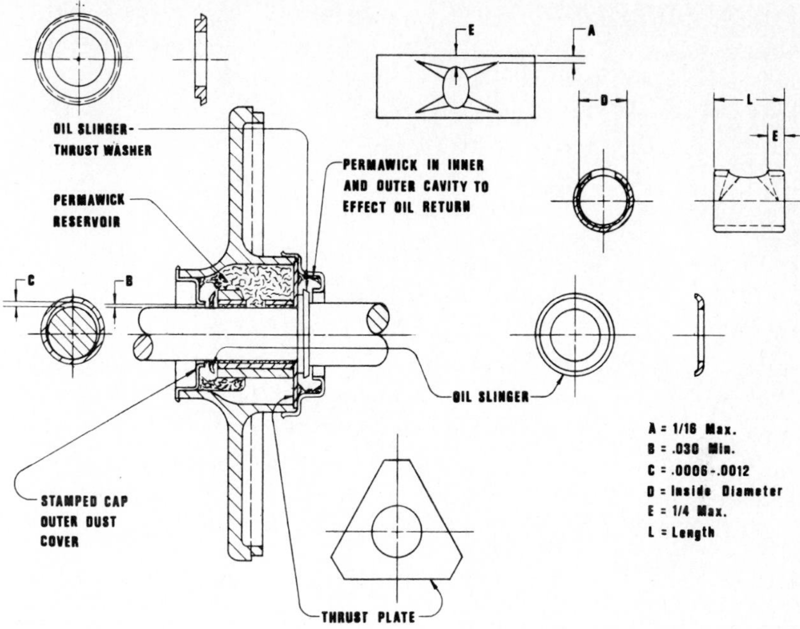

Fig. 17-3 Permawick-lubricated sleeve-bearing construction. *(Permawick Company.)*

mounted fan and blower motors of all sizes, and a porous iron-graphite bearing in laundry motors.

17-2 Sleeve Bearings: Service and Maintenance. Most sleeve-bearing motors are provided with means for reoiling. Frequency of reoiling varies a great deal, and it is generally best to obtain the recommendations of the motor manufacturer. Most sleeve-bearing motors will run for a long time without reoiling, of the order of 1 to 3 years. The motor should be so installed that the pressure of the belt pull does not come against the bearing window, lest excessive bearing wear result.

It is important not to pull the belt up too tight because excessive bearing pressures will develop, causing abnormal wear, possible "seizing," and early failure of the bearing. In general, the belt should be made only tight enough to prevent slipping. It is easier than commonly supposed to overload a motor severely and to cause premature bearing failure by excessive belt tension. A wattmeter measuring the input to the motor can be helpful; the belt tension can be reduced until it begins slipping and the minimum watts noted, after which the belt can be tightened until the watt input starts to increase noticeably.

When replacing a sleeve bearing, unless it is the oil-impregnated

Fig. 17-4 Sectional view of an all-angle sleeve bearing using a Permawick lubrication system. *(Reliance Electric Company.)*

porous type, it should be replaced with one that has a press fit in the housing of about 0.0005 in (0.01 mm). Usually the bore is slightly undersized and has to be reamed. This operation generally is best done by means of a line reamer. The shaft journal should be trued and smoothed if necessary and then carefully measured for size. A line reamer should be chosen to give a bearing clearance of 0.0006 (0.015 mm) to 0.0012 in (0.03 mm) (total diametral clearance). Porous oil-impregnated bearings may require a slightly larger clearance. When the end shields are bolted onto the motor for the line-reaming process, the rabbet fits should be carefully cleaned to make certain there is no dirt in them, as otherwise the new bearings will be reamed out of line. Porous bearings should be bought to size and not reamed.

Similarly, when reassembling a motor, it is also important that the end-shield fits be clean, as otherwise the bearings will be out of line and will bind. It is also important that there be a little free end play; if not, the bearings might seize at the thrust surfaces. If end-bumping noise is to be held to a minimum, this free end play must be small: say, 0.002 to 0.010 in (0.05 to 0.25 mm). If the motor is cushioned against end bumping, a free end play of 1/32 in (0.8 mm) may not cause objectionable noise.

If the motor uses a wool wicking, it may be relubricated with a light grade of machine oil, popularly sold in gasoline filling stations as household oil. Only petroleum oils should be used, and lubricants containing any animal or vegetable oils should be carefully avoided.

If the end shields were packed with Permawick, it is best to obtain recommendations from the motor manufacturer for replacing the Permawick material itself or for replenishing the oil, for there are different grades of the Permawick fiber and different oils are used in different cases.

17-3 Ball Bearings: Construction. Ball-bearing fractional horsepower motors are generally used for those applications in which heavy axial and radial thrust loads are encountered. Many types of ball bearings have been developed to fulfill the requirements of a wide variety of applications. For most fractional horsepower-motor applications, the single-row radial bearing is adequate and, therefore, the most popular. Such bearings are capable of handling the radial and axial-thrust loads usually encountered. For operating speeds of 3450 rpm and below, grease-lubricated ball bearings are generally used. Bearings now commonly use either a shield or a felt seal to retain the lubricant inserted by the manufacturer and to exclude dirt and other foreign matter. A *shield* is a thin steel disk affixed to the outer race and arranged to cover all the space between the inner and outer races; it fits into a labyrinth or semilabyrinth on the inner race with very narrow clearances but no rubbing contact. Either one or both sides of the bearing may be shielded. A bearing with a single shield is illustrated in Fig. 17-5. Shielded bearings are the same length (along the shaft) as unshielded bearings and are generally interchangeable with plain bearings. Sealed bearings generally use a felt seal of some sort, which retains the lubri-

Fig. 17-5 Sectional view of the front end of a motor showing a single-shielded ball bearing and both members of the starting switch. *(Century Electric Company.)*

cant and excludes dirt better than shields do. Because of the space occupied by the felt seal, sealed bearings are usually longer than either plain or shielded bearings; hence, sealed bearings are not generally interchangeable with either plain or shielded bearings. The additional length required for the seal may be added only to the inner race, or it may be added to both races. Seals may be provided on one or both sides of the bearing.

Shaft journals are ground to such a diameter as to give a light press fit between the inner race and the shaft. *When pressing a bearing on the shaft, the pressure must always be applied to the inner race directly, never to the outer race, lest the balls or the races, or both, be damaged.* The outer race fits into the recess in the end shield snugly, but loosely enough that only light hand pressure is required to insert it. For general-purpose motors, an opposed-mounting construction is extensively used. With this construction, the inner races are secured to the shaft by light press fits, and the outer races are free to slide axially in the end shields; rearward end play is limited by a shoulder in the rear end shield, and forward end play by a similar shoulder in the front end shield. This end play is commonly of the order of $\frac{1}{16}$ in (1.6 mm), but as a rule it is taken up by a steel finger spring inserted in the bearing cavity of one end shield. This spring presses against the outer race, preloads both bearings, and takes up all the end play. In general, this spring is strong enough that it takes a force of several pounds to move the shaft against it. In Fig. 17-5, a sectional view of the front end of a motor using opposed-mounting construction is illustrated. The bearing is single-shielded (on the inside of the motor) and is preloaded with spring washers to take end play out of the bearing and to contribute to quieter operation.

If axial motion of the shaft cannot be permitted, a locked-bearing construction is used. Either the front or the rear bearing can be locked, but both must not be, because differential expansion between the rotor and the housing can set up destructive axial forces. The shaft is threaded, and the inner race is locked in position by means of a thin nut and lock washer; the outer race then has to be clamped securely in the end shield by some suitable means. With locked-bearing construction, there is no endwise movement of the shaft except the small movements due to thermal expansion and the slight axial freedom in the bearing due to clearances between the balls and the races. Tight fit-up bearings have less of this freedom than loose fit-up bearings.

Ball-bearing motors are capable of withstanding considerable endwise thrusts and are generally suitable for operation in any position. The amount of thrust load a particular motor can handle safely should be obtained directly from the recommendations of the motor manufacturer or from the ball-bearing manufacturer.

Some motors use noise-damped ball bearings. A thin rubber ring is bonded to the outside of the outer race, and a steel ring is bonded to the outside of the rubber ring. Since any compression of the rubber tends to make the rotor eccentric in the gap, this construction is better suited to coupled than to belted loads. Pros and cons of such a construction are discussed by Stark.[2]

Two-pole fractional horsepower motors used on swimming pool pumps often have a ball bearing in the drive end and a sleeve bearing in the opposite end. The pulsating axial thrust imposed by the pump impeller creates a pulsating radial thrust in the opposite end of the motor, which will enlarge the clearance between the outer race of a ball bearing and the end-shield cavity but will not cause the same problem with a sleeve bearing.

Die-cast aluminum end-shield bearing cavities are constructed in one of two ways: (1) a steel ring that is cast in place and machined to size serves as the seat for the outer race of the bearing or (2) the cast-aluminum bearing seat is work-hardened, or "bearingized."

In some motor constructions the rotor and stator are purposely misaligned and with about 0.010- to 0.015-in free end play to cause magnetic centering of the rotor in the stator when it comes up to speed. The slight axial movement is used to properly locate the driven load, such as an impeller in a centrifugal pump, or to impose the axial thrust in a vertical direction.

17-4 Ball Bearings: Service and Maintenance. Lubrication of ball bearings is most important. If the bearings are open at one or both ends, they should be lubricated annually; moreover, it is a desirable practice to wash the bearings and bearing housings thoroughly approximately every 2 years, with alcohol or carbon tetrachloride. (If the motor has a commutator and brushes, take care that the carbon tetrachloride does not come into contact with either.) The bearings themselves and all grease used in them must be scrupulously clean; it is even more important to keep all dirt and foreign matter out of ball bearings than it is to keep it out of sleeve bearings. For a general-purpose lubricant, a neutral soda-base soap grease is recommended in preference to other kinds. Ordinary cup greases are frequently compounded with a calcium soap and have a melting point too close to motor operating temperatures. Greases compounded from very heavy or very light oils should be avoided. Viscosities between 200 and 500 Saybolt at 100°F give good results. *Ball bearings should not be overgreased.*

Ball bearings that are sealed at both ends generally require less attention. The claim that these have enough lubricant sealed in them for the life of the bearing can be somewhat ambiguous, for obviously the bearing would not last long after all the oil had evaporated from the grease!

Operating temperature is a major factor in the life of any ball bearing. At a bearing temperature of 40°C (104°F), the life of sealed ball bearings operating 24 h per day may be from 3 to 5 years or more. However, the life halves for each 10° (centigrade) increase in operating temperature. Application of this rule indicates that if the bearing operates at a temperature of 80°C (176°F), the life on a 24-h-day basis may be reduced to as little as 2 or 3 months. Opinion is divided as to whether it is desirable to pack grease around the outside of a shielded bearing. But one rule is certain: The higher the operating temperature, the more frequent must be the attention to lubrication.

When removing ball bearings from a shaft, it is desirable to press against the inner race; if pressure is applied to the outer race to remove the bearing, the latter must not be used again in the motor because the races are likely to become ball-dented during their removal.

STARTING SWITCHES

Some form of starting arrangement is necessary to all single-phase induction motors, as explained in Chap. 2. Split-phase and capacitor-start motors, by far the most popular types in fractional horsepower sizes, use an auxiliary winding which has to be disconnected when the motor is up to speed, as discussed in Arts. 4-4 and 5-5. The auxiliary winding is disconnected by means of a starting switch, which may be a centrifugal switch, a magnetic relay, or a thermal relay. By a wide margin, the centrifugal switch is the most popular starting device used in fractional horsepower motors.

17-5 Centrifugal Starting Switch: General. A centrifugal switch consists of two basic parts: a stationary member and a rotating member, sometimes called an actuator. The rotating member is a centrifugal device, mounted on the motor shaft; as the motor comes up to speed, centrifugal force causes an element to move, usually in an axial direction, and this motion actuates the circuit-interrupting means provided on the stator member. All of the electric contacts and current-carrying parts are part of the stator member. To avoid wear due to prolonged rubbing between the stator and rotor members, the centrifugal switch is designed so that rubbing takes place only during the starting period. A modern centrifugal starting switch usually has a life of over a million cycles.

(1) Necessity for snap action. Starting switches must be snap-acting; that is, they must break the circuit quickly and positively and must not flutter. Fluttering or prolonged arcing of the switch can cause excessive burning and pitting of the contacts; a serious loss in torque, so serious that the motor may not be able to bring its load up to speed; and damaging overvoltages on the capacitor, if one is used. This snap action can be built into either the rotor or the stator member. Today's switches generally achieve the snap action by incorporating some form of toggle action in the rotor member, although for years, Westinghouse put it into the stator member. (For example, see types CAH and FT in Fig. 18-11 that appeared in an earlier edition of this book by Veinott.[3])

(2) Switching arrangements. Most starting switches simply open a single circuit; such is the case for the usual split-phase or capacitor-start motor. A back contact was required in some types of two-value capacitor motors (Fig. 6-5). Additional circuits or contacts may be required for two-speed pole-changing motors (Arts. 4-23 to 4-26, 4-28 and 5-19).

(3) Operating speeds. Opening speed of a centrifugal switch should generally be around 75 to 80 percent of synchronous speed, for reasons illustrated in the typical speed-torque curves of Figs. 4-3, 4-4, and 5-4. The reclosing speed for a capacitor-start motor should generally be below half speed to avoid problems of regeneration (Art. 5-27).

(4) Switches for pole-changing motors. Starting switches for pole-changing (multispeed) motors have special requirements for both operating speeds and circuit arrangements and connections. Some of these requirements are discussed in Chaps. 4 and 5.

17-6 Emerson Starting Switch. The starting switch shown in Fig. 17-6 is representative of the many styles in use today in fractional horsepower motors. As shown in this figure, die-cast zinc weights *a*, which

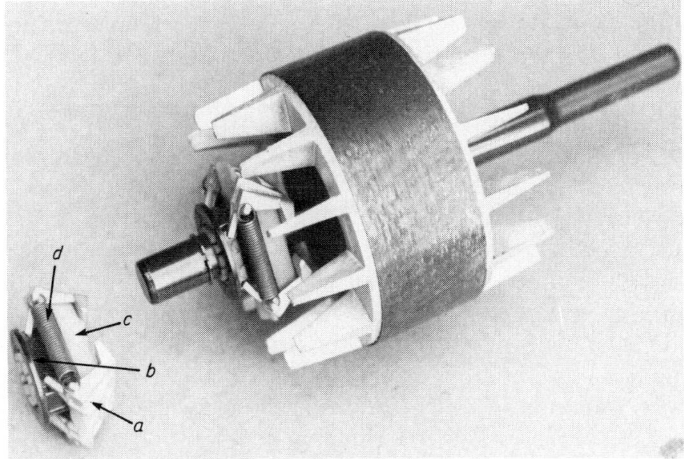

Fig. 17-6 A rotating centrifugal starting switch member. *a*, Die-cast zinc weights; *b*, phenolic spool; *c*, plastic bracket, *d*, springs. *(Emerson Motor Division.)*

fit into molded grooves in the phenolic spool *b*, are supported by the plastic bracket *c* and held in place by two springs *d*. The plastic bracket is positioned along the length of the shaft by tangs which snap into a shaft groove; it has slight freedom of rotation relative to the shaft. At rest, the spool is positioned away from the rotor. Running, centrifugal force causes the weights to move radially outward, snapping the spool to its alternative position toward the rotor and away from the stationary member of the starting switch. At standstill, the spool presses against a contact arm of the stationary switch, closing the starting circuit. When the spool snaps toward the rotor, it releases the pressure that holds the contact arm closed, allowing the natural spring action of the latter to open the contacts.

Rotating members are designed with various weights and spring tensions to vary the throw-out speed, depending on the operating speed of the motor and the number of speeds. Stationary members are available in an almost limitless variety of sizes and combinations. Some have only a single contact arm, which bears against only one side of the actuator. Some have two arms, which bear on diametrically opposite sides of the actuator, which may be connected in series or parallel or may be connected to different circuits. Some of the stationary members have enclosed contacts to keep dust and dirt away from the contacts and most stationary members are designed as an integral part of the motor terminal board.

17-7 Other Starting Switches. Many forms of centrifugal starting switches have been developed and used. One form, developed by the Ohio Electric Company, had a stationary member consisting of two stationary slip rings, secured to the front end shield and connected in series with the auxiliary winding; the rotor carried a single brush which spanned the two rings, closing the circuit when the motor was at standstill; as the rotor came up to speed, centrifugal force caused the brush to move radially outward, away from the slip rings, thereby opening the starting-winding circuit. Another quite different arrangement was the skillet type, developed by Delco; the stationary member carried a skillet-shaped arm which bore the stationary switch contacts; the governor weight was washer-shaped and slid along a radial switch pin, against a coil-restraining spring; in earlier models, the switch pin was mounted in a hole in the shaft, but in later models it is supported by a boss or projection which is cast integral with the aluminum end ring of the rotor. Westinghouse has used many forms of starting switches. A number of motor manufacturers still use their own design of centrifugal switches; most of these designs use a phenolic spool, which is arranged to slide along the shaft to actuate the stationary member, as described in Art. 17-6. Another such switch is illustrated in Fig. 17-5. Many of the previously mentioned switches are discussed in greater detail in Chap. XVIII of an earlier edition of this book.[3]

Torq Engineered Products, Inc., of Bedford, Ohio, makes a rotating member under the trade name Syncro-Snap. Actuating weights are riveted to the periphery of a conically formed disk of spring steel, and the spring-steel disk is riveted to a retaining plate. Fingers formed on the disk carry the insulating spool. When the steel spring is snapped to its opposite conformation by the weights, the insulating spool is drawn toward the retaining plate and away from the stationary member of the starting switch. A more detailed description of the Syncro-Snap is given in an earlier edition by Veinott.[4]

Thermally actuated starting switches have been used. The principle involved here is to use the heat generated by the auxiliary-winding current to open the auxiliary-winding circuit, and to use the heat generated by main-winding current to keep the auxiliary-winding circuit open so long as the motor is running; after the motor stops because of removal of excitation, the thermal relay cools, its contacts reset, and the motor is ready to start again. An inherent limitation of this arrangement is that the motor cannot be restarted immediately after it is stopped, for it is necessary to allow enough time for the thermal relay to cool and reset.

17-8 Service and Maintenance of Centrifugal Switches. Whenever a fractional horsepower motor is serviced, for whatever reason, it is a good idea to examine and inspect the centrifugal starting switch, if the motor has one. When possible, service personnel should obtain whatever instructions and advice they can from the motor manufacturer and should be guided by them. In the absence of definite instructions to the contrary, the information that follows may be helpful.

(1) Before dismantling the motor. Check and measure the actual amount of free end play of the assembled motor, as this must be taken into account later in checking the action of the starting switch. Keep the same end-play washers in both front and back end shields. If the motor will start and come up to speed, it is a good idea to start it up several times, observing the action of the switch, to see whether it is opening and closing in a normal manner.

(2) General inspection. Dismantle the motor, taking care not to lose the end-play washers. Remove the rotor from the front end shield and inspect the stator member to see that it works freely and that the contacts are neither burned nor pitted nor likely to stick. Examine it also for any evidence of loss of temper of the spring, deformation, or anything that limits normal contact pressure in the closed position. A defective stator member should be replaced by a new one. Now, inspect the action of the rotating member.

CAUTION: Do not push against the spool-type slider, but pull the actuating weights out as if they were being acted upon by centrifugal force; with many spool-type rotating members, it is possible to damage the switch by pressing in on the spool slider, for considerable force is often necessary to start the spool. The best technique is to pull out the actuating weights while pressing lightly against the spool; once weights are out, even partially, it usually does no harm and often is easier to hold the switch in the running position by holding the spool depressed. Work the switch several times to see that it works freely, and that there is no binding or sticking. If there is a close clearance between the spool and the shaft, make sure that there is no dust, dirt, or corrosion product that can impede the intended free movement of the spool slider along the shaft. If any part of the rotating element, such as a spring or spool, needs replacement, replace the entire rotating assembly; *do not replace only a part of the rotating assembly.*

(3) Checking axial location of rotating member. It is a good idea to check to see that the switch is properly located axially, especially if a new rotating member has been installed. The instructions that follow are for spool-type switches that have a slider moving along the shaft. First check the location with the rotor in the *extreme near* position. To do this, assemble the rotor in the front end shield, using only the end-play washers that were in the front end before disassembly. The switch should be tested in two positions, closed and open. With the rotor shaft preferably in a vertical position in order to ensure its location, turn the rotor slowly by hand for two or three revolutions, while observing the stator member to see that a solid pressure is maintained at the switch contacts; if an ohmmeter is available, it is helpful to connect it across the switch contacts and to watch that the observed resistance does not vary significantly while the rotor is turned. Next, check the running axial clearance between the slider and the stator switch element. To do this, move the rotating member to its running position by pulling the actuating weights outward while lightly pressing against the spool or slider (Do not do it all by pressing against the slider, for you could damage the switch!), and hold or lock it temporarily in this position. Now, measure the axial clearance between the slider and the stator member by use of a scale, feeler, or other appropriate gauge. Now, if these checks give satisfactory results, remove the rotor from the end shield and add end-play washers having a total thickness equal to the total end play, measured before the motor was disassembled, and replace the rotor; this procedure puts the rotor in the *extreme far* position. Now, repeat the test, checking for solid contact in the closed position; of course, if the running clearance was adequate in the extreme near position, it will also be ample in the extreme far position. *To recapitulate what is important: in the extreme near position of the rotor, make sure that the rotating member, in its starting position, does not press so heavily against the stator member as to cause damage to the latter, and that there is sufficient clearance in the running position. In the extreme far position of the rotor, make sure that the rotating member exerts ample pressure on the stationary member to ensure a good, solid circuit through the switch contacts.*

(4) Check of operating speeds. Switch-operating speeds are best checked in the completely assembled motor. A convenient method for doing this is to connect the motor to an adjustable low-voltage source, such as a Variac. Provide means for applying a light friction load on the shaft. Connect a suitable ammeter in the motor circuit. With a low voltage applied, of the order of one-fourth rated voltage, and a tachometer held against the shaft extension, let the motor slowly come up in speed until the switch opens; the contact opening should be clean and decisive and may be noted visually or indicated by a positive movement of the needle of the ammeter, without hanging. Let the motor coast and note the speed at which the switch recloses. This operation should be repeated several times to see that the operating speeds are consistent: first with the shaft at one limit of axial end play and then at the other limit. For ususal single-speed motors, the opening speed should be between 70 and 80 percent of synchronous; the reclosing speed should be, for capacitor-type motors, less than half of synchronous but is less important for split-phase motors, except as an indication as to the probable health of the switch. If the motor is a two-speed motor that is arranged to start on the high-speed connection, the opening speed of the starting switch must be below the lower synchronous speed. Otherwise, if the motor is set for the low-speed condition when it is started, it will come up to a speed above synchronous and then, when the starting switch operates, there will be a sudden braking of the motor, accompanied by an unpleasant jerk.

(5) Replacement of a stator member. In general, if any part of a stator switch assembly is defective it is recommended that it be replaced in its entirety by a new one obtained from the motor manufacturer. Care should be taken in installing the new member, especially when soldering the leads to it. Most such switches use phosphor bronze, beryllium copper, brass, and steel for various parts, which are generally assembled in phenolic plates or bases. Riveted joints must not be allowed to become overheated because they may loosen or damage the phenolic material. Acid solder or fluxes are to be avoided. Rosin flux of a type that will spread and splatter over the surface of the base and onto contacts, leaving a gummy deposit, is to be avoided, since dirt will accumulate on such deposits. Rosin fluxes of various brands, which leave remaining excess flux in a passive hard insulating state, preventing the just-mentioned condition, are available. Soldering operations are simpler when the switches have parts tinned where connections are to be soldered.

Many manufacturers today use stator winding leads with quick-connect terminals for connecting to the stationary starting switch member. This makes replacement of the stationary member relatively simple, but caution should be used to assure proper connections.

17-9 Magnetic Starting Relays. It was pointed out in Art. 5-20 that magnetic starting relays are sometimes used instead of centrifugal switches. It was also noted there that starting relays may be either current-operated or voltage-operated, and advantages and disadvantages of both kinds were therein discussed. A typical relay, designed for motor-starting service, is illustrated in Fig. 17-7. A sectional view of this relay is shown in Fig. 17-8. This relay utilizes a steel armature centered in a solenoid coil and a bridging-type contact arm which is closed by movement of the armature. Weight of the armature parts and number of turns in the coil are closely controlled at the factory in order to hold pick-up and drop-out currents to close limits. All moving parts are enclosed in a phenolic case, and the unit is provided with "quick-connect" terminals, as shown in the photograph. This design is available with pick-up ratings ranging from 6.3 to 35.7 A; drop-out current is

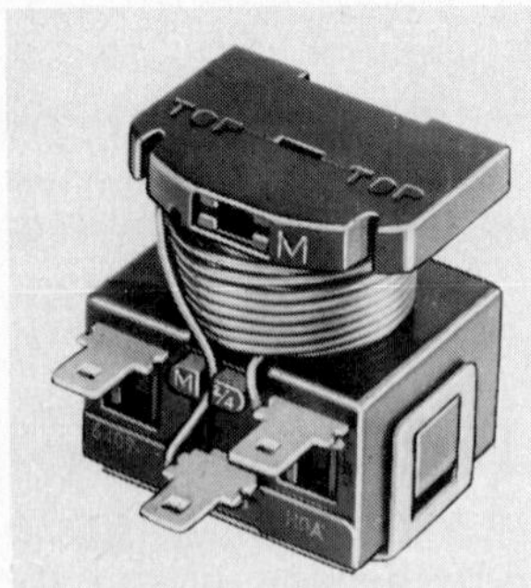

Fig. 17-7 A current-operated motor-starting relay. Models are available with pick-up currents from 6 to 36 A. The unit is gravity-opened and must be mounted vertically as shown. *(Texas Instruments Incorporated.)*

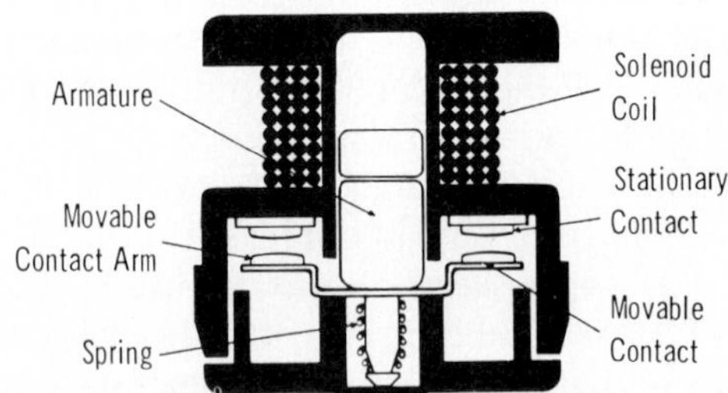

Fig. 17-8 Cross-sectional view of the motor-starting relay of Fig. 17-7 *(Texas Instruments Incorporated.)*

approximately 82 percent of pick-up current. Other designs are available with pick-up currents down to 1 A. The unit is also available as a part of a combined unit which includes a motor overload protector. This particular unit is gravity-opened and must, therefore, always be mounted in a vertical position with the coil up. (Not all relays are gravity-operated, however.) The unit illustrated is used extensively for food waste disposers, hermetic refrigeration compressors, dishwashers, and pumps.

ELECTRONIC STARTING SWITCHES

Electronic switches, which have no moving parts to wear, have long been sought by the electric motor industry. Much research has been conducted, and many patents have been granted on just such devices. Two such switches are now discussed.

17-10 Franklin's Reed-Triac Switch. This device utilizes only two components, a reed switch and a triac, housed together in a small package that either can be mounted inside the motor or can be mounted separately. Connections are shown schematically in Fig. 17-9. The triac serves as the basic switching device, taking the starting winding in and out of the circuit. It is controlled through interaction with the reed switch actuated by a coil in series with the power line. The triac is maintained in the conducting mode for the desired speed interval by means of trigger pulses applied to its gate by the reed switch's opening and closing 120 times per second in a 60-Hz motor. At standstill, the starting-winding current of a capacitor-start motor leads the line current (see Fig. 5-3), but as the motor comes up to speed this phase difference decreases and even reverses around 65 to 80 percent of synchronous

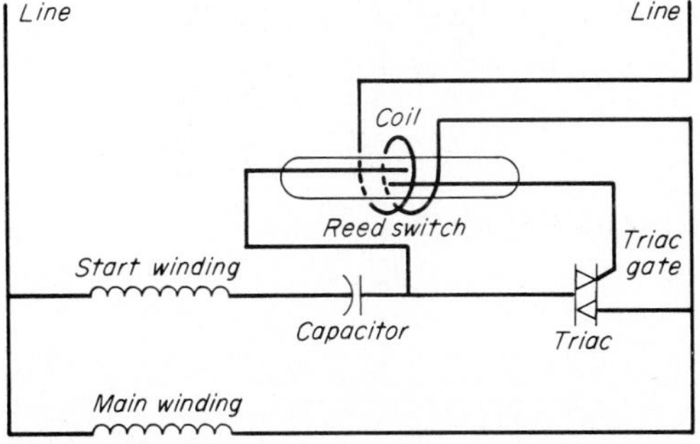

Fig. 17-9 Schematic diagram of a Reed-Triac starting switch. *(Franklin Electric Company.)*

speed, with this shift in phase angles independent of voltage. This characteristic is used to advantage to calibrate the reed switch so that it is conducting up to a more or less predetermined speed and nonconducting above that speed.

If after the triac is opened a heavy overload causes the motor to slow down to, say, half speed, the line current increases enough to close the contacts of the reed switch, firing the triac and thus energizing the starting winding.

17-11 Century Solid-State Switch. A solid-state starting switch is illustrated in Fig. 17-10. This switch employs a Hall-effect device that senses the actual motor speed by means of a magnet wheel mounted on the shaft. The speed signal from the Hall-effect sensor is fed into a custom integrated circuit which contains over 1200 transistors on a single silicon chip that incorporates both digital and analog circuitry in an area of less than 0.007 in^2.

Fig. 17-10 A solid-state switch for single-phase motors mounted inside an end shield. *(Century Electric Company, Inc.)*

RIGID AND RESILIENT MOUNTINGS AND COUPLINGS

Almost limitless is the number of different mounting and drive arrangements that have been used for fractional horsepower motors. Only a few of the most common variations are discussed in the following paragraphs.

17-12 Types of Mounting. Common and still popular mounting arrangements are rigid-foot, resilient-foot, round-frame, face, and extended through-bolt mountings.

(1) Rigid-foot mounting. A foot rigidly attached to, or an integral part of, the motor frame is probably the earliest form of mounting arrangement. It is still widely used when noise is not a major factor. Another form of rigid mounting is similar to resilient mounting but does not use resilient rings; the base is clamped directly to the end-shield hubs.

(2) Resilient-foot mounting. Resilient mounting arrangements are used extensively when quiet operation is required. These are discussed in Arts. 17-13 and 17-14.

(3) Round-frame mounting. Round frame motors, i.e., motors without a foot, may be held in place by means of a band clamped around the body of the motor. One novel development, which is made by the General Electric Company and marketed under the trade name Torsion-Flex, has three formed spring-steel arms welded directly to the motor frame. These provide torsional resiliency without the need for additional damping.

(4) Face mounting. One form of face mounting, used for oil-burner motors, is shown in Fig. 19-4. The rear end shield is provided with a rabbet fit machined concentric with the shaft; the motor is bolted to the oil burner by means of bolts passing through the extended lugs in the rear end shield. In another variation of face mounting, shown in Fig. 19-6, the extended lugs are omitted, and the motor is held in place by means of bolts projecting from the driven apparatus into the four tapped holes shown in the rear end shield. This mounting arrangement is used for a number of pump applications; it is generally called "C-face mounting."

(5) Extended through-bolt mounting. Round-frame motors are often mounted by means of extended through bolts. Such an arrangement does not center the shaft so accurately as the machined rabbet fit of a face-mounted motor, but this method of mounting is often satisfactory and convenient for fan applications.

17-13 Torque Pulsations in Single-Phase Motors. Polyphase induction motors supply a steady and smooth flow of power to the shaft. Torque developed by electromagnetic action of the windings is steady, like the torque developed by a turbine. Single-phase motors supply the torque in a series of pulses, just as in a reciprocating engine. These torque pulsations, inherent in any single-phase motor, are a major source of noise in many applications, particularly when the motor is mounted rigidly to any sheet-metal cabinet or structure capable of radiating sound. Fortunately, the effect of this major source of noise can be virtually eliminated by use of an effective resilient mounting. In order to understand the source of these torque pulsations, it is first necessary to understand the pulsating nature of single-phase power.

(1) Pulsating nature of single-phase power. In Art. 1-13 the nature of alternating current was pointed out; the current at any instant in time was shown to vary in accordance with the sine wave in Fig. 1-2. A similar curve of instantaneous amperes versus time is given in Fig. 17-11.

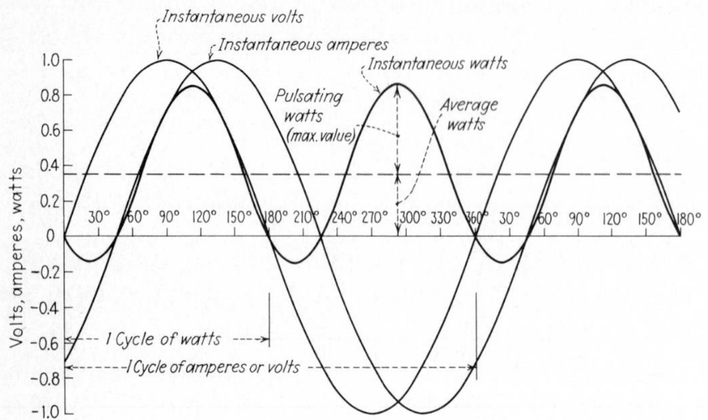

Fig. 17-11 Voltage, current, and power relations in a single-phase circuit of 70.7 percent lagging power factor.

Likewise, a curve of instantaneous volts versus time, also a sine wave, is shown. In this figure, the current wave is shown displaced 45 electrical degrees in time behind the voltage, which is the case when the power factor is 70.7 percent. The instantaneous power—or watts—in a single-phase circuit is equal to the product of the voltage at that instant by the current at that same instant. Therefore, to obtain the instantaneous watts at a number of points in time, we multiply the instantaneous volts by the instantaneous amperes, taking care to observe the plus and minus signs of both current and voltage. A curve obtained in this fashion is also shown in Fig. 17-11. Studying this curve of instantaneous watts, it is interesting to note that

1. The instantaneous watts vary, as do the volts and amperes.

2. The frequency of this variation is *twice* the supply-circuit frequency; i.e., in a 60-Hz circuit, the power frequency is 120 Hz.

3. The power is not always positive but is negative at certain portions of the cycle; negative power indicates a reversal in direction of power flow; e.g., if the power input to a motor is taken as positive, the power is said to be negative when the motor is feeding power back to the line.

4. It will be noted that the power can be thought of as consisting of two components. The first component is the steady power, or average watts, which remains constant in value, as in a dc circuit; the second component is the pulsating watts, which vary sinusoidally with time at a frequency double the frequency of the applied voltage. The pulsating watts are superimposed upon the average watts.

5. Pulsating power is greater than the average power, showing that the former is of major importance.

(2) Pulsations in torque. In any single-phase motor, the torque produced by the electromagnetic forces consists of two components: (1) an *average torque* and (2) a *pulsating torque*, alternating at twice line frequency. Since the power input to the motor, by the very nature of single-phase power, varies over a wide range, the torque developed by the electromagnetic forces must also vary over a wide range. This point becomes obvious when the fundamental law of conservation of energy is considered: when the power input is zero or negative, there can be no power output, and hence no torque. Readers interested in pursuing this subject of torque pulsations further are referred to an article on the subject by Kimball and Alger[5] and also to Arts. 10-7 and 12-8 in another book by Veinott;[6] Art. 10-7 tells how to calculate this torque when only the main winding is energized, and Art. 12-8 when both windings are energized. In addition, Krause has published an interesting study of this phenomenon.[7] Although the flywheel effect of the rotor smooths out variations in speed and permits the motor to carry the load steadily, it must be remembered that the torques due to electromagnetic forces contain a strong component of pulsating torque. If the stator is rigidly attached to the driven device, this double-frequency torque pulsation and its harmonics are likely to be the principal source of noise and vibration by the motor, particularly if the motor is mounted on any structure capable of vibrating and radiating sound. Dominant pitch, for 60-Hz motors, is 120 Hz, a note about one octave below middle C on the piano. It is primarily to eliminate this 120-Hz hum that resilient mountings are used. How this is done forms the subject of the following article.

17-14 How Resilient Mounting Eliminates Noise Due to Torque Pulsations. We saw, in the preceding article, that the torque developed in a single-phase motor consists of two components: an average torque and a pulsating torque, alternating at twice line frequency. An ideal resilient mounting transmits the average torque to the stator, but absorbs the pulsating torque without transmitting any of it to the motor-mounting supports. The most common, and most effective, arrangement uses two neoprene mounting rings, one on each end of the motor. In one form of construction, each mounting ring consists of two concentric steel rings, between which the actual neoprene ring is bonded. The inner steel ring is pressed onto a machined fit on the end shield, and the outer ring is clamped in a mounting cradle. The stator is thus free to move, but the principal freedom of motion is about the center of the shaft; motion about the center of the shaft stresses the rubber in shear, the most effective way to use rubber to isolate noise and vibration, because the modulus of elasticity of rubber is much less in shear than in compression or in tension. Since the whole motor is free to vibrate about the center of the shaft, the double-frequency torque pulsation is not transmitted to the base. For best results, the center of gravity of the movable stator should coincide with the center of the shaft; a capacitor, or any other substantial weight, mounted on the frame of the motor brings the center of gravity of the stator slightly above the center line of the shaft, causing a slight translational force to be exerted on the cradle base, because the stator tries to oscillate around its own center of gravity. However, a capacitor is usually light enough that the effect is small. One should not, however, mount a heavy weight on the stator of a resiliently mounted motor if one wants it to be quiet.

Variations on the neoprene-ring mounting are numerous. Sometimes the inner steel ring is omitted entirely; slipping is prevented by using lands, a spline, or a hexagonal shape on the nose of the end shield and molding the inside of the neoprene ring to fit. Sometimes, too, the outer steel ring is not bonded to the neoprene; it is held from slipping by friction. Often the neoprene is molded with hollows in it, to provide greater torsional resiliency. Actual use of neoprene rings is illustrated in Fig. 17-1. A mounting arrangement using the neoprene rings is commonly referred to as rubber mounting, because these rings were once molded from natural rubber.

Polyphase and dc motors do not have the same inherent double-frequency torque pulsation as single-phase motors, but they are frequently rubber-mounted because the companion single-phase motor is rubber-mounted. Resilient mounting does help, though it is not so necessary as it is for the single-phase motor. A permanent-split capacitor motor tends to have less torque pulsation than a single-phase motor without the capacitor, but the capacitor cannot entirely dispense with the need for resilient mounting.

17-15 Flexible Couplings. Flexible couplings may be used either to allow for misalignment between shafts or to provide torsional flexibility between the motor and its load. A coupling may take care of misalignment and still provide no torsional flexibility, but the latter feature is quite essential in many applications. It has been brought out that single-phase motors have very pronounced double-frequency torque pulsations set up by the electromagnetic forces in the motor. Resilient mounting isolates the effect of this pulsating force from being transmitted through the *stator* to the frame of the driven device. However, it must be remembered that the pulsating torque is exerted on the *rotor* as well as on the stator. Therefore, 60-Hz motors will have a 120-Hz vibration in the rotor as well as in the stator. If the motor is belted to the load, the belt will usually isolate this vibration from the driven load. However, if no torsional flexibility is provided, this 120-Hz vibration can be transmitted to the driven member of the load and is, in some cases, objectionable. One example is a large fan mounted directly on the motor shaft. When, on one application, it was found that a large fan rigidly mounted on the motor shaft emitted an objectionable 120-Hz hum, a special hub was designed to provide torsional flexibility; the construction used for this purpose is shown in Fig. 17-12.

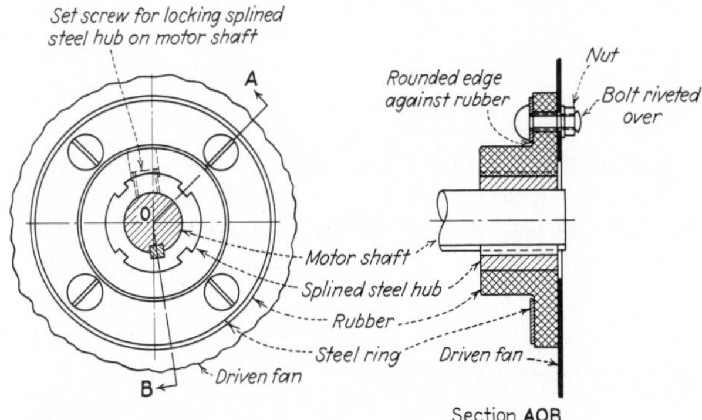

Fig. 17-12 Construction for resiliently mounting a fan on a motor shaft. *(Carrier Corporation.)*

Another interesting application is that of hot-water circulating pumps used for residences. Couplings used to connect the pump impeller to the motor are provided with torsional flexibility in order to prevent the 120-Hz hum from getting into the hot-water system and setting up objectionable noises in the radiators.

No amount of elastic suspension of the stator will prevent transmission of the torque pulsation from the rotor to the rotating member of the driven device; only a coupling device with torsional flexibility will help in this respect.

THERMAL PROTECTORS

Need for some form of thermal protection for fractional horsepower electric motors has long been recognized, and a wide variety of devices to achieve that end have been developed and used. First to be used were current-responsive thermal relays, located remotely from the motor; but, in the early thirties, thermal devices responsive to both current and temperature began to be used. Nowadays most fractional horsepower motors are being supplied with a built-in thermal protector, sensitive to both current and temperature. One important reason for such protectors is the prevention of fires due to motor burnouts, whatever the reason for the burnout conditions. Another important function of the protector is to prevent damage to or burnout of the motor during abnormal conditions that may later be remedied automatically or by human intervention. An example of the first case is to protect the motor from overheating during a temporary period of prolonged undervoltage; an example of the second case would be a motor stalled by a jamming of the driven device which was later cleared by human intervention.

17-16 General: Definitions. Definitions of some commonly used terms follow:

Thermal protector. A protective device for assembly as an integral part of the machine, which, when properly applied, protects the machine against dangerous overheating due to overload and, in a motor, failure to start [National Electrical Manufacturers Association (NEMA) definition]. The words "thermally protected" appearing on the nameplate of a motor indicate that the motor is provided with a thermal protector (NEMA definition).

Inherent-overheating protective device. A term that has been widely used to designate a device, built integral with the motor, which is responsive to motor temperature *and to motor current.* The term has subsequently been so loosely used that it is now neither precise in meaning nor so popular in use as it once was.

Automatic-reset thermal protector. A thermal protector that resets automatically after its parts have cooled sufficiently.

Manual-reset thermal protector. A thermal protector that, once it has tripped, stays open until it is reset manually, usually by means of an external reset button. Normally, the device is "trip-free"; that is, it will trip automatically regardless of the position of the reset button and, further, it cannot be reset until the protector parts have cooled somewhat.

Thermally protected motors and the way they are marked are discussed in Art. 1-4. Many connection diagrams for such motors are discussed in Chaps. 4 and 5. In the following paragraphs we shall discuss the more popular devices currently being used, and we will provide a qualitative discussion of their principles of operation, passing by a whole host of older and now less popular devices, many of which were discussed in *Fractional Horsepower Electric Motors.*[3] We shall start by discussing what the Underwriters Laboratories (UL) expects of such devices.

17-17 Underwriters Laboratories Standards for Thermal Protectors. What is expected of a properly applied thermal protector in a fractional-horsepower electric motor is very well spelled out in UL 547,[8] from which most of the material in this article is extracted.

(1) Scope. The scope of the standard is limited to dc and ac single-phase motors rated 15 hp or less and 600 V or less. It does not include devices that require an external contactor or other component; nor does it include requirements for hazardous locations.

(2) Design and construction requirements. Design and construction requirements provide the following: no operating part other than the reset button shall be exposed, no uninsulated live parts shall be exposed, manual-reset protectors shall be trip-free, it shall be difficult to substitute a protector or thermal element of higher rating, and the protector must be resistant to corrosion. Minimum electrical spacings through air and over surface are also specified.

(3) Protector performance tests. Protector performance tests for a 110- or 115-V motor are taken at 120 V; for a 220- or 230-V motor, they are taken at 240 V. For any other voltage rating, the tests are taken at 100 to 105 percent of rated voltage.

(4) Temperature limits. When the motor is running with the maximum load it can carry without causing the protector to trip, the total temperature shall be not more than 140°C (284°F) for a motor employing Class A insulation, or more than 165°C (329°F) for a motor employing Class B insulation. Locked-rotor temperature tests are taken at full voltage, with the protector cycling. Temperatures are measured by iron-constantan thermocouples of no. 30 American wire gauge (Awg), applied to the winding, and a potentiometer type of instrument. The tests are conducted for a period of 72 h. The maximum temperatures permitted by the standard are given in Table 17-2. If a manual-reset protector is being tested, it is operated for 10 cycles, with the protector being reset as quickly as possible after it has opened the circuit.

(5) Locked-rotor endurance test. An automatic-reset protector shall operate for 15 days with the motor for which it is designed, without

TABLE 17-2 Limiting Temperatures during Locked-Rotor Cycling Tests on Thermally Protected Motors

	Maximum temperature		Maximum average temperature, °C (°F)	
Types of protector	Class A insulation	Class B insulation	Class A insulation	Class B insulation
Automatically reset:				
During first hour of operation	200 (392)	225 (437)		
After first hour of operation	175 (347)	200 (392)	150 (302)	175 (347)
Manually reset:	200 (392)	225 (437)		
Thermal cutoff:				
Opens during first hour of test	200 (392)	225 (437)		
Opens after first hour of test	150 (302)	175 (347)		

Note: Temperature limits have not been established by Underwriters Laboratories for Class F and Class H insulation systems.

Source: From *Thermal Protectors for Motors,* UL 547, Underwriters Laboratories Inc., Northbrook, Ill., 1980.

permanent injury to the motor or excessive deterioration of the insulation. However, it may permanently open the circuit prior to the expiration of that period if it is specifically designed to do so, and if testing of three samples shows that it will do so consistently and reliably without grounding to the motor frame, injury to the motor, or any evidence of fire hazard. A manual-reset protector must open the circuit 50 times without damage to itself or permanent injury to the motor; for this test, the protector is reset as soon after each opening as it can be made to do so.

The preceding give a fair idea of what is expected of a thermal protector, but for the latest complete details of UL requirements, reference should be made to the latest issue of the UL standard.[8]

17-18 Essential Parts of a Thermal Protector. Although many different types of thermal protectors have been developed and used over the years, there is now no doubt that the most popular and most widely used type today is the disk type, produced by Texas Instruments Incorporated, who report that over 400 million of them have been put into fractional horsepower motors. A sectional drawing of one form of automatic- and manual-reset disk-type protector is shown in Fig. 17-13, where the essential parts are labeled:

(1) Bimetal disk. This is the only moving part, a bimetallic disk,

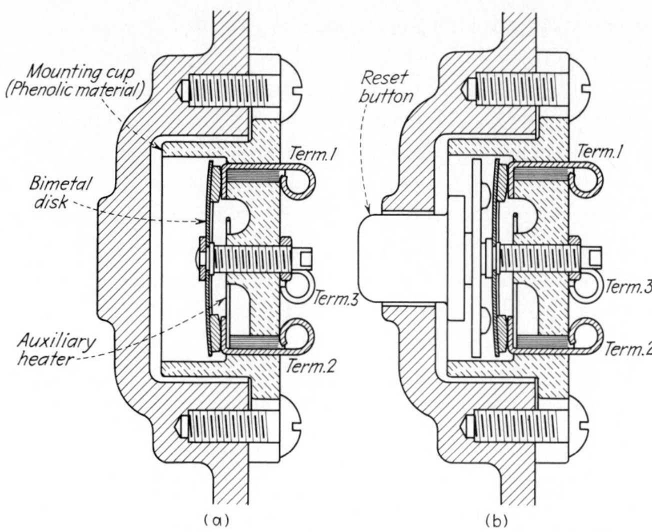

Fig. 17-13 Sectional views of typical disk-type thermal protectors. (*a*) Automatic reset; (*b*) manual reset. *(Texas Instruments Incorporated.)*

invented by J. A. Spencer. The disk has a dished shape, with the high-expansive side of the bimetal on the concave side of the disk at normal temperature. As the disk is heated, the bimetal on the inside of the disk tries to expand but is restrained from doing so by the concave shape of the disk, for the outer side, being of low-expansive metal, does not expand similarly; as the temperature of the disk is increased, stresses are set up in the disk, tending to buckle it; as soon as the temperature has reached a predetermined value, the disk snaps suddenly over to its other position. When the disk cools down to a definite temperature, it snaps back to its original shape. It is to be noted that this disk operates when, and only when, it reaches a certain fixed temperature; depending on application requirements this fixed temperature setting may be varied over a range of 60°C or greater.

(2) Auxiliary heater. This is a small preformed heating element, of an alloy of resistance wire such as nichrome, placed close to the bimetallic disk.

(3) Mounting cup or base. All parts are mounted on a phenolic cup which, in the illustration, is provided with mounting ears for attaching the protector to a recessed surface, such as an end shield of a motor. The bimetal disk is held in place by a single central screw, shown in the figure; the opening temperature of the disk is adjusted by this screw, which is then locked in place and sealed. The mounting cup also positions and supports the terminals, which pass through it as shown and connect to the silver stationary contacts inside the cup. The silver moving terminals are welded to the disk as shown in the drawing.

(4) Reset button. Figure 17-13*b* shows a reset button, made of bright red nylon. It is spring-biased away from the disk. When pressed inward, it pushes a metal plate, carrying two or more coined dimples, against the outside of the disk, tending to reset the latter. Now, the disk of a manual-reset protector is normally designed so that, once it has tripped, it goes so far over center that it will not reset at any ordinary room temperature likely to be encountered. However, by pressing against the outer side of the disk, the resetting temperature can be increased. The motion of the reset button is limited mechanically so that it always stops short of pressing the disk over center and closing it. Actually the travel of the reset button is set so that when it is against its inward limit of travel (as it would be if clamped, for example), the resetting temperature is raised to a value normal to that of an automatic-reset protector; hence this manual-reset protector would cycle as if it were an automatic-reset protector. Thus, the protector shown is a trip-free protector.

17-19 Elementary Principles of Operation of Thermal Protectors. For the purpose of understanding the elementary principles of operation of a thermal protector, let us consider one of the types shown in Fig. 17-13, mounted in the motor end shield. Let us first consider what happens under locked-rotor conditions, and then under running conditions. Finally, let us make some comparisons with overcurrent devices.

(1) Locked-rotor conditions. Under locked-rotor conditions, the windings of a split-phase or capacitor-start motor heat very rapidly, especially those of the former. The auxiliary winding of a split-phase motor may approach dangerous temperatures in 7 s or less, and those of a capacitor-start motor in perhaps 20 to 40 s, depending on the design. During this time, the temperature of the end shield does not change sensibly. In short, the windings can burn out while the end shield is still cool. Hence, if the protector mounted in the end shield responded only to temperature, it would not protect the motor windings from burnout under locked-rotor conditions. For this reason, motor current is passed through the auxiliary heater, and the heat generated therein radiates directly to the disk, causing it to heat up and trip in a time sufficiently short to protect the motor windings. Since the bimetallic disk itself carries current and is of finite resistance, heat is generated directly in the disk itself, accelerating its tripping. Thus, under locked-rotor conditions, the protector is caused to trip by heat generated within the protector itself, not by heat received from the motor.

Now, what happens after the first tripping of the protector? The end shield is cool and starts immediately to absorb heat from the disk and heater of the protector, both of which cool rapidly and cause the disk to reset, energizing the windings and causing current to flow through the protector again. This time, the tripping time is much shorter than for the first trip because it is necessary to heat the disk up from its reclosing to its opening temperature, a matter of 25 to 50°C, rather than from room temperature, as it did the first time. The protector cycles, at first relatively rapidly; then as the end shield heats up and approaches the reclosing temperature of the protector disk, the "off" time increases, markedly decreasing the cycling rate. All these considerations are shown graphically in Fig. 17-14; for example, the protector is seen to cycle six times in the first 15 min, but only three times in the last 15 min.

Fig. 17-14 Winding temperatures in a thermally protected motor using an automatic-reset thermal protector; rotor is blocked, and the protector is cycling automatically. Temperature readings were taken automatically by a Leeds and Northrup Micromax recording instrument. *(Reliance Electric Company.)*

(2) Running loads. The purpose of the protector is to prevent the windings from reaching a dangerous temperature, regardless of how they might arrive at this temperature. If it were possible to make the protector infinitely small and insert it inside the winding at the hottest spot, ideal protection could be obtained by a device that was only temperature-sensitive. A temperature-sensitive protector, attached to the frame or end shield, would not be sufficient, because the windings are hotter than the end shield *by a varying amount which depends mainly upon the amount of current flowing through them.* In any given motor, the temperature rise of the winding above the frame (not above ambient) tends to be proportional to the I^2R losses in them, that is, to the square of the current. If a disk-type thermal protector is mounted in the end shield, the temperature of the disk, when no current is flowing through the protector, will be the same as that of the end shield. When current flows through the protector, the disk becomes heated, and the temperature rise of the disk above the end shield is approximately proportional to the I^2R losses in the protector, that is, to the square of the current. Since the end shield will generally approximate the temperature of the frame, we can sum up the situation of the end-shield-mounted protector in the following terms:

The end-shield temperature is lower than that of the windings being protected by an amount proportional to the square of the current, but the temperature of the disk is higher than that of the end shield by an amount also proportional to the square of the current; hence, by a proper selection of protector characteristics, the temperature of the disk can be made to approximate very closely the temperature of the windings for a wide range of load currents. In such a case, the disk is set to open at the limiting winding temperature desired. These principles are discussed more fully in a paper by Veinott and Schaefer.[9]

Thus it is clear that the auxiliary heater, so necessary for locked-rotor protection, serves a useful purpose in obtaining running protection. In selecting a protector, it is sometimes found that the one that gives ideal running protection does not give adequate locked-rotor protection, and vice versa, so that a compromise has to be made. There is available such a wide variety of protectors of differing characteristics that a protector that gives protection closely approximating the ideal requirements discussed can usually be found.

(3) Thermal protector compared with overcurrent device. An external overcurrent device operates on the theory that the winding temperature depends on only the motor current, so it is set to operate at a certain current, usually of the order of 115 to 125 percent of full-load current. Now, it is not overcurrent that damages the windings, but overtemperature. Many factors other than current can elevate the winding temperature: location of the motor in a high ambient, as sometimes occurs in appliances; partial or complete blocking of the supply of ventilating air; dust or dirt accumulations on the windings themselves, or in air passages; proximity to hot surfaces or bodies; operation on overvoltage, etc. The thermal protector, located inside the motor, operates on motor temperature as well as motor current. It protects against dangerous temperatures at small currents, at which an overcurrent protector would not trip, and permits the motor to carry more overload at low ambient temperatures, which it can safely do without overheating. In short, a properly applied thermal protector affords more positive protection and permits fuller utilization of overload capabilities of the motor than does the remote overcurrent protector.

17-20 Thermal Protectors for Single-Phase Motors. Thermal protectors are available in a very wide assortment of designs and characteristics. Two broad classifications are end-shield-mounted and on-winding protectors.

(1) End-shield mounted protectors. Figure 17-13 shows sectional views of automatic-reset and manual-reset thermal protectors designed for end-shield mounting. Other varieties, sizes, and kinds of automatic-reset and manual-reset protectors are illustrated in Fig. 17-15. The curved-body and rectangular types, when provided with a bracket, and the eared-base type are generally used in motor end shields; the round-base type is often mounted directly on the dome of refrigerator compressors. Curved-body protectors are also suitable for inserting between the winding end turns and the frame. Texas Instruments protectors come in five sizes of disk, ranging from ½ to 1½ in, and some have two heaters.

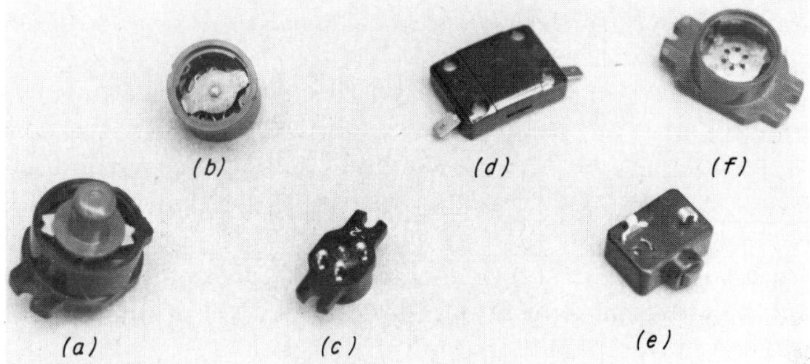

Fig. 17-15 Types and sizes of thermal protectors used in fractional horsepower electric motors. (*a*) Ear type, 1-in disk, manual reset; (*b*) round base, ¾-in disk, automatic reset; (*c*) ear type, ½-in disk, automatic reset. *(Texas Instruments, Incorporated.)* (*d*) Curved body, Model 20M, automatic reset; (*e*) rectangular, Model 30M, manual reset; (*f*) ear type, Model 14M, automatic reset. *(Therm-O-Disc Incorporated.)*

(2) On-winding protectors. As motors have progressively been packed into less and less space and end-shield material has been decreased, as illustrated in Fig. 4-24, temperature-protection problems and the difficulty of mounting the protector have increased. To meet these demands, on-winding protectors with heaters were developed. The general appearance of these devices is illustrated in Fig. 17-16. The cases are live and, in practice, the devices are supplied with a 0.005-in or an 0.008-in Mylar insulating sleeve that extends ³⁄₁₆ in beyond the case; they are normally furnished with leads.

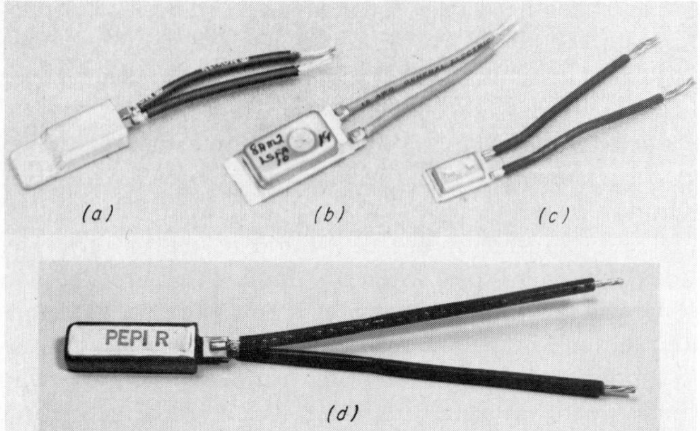

Fig. 17-16 On-winding-type thermal protectors. (*a*) 2AM protector with single heater; (*b*) 8AM protector with single heater; (*c*) 7AM protector with no heater. *(Texas Instruments Incorporated.)* (*d*) Model F with no heater. *(Portage Electric Products, Inc.)*

Smaller versions have been developed for use with shaded-pole and permanent-split capacitor motors. In general, these are two-lead devices with no auxiliary heater (the bimetal blade carries current and has some heater effect in itself), and protective Mylar sleeves are used because the case is live. Examples are shown in Fig. 17-16. A sectional view of such a protector is shown in Fig. 17-17. Texas Instruments, Incorporated, reports that over 500 million of its 7AM protector have been put into use in fractional horsepower motors.

The principles through which they protect the motor windings are essentially the same as those discussed in Art. 17-19, except that less heater effect is needed than for end-shield-mounted protectors because

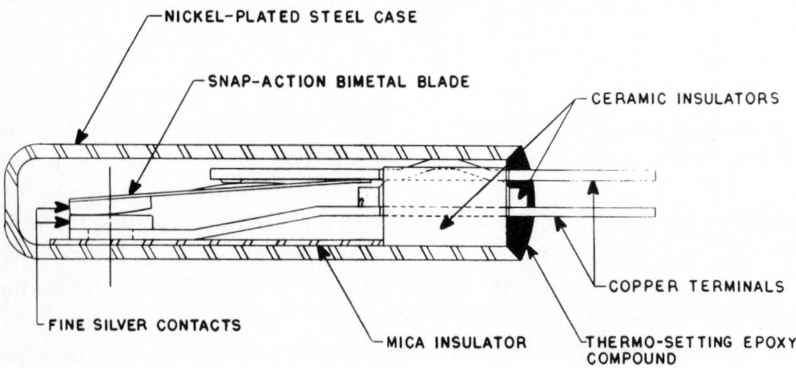

Fig. 17-17 Cross-sectional view of an on-winding thermal protector for fractional horsepower motors. *(Portage Electric Products, Inc.)*

they are located in a closer heat-receiving relationship with the windings they are protecting; it is essential that the protector be in contact with both the main and auxiliary windings. In general, contact ratings are available for almost all sizes and ratings of fractional horsepower motors.

Unfortunately, this construction does not lend itself to manual resetting. Partly for this reason, end-shield-mounted protectors are still used where on-winding protectors might otherwise be used.

(3) Thermal cutoffs. Some subfractional horsepower motors use a thermal cutoff, commonly referred to as a one-shot protector, to protect the windings from overheating. Figure 17-18 shows the cross section of a thermal cutoff; the thermal pellet is electrically nonconductive; under normal operating temperatures, the solid pellet compresses a spring which holds the star contact against the contact lead. When a fault temperature is reached, the pellet melts, the trip spring moves the star contact away from the contact lead, and an open circuit results. Since the case is live, a Mylar sleeve is placed over its length.

A thermal cutoff is sensitive to temperature only, and, unlike other

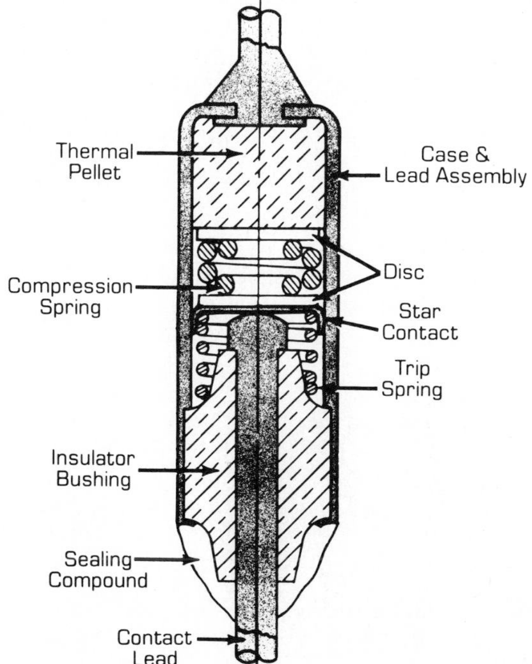

Fig. 17-18 Cross-sectional view of a thermal cutoff for subfractional horsepower motors. *(Micro Devices Product Group, Therm-O-Disc Incorporated.)*

protectors, the opening temperature does not derate with current flow. Maximum winding temperatures, when using a thermal cutoff as specified in UL 547[5], are as given in Table 17-2; in addition, the thermal cutoff must meet the requirements in UL 1020.[10]

Micro Devices Products Group, Therm-o-Disc, markets such a device under the trade name Microtemp; rated opening temperatures range from 60 to 240°C.

17-21 Thermal Protectors for Three-Phase Motors. Thermal protectors for three-phase motors are illustrated in Fig. 17-19. These are similar to the end-shield-mounted protectors discussed earlier, except that they have three heaters, and three or six terminals, depending upon whether they are to be used with single- or dual-voltage motors. They are *neutral-opening protectors;* that is, the disk itself forms the star point. This means that they are suitable only for Y-connected motors, not for delta-connected motors. But, since most fractional horsepower motors can be designed for star connection, this poses no serious limitation. Since the star point is at ground potential in most cases, a neutral-opening protector leaves all three windings "hot" when the protector opens as one phase is left connected to each line. Connections for a dual-voltage arrangement are shown in Fig. 17-20.

GEARMOTORS

Since most machines that are driven by electric motors operate at speeds well below normal motor speeds, some form of speed reducer is usually needed. Gears are extensively used for such purposes. The speed-reducing gears may be built into the driven machine, or the gear-

Fig. 17-19 Thermal protectors for three-phase motors. These are designed to open the neutral or star point of a Y-connected motor. *(Texas Instruments Incorporated.)*

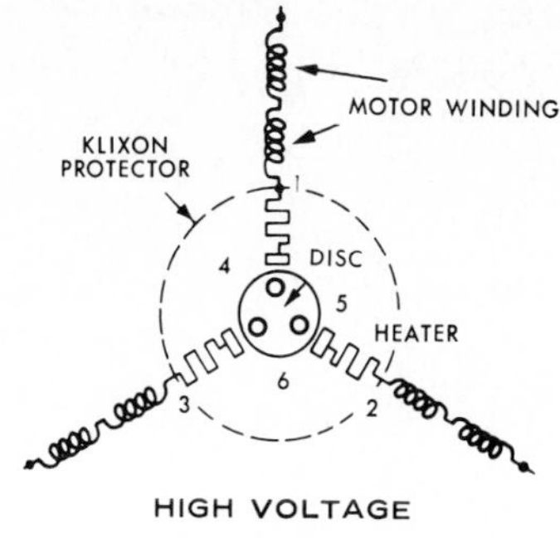

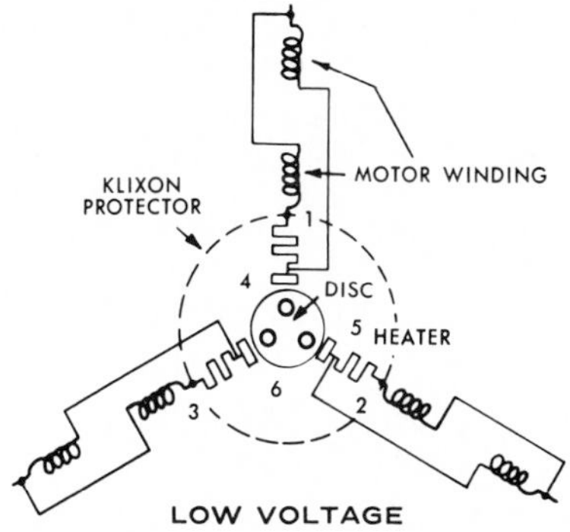

Fig. 17-20 Connections for dual-voltage neutral-opening thermal protectors. *(Texas Instruments Incorporated.)*

reduction unit may be made a part of the motor. Sometimes, the speed reducer is a separate unit. Gearmotors, having the motor and reducer built as an integral unit, are frequently used both for economy and for saving space.

17-22 Fractional Horsepower Gearmotors. There are two basic types of gearmotors commonly used in fractional horsepower sizes: (1) right-angle gearmotors, in which the output shaft is at right angles to the motor shaft and (2) parallel gearmotors, which have an output shaft parallel to the motor shaft but not generally in line with it.

(1) Right-angle gearmotors. These are the most compact and economical construction for speed reductions of 6:1 or more. For speed ratios between 5:1 and 70:1, a single worm stage is employed. Such an arrangement is illustrated in Fig. 17-21, where a portion of the gear case has been removed to show the worm and its mating gear. For speed ratios between 70:1 and 300:1, a two-stage construction is used, combining a worm stage with a parallel stage. A typical arrangement is illustrated in Fig. 17-22, where the housing has been cut away enough to show all four gears of the two stages; it will be noted that the final

Fig. 17-21 A right-angle single-stage gearmotor. Worm is cut on the steel worm shaft, then carburized and ground. The worm shaft has its two bearings and is driven by the motor shaft through a splined coupling. *(Reliance Electric Company.)*

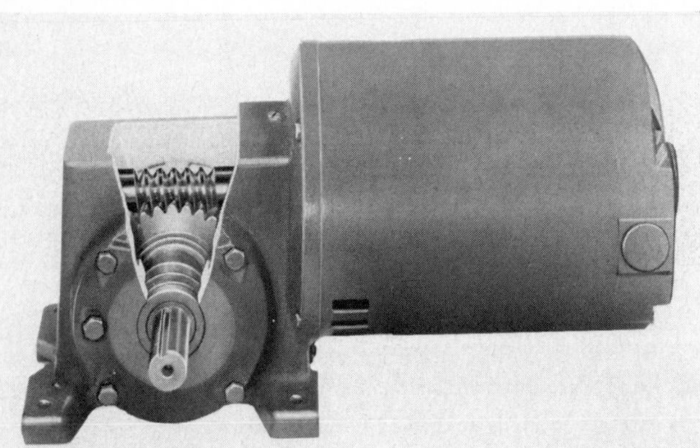

Fig. 17-22 A right-angle two-stage combination gearmotor. First stage is the same as shown in Fig. 17-21. Second stage uses helical gears cut from steel, which are then hardened and honed. *(Reliance Electric Company.)*

output stage uses helical gears. Mechanical efficiency of the gears may be as high as 90 percent for a 5:1 reduction but may be as low as 40 percent for a 70:1 reduction. A similar range of 90 percent efficiency down to 40 percent efficiency would apply to the combination units at ratios progressing upward from 70:1 to 300:1.

Lubrication is very important. The gear case should be flushed out and a fresh supply of the lubricant recommended by the manufacturer added every 750 h of operation.

(2) Parallel gearmotors. These units are more efficient but less compact than right-angle units. They sometimes adapt more readily to the driven machine because the output shaft is in line with, or at least parallel to, the motor shaft. They may be built with one, two, or three stages, commonly known as single parallel, double parallel, or triple parallel. Generally there is a loss of about 3 percent in each stage, so that the efficiencies are, approximately, 97 percent for single parallel, 94 percent for double parallel, or 91 percent for triple parallel. The gears can be spur, helical, or herringbone. Spur gears are generally not used, except in the slowest speed stage of a triple parallel unit, because they are inherently noisier at high speeds. Helical gears are rapidly replacing herringbone gears in the lower horsepower gearmotors because they can be made with greater accuracy at a reasonable cost.

Lubrication is also important for parallel units. In most cases, any good grade of motor oil or turbine oil is satisfactory. Under normal operating conditions, an oil change should be made at 1500-h intervals of operation. With single parallel gearmotors, the mounting feet may be a part of the motor, leaving the gear unit overhanging, or a part of the gear housing, with the motor overhanging. A typical example is illustrated in Fig. 17-23.

17-23 Gear Reducers for Fractional Horsepower Motors. Sometimes convenience is best served by using a separate gear-reducer unit. This

Fig. 17-23 A parallel gearmotor. Gear ratios are from 2.25:1 to 129.7:1. The gear case has its own feet, and the motor is bolted to it. *(Reliance Electric Company.)*

gives the user a wider range of choice in selecting the driving motor. The Reliance Electric Company offers a line of gear reducers for fractional horsepower motors in parallel-shaft and right-angle types, with provision for flange-mounting, C-face mounting, and integral rigid base. Some designs are suitable for vertical input when horizontal space is limited. Gear ratios vary from 2.25:1 to 336:1.

17-24 Subfractional Horsepower Gearmotors. A number of subfractional horsepower gearmotors were discussed and shown in Chaps. 9, 10, and 12. There are, perhaps, three basic types of such motors: (1) right-angle gearmotors, (2) parallel gearmotors, and (3) planetary gearmotors.

(1) Right-angle gearmotors. A right-angle gearmotor with two worm-gear stages is illustrated in Fig. 17-24. Single-reduction gearing may be had for ratios from 2:1 to as high as 60:1; with a 1725-rpm motor, this means output speeds of 900 to 30 rpm. Double-reduction gearing is used for ratios from 36:1 to 2880:1, giving, for 1725-rpm motors, output speeds from 50 to 0.6 rpm. Universal motors operate at higher speeds.

CAUTION: Most worm-gear reducers are irreversible, i.e., the motor cannot be turned from the slow-speed shaft. Care must be taken not to connect high-inertia loads directly to the output shaft, because the

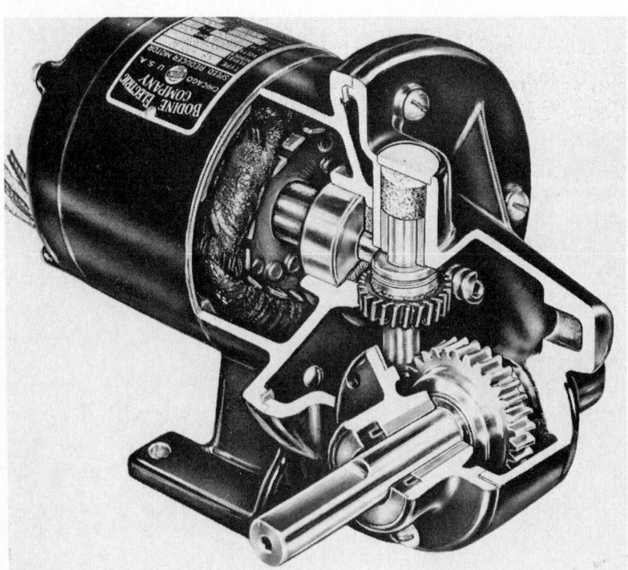

Fig. 17-24 Fractional horsepower motor with double-reduction worm gear. May be used with shunt, universal, induction, or synchronous motors. *(Bodine Electric Company.)*

momentum of the load during stopping or starting may damage the gears. Many of these motors, particularly double-reduction universal motors, can deliver more torque than the gears can handle continuously. Care should, therefore, be exercised to avoid applying these motors to devices that can become jammed—lest the gears be damaged. For such applications a safety clutch or shear pin between the motor and its load is recommended.

(2) Parallel gearmotors. A compact-design four-parallel gearmotor is illustrated in Fig. 17-25. The spur gears are of case-hardened steel; the primary stage uses helical gears of heavy-duty phenolic for quietness. Grease-lubricated ball bearings are used on the motor shaft. Needle bearings are used on the final output shaft to obtain a high overhung-load capacity. Two screw plugs in the gear housing provide means for relubrication and maintenance of a proper lubricant level. The motors are normally shipped without oil in the gear case.

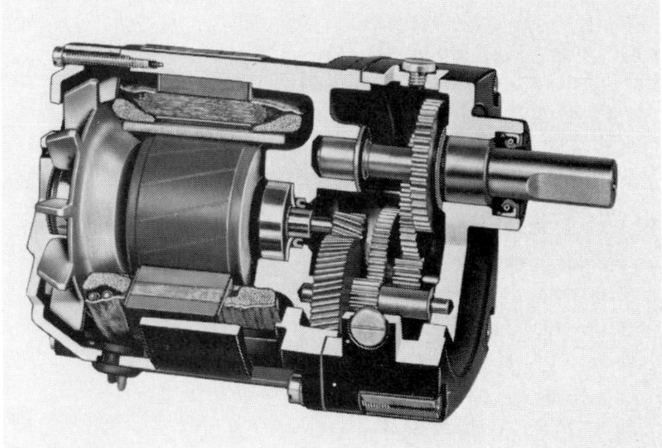

Fig. 17-25 A four-stage parallel gearmotor of compact design. The primary stage is helical and of heavy-duty phenolic for quietness. Other gears are of case-hardened steel. Thirteen standard gear ratios for 60-Hz motors offer output speeds of 300 to 2 rpm. Output torque ratings are from 1.5 to 25 lb-in. *(Bodine Electric Company.)*

(3) Planetary gearmotors. For very high gear ratios, especially in very small motors, planetary gears are often used. One such motor is illustrated in Fig. 12-6; as stated in the caption, this motor is available in ratios up to 21,808:1. Even higher ratios can be obtained.

References

1. Bearings and Lubricants, Mechanical Drives Reference Issue, *Mach. Des.*, Aug. 23, 1984.
2. Stark, J. H.: "Elastic Girdles for Ball Bearings," *Prod. Eng.*, September 1951, pp. 188–193.

3. Veinott, C. G.: *Fractional Horsepower Electric Motors,* 2d ed., McGraw-Hill Book Company, New York, 1948.
4. ———: *Fractional- and Subfractional-horsepower Electric Motors,* McGraw-Hill Book Company, 3d ed. New York, 1970.
5. Kimball, A. L., and P. L. Alger: "Single-Phase Motor Torque Pulsations," *AIEE Trans.* June 1924, p. 730.
6. Veinott, C. G.: *Theory and Design of Small Induction Motors,* McGraw-Hill Book Company, New York, 1959. (Out of print, but available from University Microfilms, P.O. Box 1647, Ann Arbor, Mich. 48601.)
7. Krause, P. C.: "Simulation of Unsymmetrical 2-Phase Induction Machines," *IEEE Trans. Power Appar. Syst.*, vol. 84, 1965, pp. 1025–1037.
8. *Thermal Protectors for Motors,* UL 547, Underwriters Laboratories Inc., 333 Pfingsten Rd., Northbrook, IL 60062, 1980.
9. Veinott, C. G., and L. L. Schaefer: "Fundamental Theory of Inherent Overheating Protection under Running Overload Conditions," *AIEE Trans.*, vol. 68, 1949, p. 266.
10. *Thermal Cutoffs for Use in Electrical Appliances and Components,* UL 1020, Underwriters Laboratories, Inc., 333 Pfingsten Rd., Northbrook IL, 60062, 1983.

Bibliography

Agnoff, Charles: "Multipurpose Bearings," *Mach. Des.*, Aug. 31, 1967.

Bremer, E. B., and C. G. Veinott: "Refrigerator Motors Need Thermal Protection," *Electr. Refrig. News,* Oct. 11, 1933.

Danowski, Gene: "Selecting Speed Reducers," *Mach. Des.* Aug. 23, 1984.

National Electrical Manufacturers Association: *Renewal Parts for Motors and Generators (Performance, Selection, and Maintenance),* Standards Publication no. RP1-1981, 2101 L Street N.W., Washington, DC 20037, 1981.

Pesek, S. J., and W. A. Weinkamer: "Strip-Type Bearings," *Mach. Des.*, March 10, 1965, pp. 35–39.

Potter, C. P.: "The Inherent Overheating Protection of Single-Phase Motors," *AIEE Trans.*, vol. 60, 1941, pp. 993–996.

CHAPTER

18

TEST EQUIPMENT AND PROCEDURES

Tests, properly taken and properly evaluated, may be used as a quality-control check on a motor to verify whether the motor was correctly made, whether in a factory or in a service shop. The design engineer runs a number of tests to make a thorough evaluation of the design. The application engineer runs tests to help the engineer decide what motor to use for any particular application. Space does not permit us to treat this subject comprehensively; it is covered in considerable detail in Institute of Electrical and Electronics Engineers (IEEE) publications. However, we shall treat the subject briefly and refer the reader to more detailed publications.

METERS AND EQUIPMENT

18-1 Power Source. For accurate and reliable testing it is absolutely essential to use a good source of adjustable-voltage power that holds reasonably steady during the tests.

(1) For ac motors. An adjustable autotransformer, connected to a source of voltage of good wave form, is recommended. Figure 18-1

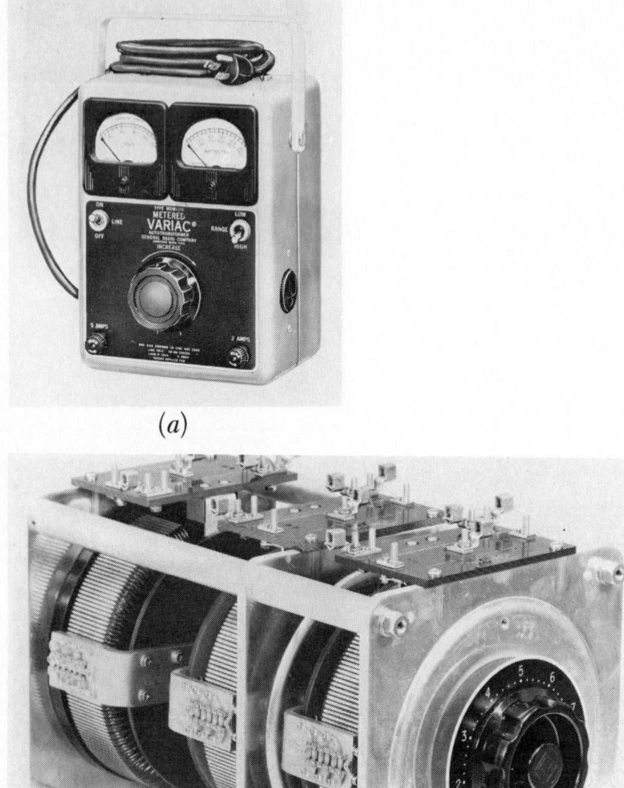

(*a*)

(*b*)

Fig. 18-1 Adjustable autotransformers for controlling ac voltage. (*a*) A Variac autotransformer used for single-phase applications that consists of a variable autotransformer and two meters built into a portable carrying case; (*b*) a ganged assembly of three Variac autotransformers used for three-phase circuits; (*c*) type 1581-A Variac automatic voltage regulator. (*Technipower.*)

illustrates typical autotransformers used for such purposes; they are marketed by the Technipower of Danbury, Connecticut, under their registered trade name Variac. The Variac autotransformer consists of a single-layer winding on a toroidal silicon-steel core. As the control knob is rotated, a graphitic brush traverses the winding, tapping a portion of

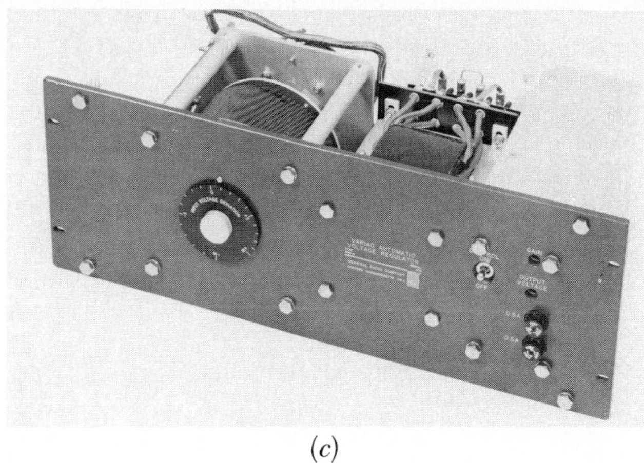

(*c*)

Fig. 18-1 (*Continued*)

the total voltage across the winding. The volts per turn range from 0.3 up to 1.0, depending on the size of the unit. Uncased models are offered for panel mounting, and cased models with carrying handle, cord, and built-in overload protector are available for portable use. Portable models are also available with one, two, or three meters built integrally. For polyphase circuits, two or three units are ganged and operated by the same control knob, as shown in Fig. 18-1*b*. An automatic voltage regulator is shown in Fig. 18-1*c;* a solid-state control circuit and servomotor automatically position the autotransformer brush to hold the output voltage constant. Inverters, which generally have a poor wave form, are not recommended for general testing of ac motors.

(2) For dc motors. For conventional dc motors, use of a motor-generator set with field control is recommended; batteries, of course, may also be used. Use of rectified alternating current is not recommended unless the motor being tested has been designed to be used with such a source. (See Arts. 13-2 to 13-7.)

18-2 Measurement of Voltage. For ac circuits, use of commercial, low-loss, dynamometer-type instruments is suggested. For 115- and 230-V circuits, scales needed are 0 to 150 and 0 to 300, respectively. Such meters are usually readable down to about half voltage. Generally they are calibrated for 60 Hz; if readings on 400 Hz are to be taken, specially calibrated meters should be used.

For dc circuits, dc voltmeters of the d'Arsonval moving-coil type are recommended. For measuring form factor, the voltage needs to be measured with both ac and dc meters (Art. 13-4). But do not use a dc voltmeter on an ac circuit!

For more details on various kinds of electrical measurements, see the relevant IEEE master test code.[1,*]

18-3 Measurement of Current

(1) Ac motors. For measuring 60-Hz currents commercial ammeters, preferably of the dynamometer type, are usually suitable; for 400-Hz circuits, specially calibrated meters are usually needed. Maximum scale values generally needed for tests on fractional horsepower motors are the following: for full-load currents, 15, and for locked rotor currents, 65. For testing subfractional horsepower motors, much lower scales are needed.

Multiscale meters are available and are recommended when a wide range of horsepower ratings are to be tested. Often, a single-scale meter and current transformer are used to give a wide range of currents that can be measured.

(2) Dc motors. Dc ammeters usually use a d'Arsonval moving-coil element; often they use external shunts so that a wide range of currents can be measured by the same instrument. Such an instrument reads average current. When the current is supplied from a rectified ac source, the current may have considerable ripple; in such a case, the current should usually be measured with both a dc ammeter that reads average current and an ac ammeter that reads root-mean-square (rms) current: the ratio of these two readings is the *form factor* of the current.

18-4 Measurement of Power Input (Watts). Low-loss meters are recommended; even so, corrections should be made for these losses, as pointed out in Art. 18-11. Potential and current scales should correspond to the scales of the voltmeter and ammeter being used in the test. For checking the power factor of electrolytic capacitors (Art. 8-4), low-power-factor meters are essential.

Polyphase wattmeters are essentially two single-phase instruments with both moving elements mounted on the same shaft. They are not necessary for measuring polyphase power, which can always be measured with two single-phase wattmeters; some find them more convenient, however. For dc power measurements, wattmeters are unnecessary.

18-5 Measurement of Power Factor. It is recommended that the power factor be computed from the readings of volts, amperes, and watts, rather than from a direct reading on a power-factor meter. In a single-phase circuit,

*For numbered references, see References at end of this chapter.

$$\text{Power factor} = \frac{\text{watts}}{\text{volts} \times \text{amperes}} \tag{18-1}$$

In a two-phase circuit,

$$\text{Power factor} = \frac{\text{total watts}}{2 \times \text{volts} \times \text{amperes}} \tag{18-2}$$

In a three-phase circuit,

$$\text{Power factor} = \frac{\text{total watts}}{\sqrt{3} \times \text{line-to-line volts} \times \text{line amperes}} \tag{18-3}$$

18-6 Measurement of Frequency. For usual tests, it is not necessary to measure the frequency of a commercial 60-Hz circuit, because the frequency is usually held very closely. A vibrating-reed instrument can give an approximate indication of frequency. More accurate results can be obtained with an indicating instrument with the normal-frequency point in the middle of the scale and calibrated to hundredths of a hertz. An alternative method is to measure the speed of a small synchronous motor, operated on the same circuit.

18-7 Measurement of Speed. Close and accurate measurement of speed is usually essential to good test results. Several instruments and methods are available.

(1) Revolution counter and stopwatch. A revolution counter is used to count the revolutions in a period of time, measured by a stopwatch.

(2) Speed indicator (integrating tachometer). This instrument combines a revolution counter and stopwatch in a single case. Pressing the button automatically starts the counter and the stopwatch, which automatically stops the counter at the end of a predetermined interval of time, of the order of 3 to 5 s. Such an instrument has been called an integrating tachometer.

(3) Centrifugal tachometer (horn). As the name implies, this instrument has a centrifugal mechanism and gives an instantaneous reading of speed.

(4) Eddy-current-drag tachometer (Smith). This instrument indicates speed by measuring the force exerted on a drag cup by a rotating permanent magnet.

(5) Stroboscopic tachometer. By this means, a rotating pattern is illuminated by periodic flashes of light of very short duration (of the order of 10 to 50 μs). The frequency of the flashes is adjusted until the pattern appears to stand still; rpm is read directly from a scale on the instrument. This method is particularly valuable for measuring speed of subfractional horsepower motors, many of which would be overloaded by a conventional tachometer.

(6) Photoelectric tachometer. This instrument emits an infrared beam that senses the rotation of a marker-taped surface; each rotation of the marker through the beam is counted, then indicated on a liquid crystal display. Figure 18-2 shows a digital photoelectric hand tachometer made by Biddle Instruments, Blue Bell, Pennsylvania. The system counts for 0.6 s, then displays the rpm at 0.6-s intervals; when the button is released, the displayed number remains for about 3 min. Their catalog no. 359982 has a range of 100 to 200,000 rpm; for speeds below 100 rpm, four equally spaced marker tapes are placed on the rotating surface and the reading is divided by 4.

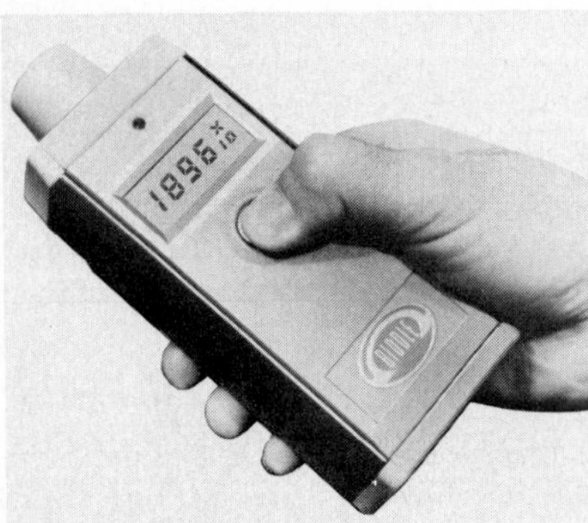

Fig. 18-2 A digital photoelectric hand tachometer showing operating pushbutton, target LED, and liquid crystal display indicating 18,960 rpm. *(Biddle Instruments.)*

(7) General. Tachometers should be checked frequently; a small synchronous motor is most convenient for this purpose. Stroboscopic tachometers often have a vibrating reed for calibration purposes; the timing of the flashes is adjusted to make the reed appear to stand still. Rubber tachometer tips must be kept free of oil and grease so that they will not slip; dipping them into powdered chalk occasionally is helpful.

18-8 Measurement of Slip. Slip of an induction motor is the difference between the actual speed and the synchronous speed. When the slip is less than 100 rpm, it is usually more accurate to measure it directly than it is to measure the speed and calculate the slip. A convenient way to do this is to use a stroboscopic tachometer, adjusted to flash at line frequency, to view a pattern on the shaft of the motor; the apparent backward rpm of the pattern is counted, giving the slip revolutions directly. The flashing light source can be either a neon bulb or an arc lamp. Slips in excess of 100 rpm become difficult to count; they can be determined by subtracting the actual measured speed from the synchronous speed.

18-9 Measurement of Resistance

(1) Resistance of windings. For measurements of resistance of windings which are 1 Ω or more, a Wheatstone bridge is recommended. Kelvin double bridges are recommended if the resistance is less than 1 Ω. A Wheatstone bridge has two leads, and the resistance of the bridge leads must be subtracted from the reading. The Kelvin double bridge has two current leads and two potential leads; the latter are applied directly to the winding being measured, and the resistance of the bridge leads is not a factor. Multitesters, which have a liquid crystal display, offer a convenient means for quickly measuring the approximate resistance of a winding, particularly that of an armature, but bridge methods are considered more accurate and reliable. Further information on the techniques of resistance measurement is given in an IEEE test code.[2]

Regardless of how winding resistance is measured, the room temperature should be noted and the resistance corrected to 25°C. [See Art. 18-16 (1).]

(2) Insulation resistance. Insulation resistance readings are generally of less interest to users of small machines than to users of very large machines. Insulation resistance can be measured by means of a megger, or by other means, described by the IEEE.[3]

18-10 Measurement of Temperature. The temperature of fractional horsepower motors is measured (1) by thermocouple, (2) by rise of resistance, and (3) by thermometer. The former two are the most popular and convenient; the thermometer method has all but been eliminated. Thermocouples can be cemented or attached to the windings by modeling clay or other means. The temperature of the thermocouples can be indicated directly by a potentiometer calibrated for the kind of thermocouple used.

Termperature rise by resistance is taken by measuring the cold resistance of the winding, and then the hot resistance of the winding, measured *immediately* at the conclusion of the heat run, before the windings have time to cool sensibly. The temperature rise of copper windings is computed by

$$T = \left(\frac{R_h}{r_t} - 1\right)(235 + t) \qquad (18\text{-}4)$$

where T = total temperature rise in °C above temperature t
R_h = hot resistance
r_t = cold resistance at temperature t
t = temperature of winding in °C when r_t was measured

If the winding is aluminum, substitute 225 for 235 in Eq. (18-4).

More information on temperature measurements in rotating

machines can be found in an IEEE master test code,[4] supplemented by additional information in the test codes for specific kinds of machines.

Seely has developed an ingenious method which makes it possible to measure the resistance of an ac winding while the motor is running, thus permitting measurement of the winding temperature by the resistance method, without shutting down the motor.[5] He does this by means of a potential transformer that keeps the ac line voltage off the bridge and by a capacitor that keeps the bridge from measuring the resistance of the ac power supply.

For measurement of temperature under rapidly rising conditions, such as with a locked rotor, the thermocouple and potentiometer method is recommended; either the thermocouple or resistance method may be used for slowly rising temperatures.

18-11 Meter Loss Corrections. A wattmeter can be connected in either of the two ways shown in Fig. 18-3. With either connection, part of the internal losses in the wattmeter will be read by the wattmeter:

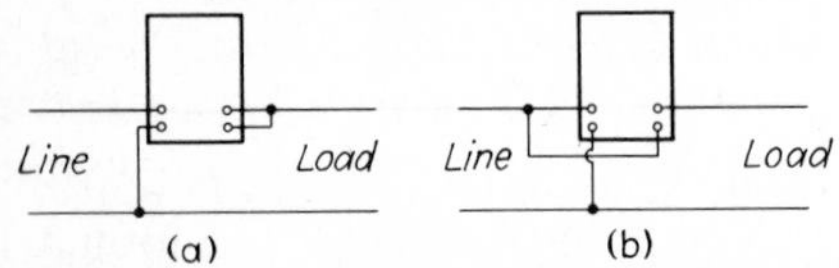

Fig. 18-3 Two ways of connecting a wattmeter in a single-phase circuit. (*a*) Wattmeter reading includes losses in potential coil; (*b*) wattmeter reading includes losses in current coil.

in Fig. 18-3*a*, the loss in the potential coil is included; in Fig. 18-3*b*, the loss in the current coil is included. In neither case does the wattmeter give the true reading. In general, it is recommended that the connections of Fig. 18-3*a* be used and that the loss in the potential coil be subtracted from the wattmeter reading to obtain the power absorbed by the load.

The loss in the potential coil is given by

$$\text{Loss} = \frac{E^2}{R_p} \tag{18-5}$$

where E = actual voltage being measured

R_p = total resistance in the potential-coil circuit, including that of an external multiplier, if used

It is possible to read the loss on the wattmeter itself by exciting line connections shown in Fig. 18-3*a*, leaving open the load connections.

Some wattmeters are provided with internal compensation for the loss in the potential coil; such an instrument must be connected as in Fig. 18-3*a*, with no correction made for potential-coil losses.

18-12 Meter Connections for Single-Phase Circuits. Recommended connections for testing single-phase motors are shown in Fig. 18-4. With this connection, it is necessary to subtract the loss in the potential coil of the wattmeter (unless the wattmeter has internal compensation therefor), and also the loss in the voltmeter, from the actual wattmeter reading in order to obtain true watts. The potential-circuit loss can be read directly on the wattmeter by leaving the load and ammeter short-circuiting switches open and closing the line and potential switches. The potential switch shown should be opened when currents are being read. The ammeter short-circuiting switch is to protect the current coils of the ammeter and wattmeter; it should be left closed at all times except when reading amperes or watts.

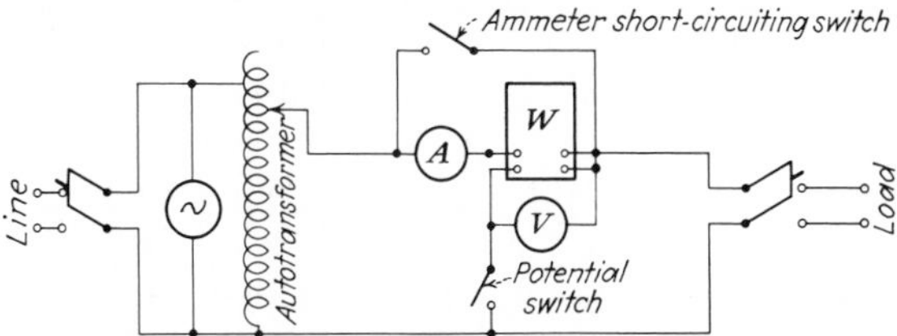

Fig. 18-4 Instrument connections for metering single-phase power.

18-13 Meter Connections for Polyphase Circuits

(1) Three-phase circuits. Recommended connections for testing three-phase motors are shown in Fig. 18-5. In this figure W_1 and W_2 may be two single-phase wattmeters, or the two parts of a polyphase wattmeter. Although the figure shows two wattmeters, three voltmeters, and three ammeters, it is more customary to use only a single instrument of

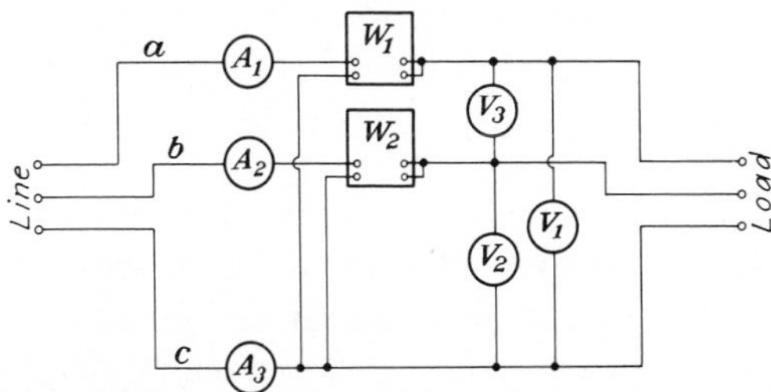

Fig. 18-5 Instrument connections for metering three-phase circuits.

each kind, switching them around by special switches arranged for this purpose. When testing motors, these two wattmeters will never give the same reading; when the power factor of the motor is less than 50 percent, the smaller wattmeter will read negatively. The problem, in any given set of readings, is to determine the sign of the smaller wattmeter reading.

(2) Determination of the sign of the smaller wattmeter reading. Provision has to be made to reverse the connections to the potential coil of the wattmeter that reads the lesser. The problem is to know which position of this reversing switch indicates a positive sign and which indicates a negative sign. There are several ways to determine this. Run the motor at no load and open line *a* momentarily; in this case, the motor will be running on single-phase power, and so this wattmeter will have to read positively. The position of the reversing switch that makes the wattmeter read positively should be marked as the positive position; the other, of course, is the negative position. To check this, run the motor at no load on all three phases; W_2 should be negative, as indicated by the negative position of the switch. As the motor load is increased, the reading of W_2 should decrease and, finally, reverse, so that the reversing switch has to be thrown over to the positive position. When the sign of the smaller reading is negative, it is subtracted from the larger reading; when it is positive, it is added. If a permanent or semipermanent test setup is made, the two positions of the potential-coil reversing switch should have the (+) and (−) positions marked.

If the test setup has been torn down and there is reason to doubt whether the sign of the smaller reading is correct, a simple test is afforded by the following relations:

If $W_1 < 0.866\ EI$, then W_2 is −.
If $W_1 > 0.866\ EI$, then W_2 is +.

(3) Meter losses in three-phase circuits. The best way to make the meter-loss corrections is to total the wattmeter readings algebraically and then to subtract from this total the sum of the losses in the potential circuits of both wattmeters. This is probably safer than correcting each wattmeter reading individually, for when the sign of the smaller wattmeter reading is negative, the potential loss is numerically added to the reading.

18-14 Measurement of Mechanical Output. There are a number of ways to measure the mechanical output of a fractional horsepower motor, some of which are described next. The first three are discussed in much greater detail in an IEEE test procedure.[6]

(1) Prony brake. For many test purposes, the prony brake offers a convenient and economical way of loading a motor. A typical brake is shown in Fig. 18-6. The diameter of the pulley wheel is not important, but the *length of the arm is.* Accurate scales preferably of the gravity-balance type, should be used. When determining the tare, the thumb nuts should be loosened and the arm must be loose on the pulley. When load readings are being taken, there must be no hand pressure on the wing nuts.

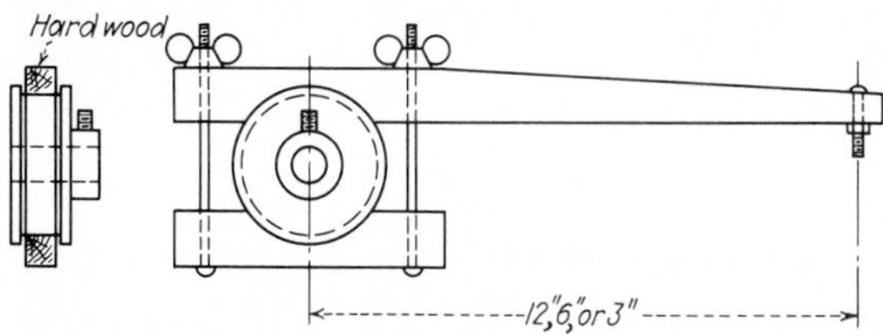

Fig. 18-6 A prony brake for testing small motors.

(2) Dynamometer. A dynamometer is illustrated in Fig. 18-7. It is a dc machine, capable of both armature and field control. By connecting load resistors across the armature, the unit can absorb power whether

Fig. 18-7 A dynamometer for testing fractional horsepower electric motors. The dynamometer unit (left) is mounted in cradle bearings, and a torque arm, attached to the back side, is hooked to the scales, which therefore measure the torque developed by the dynamometer. The motor being tested (right) is mounted on a platform, the height of which is adjustable to facilitate shaft alignment; the test motor drives the dynamometer through the flexible coupling shown. *(Reliance Electric Company.)*

separately or self-excited, though it is generally more stable when it is separately excited. If paralleled with a dc bus, it can pump power back into the dc bus (an especially desirable feature when taking heat runs), and by proper adjustment of the field current can be made to develop either motoring or generating torque, a necessary feature when speed-torque curves are taken. Dynamometer windage loss does not show up in the developed torque and so, for accuracy, a correction for it needs to be made.

A hysteresis-type dynamometer with the cover removed is shown in Fig. 18-8; it is made by Magtrol, Incorporated, Buffalo, New York. Also shown is Magtrol's standard solid-state power supply for providing from 0 to 28 dc volts to the dynamometer and a three-way digital readout which displays speed, torque, and power in either watts or horsepower. A display hold button holds the entire display indefinitely. Dynamometer torque ranges of from 2.5 oz-in to 200 lb-in are available. Other readouts, including polyphase, are also available. Magtrol's model 4619 controller provides xy recording capability.

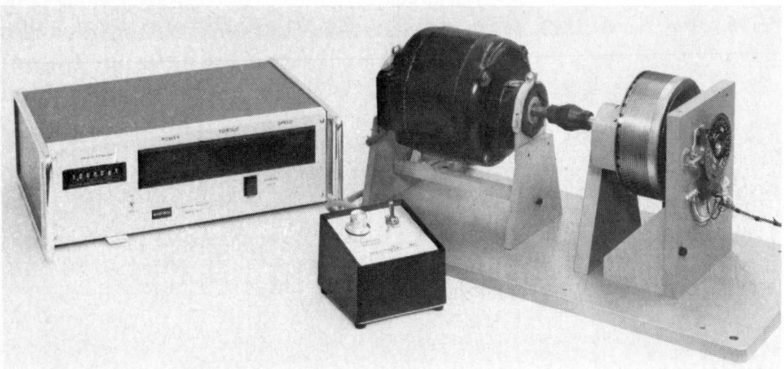

Fig. 18-8 A hysteresis-type dynamometer, a solid-state power supply for providing dc power to the dynamometer, and a digital readout which displays speed, torque, and watts. *(Magtrol, Incorporated.)*

(3) Rope and pulley. This method is particularly useful in testing subfractional horsepower motors, but it is often used for testing larger motors. A small cord, suspended from a spring scale, is wrapped around a smooth pulley on the motor shaft; it is wrapped around the pulley enough times that, when the cord is tightened by a virtually negligible pull on the free end, the scale will measure the motor's pull. A distinct curvature in the free end of the cord as it leaves the pulley is the only conclusive evidence that there is no force on the free end. The pulley face should be wide enough that all the turns required can be in a single layer, but if this is not so, at least the first two turns nearest the scale end of the cords should be in the first layer. In calculating torque, the radius is taken as the radius of the pulley plus the radius of the cord under tension. The cord should pull directly against the scale, not obliquely.

(4) Torque Table. Torque Table is a registered trademark of Lebow Products of Troy, Michigan. One of their devices is illustrated in Fig. 18-9; it is a reaction-torque-measuring structure which does not require cradle or trunnion bearings as do dynamometers. The unit consists of a plate supported from a fixed mounting base by means of four torque-sensitive flexure straps. Foil-type strain gauges are bonded to these flexures, one to each, and connected electrically to a four-arm bridge circuit, from which an electric signal, proportional to torque, is developed. The Torque Table is calibrated by hanging known weights on the calibrating arm shown in the figure; this arm is removable. When the test motor is on the Torque Table, it is mounted so that the center of the motor shaft passes through the points in space formed by the intersections of lines drawn through the four flexure straps—a necessary feature because the cradle pivots about this axis.

The motor is loaded by means of an eddy-current absorption brake. Since the Torque Table itself does not rotate, motors of any speed can be tested on it. The torque indicated by the Torque Table is the torque developed by the magnetic field, less the friction torque of the motor bearings. At a steady speed, this torque is equal to that of the shaft;

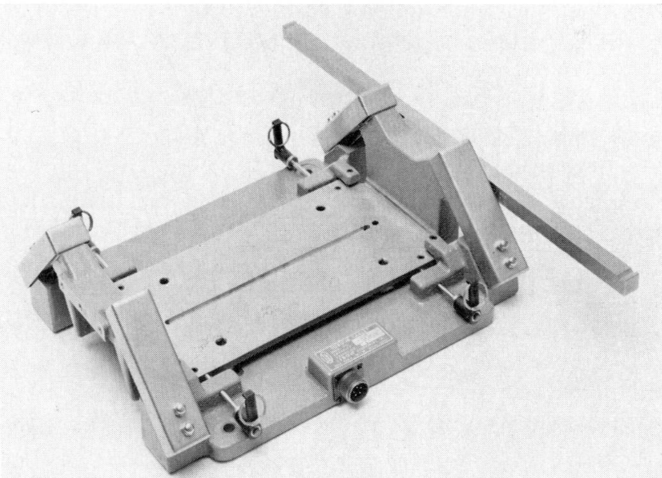

Fig. 18-9 A Torque Table, which is a device for measuring the torque reaction developed by the stator of a motor undergoing test. The unit accommodates NEMA frames 56 and 66 and has a capacity of 50 in-lb. Other similar units have capacities from 5 to 2000 in-lb. *(Lebow Products.)*

during acceleration, the Torque Table measures the torque output of the shaft plus the torque consumed in accelerating the internal inertia of the rotor.

TESTS TO TAKE: SINGLE-PHASE INDUCTION MOTORS

Small electric motors are normally given some sort of routine test before they leave the factory, and service shops should likewise test rebuilt motors. A complete engineering test is less often required, but both types of tests are described next, as well as in the IEEE test procedure.[6]

18-15 Routine Test. A routine test may consist of the following:

(1) No-load readings. After the motor has run at no load long enough for the bearings to warm up, readings of watts and amperes at rated voltage are taken. While the motor is running, it should be observed for unusual noises and for end play.

(2) Full-load readings (optional). Holding the motor output at rated value and applying rated voltage and frequency, watts, amperes, and rpm are measured.

(3) Locked-rotor readings. With rotor locked and full voltage applied, the input watts and amperes and, when possible, torque are measured. These readings should all be taken in not over 4 or 5 s, for the motor heats rapidly.

18-16 A Complete Engineering Test. Those who wish to take a complete engineering test should consult the IEEE Test Procedure for Single-phase Induction Motors.[6] The more common tests taken, with a few of the necessary precautions, are given next.

(1) Cold resistances. Resistances of both main and auxiliary windings should be measured at room temperature and corrected to a standard temperature of 25°C, which can be done by using the following equation:

$$R_{25^\circ\mathrm{C}} = \frac{260}{235 + t} r_t \tag{18-6}$$

where $R_{25^\circ\mathrm{C}}$ = resistance at 25°C
r_t = resistance at temperature t

The resistance of the auxiliary winding should not include the switch resistance. For windings of aluminum wire, substitute 225 for 235 in Eq. (18-6).

(2) Locked-rotor readings. Make the connections as shown in Fig. 18-4, impress approximate rated voltage on the main winding, and read, within 3 or 4 s, watts, amperes, and volts; measure the resistance of the winding immediately afterward. (Do not take time to adjust the voltage if it is within 5 percent of rated value.) Repeat these readings with the auxiliary winding, with capacitor, if any, in the circuit; also measure the voltages across the auxiliary winding and across the capacitor.

(3) Breakdown torques. Breakdown torque is the maximum torque the motor can carry without an abrupt drop in speed. (See Fig. 18-10.) It is taken by increasing the torque load on the motor until the motor "breaks down," that is, the speed suddenly falls off. This test needs to be taken as rapidly as possible so that the motor will not overheat during the process but not so rapidly as to introduce inertia errors into the readings.

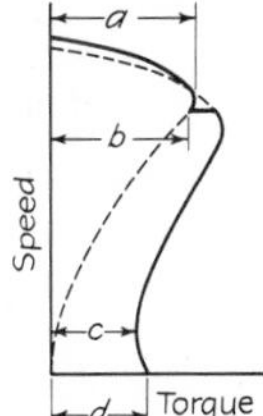

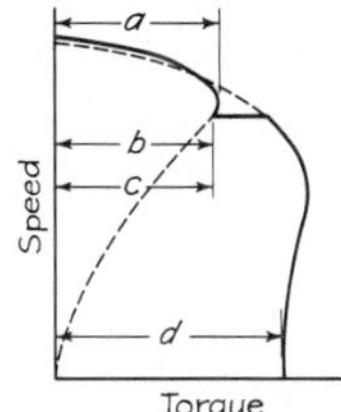

Fig. 18-10 Typical speed-torque curves of single-phase motors illustrating definitions of the various kinds of torque. *a*, Breakdown torque; *b*, switching torque; *c*, pull-up torque; *d*, locked-rotor torque.

(4) Switching and pull-up torques. The nature of switching and pull-up torques is illustrated in Fig. 18-10. Switching torque of a capacitor-start motor can be measured with the help of a prony brake. Start with a light load and increase it by tightening the wing nuts until the motor falls off in speed abruptly and drops down onto the starting connection; since the torque on the starting connection is generally greater than the breakdown torque, the motor will accelerate and transfer back to the running connection, cycling back and forth from one connection to the other. Now, decrease the torque gradually until the motor ceases cycling; that torque will be the switching torque.

As illustrated in Fig. 18-10, pull-up torque may occur at any speed, from very low to switching speed or even higher in some designs. Usually the rope and pulley is the preferred method for measuring pull-up torque, starting at standstill and allowing the motor to accelerate slowly (without pulling on the rope) until the motor is running above breakdown. The minimum value read on the scale is an indication of pull-up.

(5) Locked-rotor torque and current. Locked-rotor torque varies with rotor position, and so the actual torque developed has to be measured in a number of positions to take this effect into account. One way to do this is to use a brake arm, and turn the motor over onto the round part. Then, roll the motor slowly away from the scales, watching the latter for its minimum reading. The test must be performed rapidly, because the motor will heat too much otherwise. Line current can be observed during this test.

(6) No-load saturation: friction and windage losses. The purpose of this test is to be able to segregate losses; it permits the separate determination of the friction and windage, and also the core losses. With the motor running at no load, take a series of readings of volts, amperes, and watts, starting at 130 percent of rated voltage and varying the voltage downward until the motor loses speed rapidly or until the current starts to increase. If it is a capacitor motor, only the main winding should be energized for this test. Amperes and net watts should be plotted against voltage and the net watts curve extrapolated to zero volts to give the value of friction and windage. (See Art. 18-24.)

An alternative method of measuring friction and windage is to measure, by means of a stroboscopic tachometer and stop watch, the rate at which the speed decreases when power is suddenly removed. The weight of the rotor and its radius of gyration must be known. Then friction and windage are

$$F \text{ and } W = Kwk^2 S \frac{dS}{dt} \tag{18-7}$$

where S = speed at which deceleration rate was measured
dS/dt = rate of deceleration, rpm/s
k = radius of gyration, ft or in
w = weight of rotor, lb
K = $4{,}622 \times 10^{-7}$ if k is in feet and 32.1×10^{-7} if k is in inches

(7) Load-performance test. This test is taken to obtain performance at various loads, giving a typical set of performance curves as illustrated in Fig. 18-11. Starting at a heavy load and holding rated voltage, a set of readings of current, power, and rpm is taken; the load is reduced by steps and the readings repeated. Horsepower, efficiency, and power factor are calculated from these readings and plotted. If a capacitor-run motor is being tested, connect it up as in Fig. 18-5, except that both *a* and *b* line leads should be connected to the same side of the line. The *a* and *b* load leads should be connected to the main and auxiliary windings, respectively, and *c* to the common motor lead. In this way readings of main and auxiliary phases, independently of one another, are

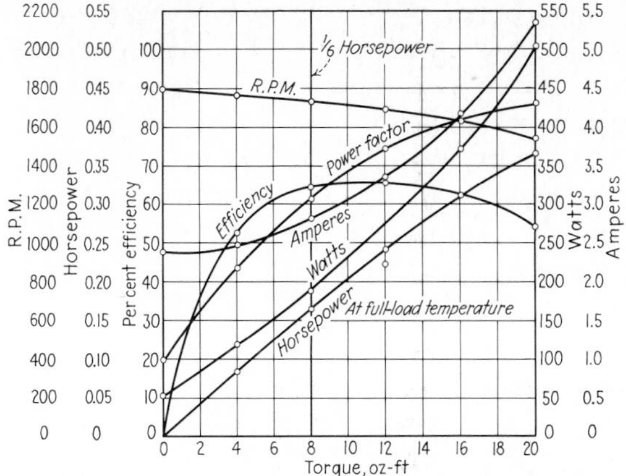

Fig. 18-11 A load-performance test on a single-phase induction motor rated at ⅙ hp, 115 V, 60 Hz, and 1725 rpm.

obtained. Capacitor voltage, and also auxiliary winding voltage, at each load point should be measured and recorded.

(8) Full-load saturation. The purpose of this test is to determine how the full-load performance of the motor is affected by changes in line voltage. Holding torque constant at rated value, take readings of input and rpm at seven different impressed voltages in steps of 10 percent; three voltages should be above, three below, and one at rated voltage. Readings are plotted against voltage.

(9) Temperature test. Temperature tests are taken to determine the temperature rise of the windings above ambient when the motor is carrying rated output. During these tests, the motor must be shielded from drafts, especially if the motor is enclosed. The motor should be mounted on wood, rather than on metal, which can, sometimes, conduct away enough heat to affect the readings. It is preferable to hold mechanical output constant, but often it is more convenient to hold the input current or watts constant. Temperatures are measured as described in Art. 18-10. Readings should generally be taken every 15 min until temperatures become constant, unless the motor carries an intermittent rating, in which case readings should be taken more frequently and for only as long as indicated by the time rating.

(10) Speed-torque tests. A speed-torque test may be taken to determine the accelerating characteristics of the motor, to look for possible cusps, or to diagnose the action of the starting switch. One method of taking such a test is to use a dynamometer; the dynamometer field should be adjustable, and the dynamometer armature should be connected to a separate dc bus, the voltage of which is adjustable. Torque readings are taken at enough points to plot the speed-torque curve. Each reading should be taken as quickly as possible, for the motor heats rapidly on this test, and it is usually necessary to pause between readings to keep the motor from overheating.

An alternate method is illustrated in Fig. 18-12. Here, the motor brings a large flywheel up to speed, while the speed torque is automatically plotted on the *xy* recorder. The complete speed-torque curve is traced during a single acceleration of the motor.

Fig. 18-12 Test setup for taking a speed-torque test. Stator is mounted on a Torque Table (Fig. 18-9), which furnishes the signal to the *y* axis of the *xy* recorder. The speed signal which activates the *X* axis is furnished by the tachometer at the far right. The flywheel in the center provides slow acceleration. *(Lebow Products.)*

(11) Winding-ratio test. This test is needed only for capacitor-run motors when it is desired to segregate the losses; if the winding ratio is known from design data, it is not necessary to take the test. Run the motor with rated voltage E_m impressed on the main winding only, and measure E'_a, the auxiliary-winding voltage (there should be no load on the shaft). Impress E_a, arbitrarily chosen as approximately 18 percent more than E'_a (the underlying idea is to operate the motor at normal flux), upon the auxiliary winding, letting the motor run at no load; then measure the voltage induced in the main winding, which we shall call E'_m. The winding ratio is

$$a = \sqrt{\frac{E'_a E_a}{E_m E'_m}} \tag{18-8}$$

where

$$a = \frac{\text{effective conductors in auxiliary winding}}{\text{effective conductors in main winding}}$$

TESTS TO TAKE: OTHER TYPES

18-17 Adjustable Varying-Speed Fan-Duty Motors. Fan-duty adjustable-speed motors should, in addition to other tests, also be checked for minimum starting volts when connected for the lowest operating speed; in general, these motors are either permanent-split capacitor or shaded-pole motors. They ought to be able to start with the rotor in any position on a voltage of not over 90 percent of rated.

18-18 Polyphase Induction Motors. A routine test on a polyphase motor comprises the same tests as on a single-phase motor (Art. 18-15). Likewise, a complete engineering test comprises the same tests as for a single-phase motor, with minor exceptions. When locked-rotor readings are taken on a three-phase motor, all three phases must be simultaneously excited. Also, a polyphase motor does not have switching torque; nor is a winding-ratio test applicable. More detailed information is available in IEEE test procedures.[7,8]

18-19 DC Motors. Testing procedures for dc motors are covered in considerable depth by IEEE.[9,10] Briefly, such tests fall into four broad categories: preliminary, performance-determination, temperature, and miscellaneous.

(1) Preliminary tests. Such tests generally include

1. Verification of terminal markings and rotation
2. Resistance measurements of all windings
3. Air-gap measurement
4. Potential drop and polarity of field coils
5. Vibration
6. Brush setting
7. Dielectric tests

(2) Performance-determination tests. These tests include

1. Magnetic saturation (no-load saturation curve)
2. Commutation
3. Speed regulation
4. Efficiency using the input-output method
5. Field, armature, brush-contact, and stray-load losses
6. Load performance from 250 to 300 percent down

(3) Temperature tests. These tests are run at rated load.

(4) Miscellaneous tests. Miscellaneous tests include

1 Audio noise
2 Telephone-influence factor
3 Voltage wave shape
4. Response
5 Winding inductance
6 Shaft currents

18-20 Universal Motors. In general, the techniques for testing universal motors are about the same as for testing dc series motors, except as these tests may be modified by higher operating speeds. When taking a performance test, it may be more convenient to take readings at selected intervals of speed, rather than torque, as is done for constant-speed motors. A complete engineering test should include performance tests at 60 Hz, 25 Hz, and on direct current.

18-21 Synchronous Motors. Reluctance-type synchronous motors, both single-phase and polyphase, can generally be tested in pretty much the same way as their induction counterparts, except that there are two additional torques of interest: pull-in and pull-out. Pull-in torque is greatly affected by the inertia of the connected load but is often measured with no external inertia; for this reason, either a brake or a rope and pulley may be used, but not a dynamometer or external flywheel. However, if the external inertia the motor has to drive is known, it may be more useful to measure the pull-in torque with this amount of external inertia connected. Pull-out torque can be measured in much the same fashion as breakdown torque is taken.

Hysteresis-type motors can be tested in essentially the same way as reluctance motors are, except that their pull-in torque is not affected by external inertia. Inductor motors have to start and synchronize in a fraction of a cycle and can handle but little external inertia.

18-22 High-Potential Test (Dielectric Test). High-potential tests are taken primarily to determine whether the windings are adequately insulated from ground and from one another and to ensure the adequacy of clearance distances. These tests are for new or like-new machines only, since the accumulation of dirt and moisture on machines in service will decrease the dielectric properties.

The high-potential test voltage should be applied successively between each electric circuit and the frame, with those windings not under test and other metal parts connected electrically to the frame. As used here, an *electric circuit* consists of all windings and other live parts which are conductively connected to one power supply when starting or running. No leads should be left unconnected. Capacitors should be left connected to their associated windings.

Table 18-1 gives high-potential test voltages, established by National Electrical Manufacturers Association (NEMA), for clean, new, dry machines tested at the factory. When tested later, the test voltage should not be over 85 percent of the values given. Repeated application of the test voltage is not recommended, for such a practice may damage otherwise good insulation.

TABLE 18-1 Dielectric Test Voltages for Fractional Horsepower Motors

	Duration of application of test voltage	
	1 min	1 s
Universal motors (rated for operation on circuits not exceeding 250 V):		
Motors rated ½ hp and larger and all motors for portable tools	2E + 1000	2.4E + 1200
All other motors	1000	1200
AC motors:		
½ hp and larger and rated for all voltages	2E + 1000	2.4E + 1200
Less than ½ hp but rated for more than 250 V	2E + 1000	2.4E + 1200
Less than ½ hp and rated for 250 V or less	1000	1200
DC motors:		
½ hp and larger and rated for all voltages; less than ½ hp but rated for more than 250 V		
Armature or field windings for use on adjustable-voltage electronic power supply	2E* + 1000	2.4E* + 1200
All other armature or field windings	2E + 1000	2.4E + 1200
Less than ½ hp and rated for 240 V or less	1000	1200

Note: Voltages given in this table are for clean, dry, new machines tested at the factory. When testing machines that have been in service for some time, the authors recommend reducing these voltages by at least 25 percent.

*AC line-to-line voltage of the power supply selected for the basis of rating.

Source: From *NEMA Motor and Generator Standards*, NEMA Publ. no. MG1-1978, National Electrical Manufacturers Association, Washington, 1978.

Figure 18-13 shows a Model 106-2.5 dielectric tester made by the Slaughter Company, Ardmore, Oklahoma. The output, as indicated on the meter, is adjustable, and insulation breakdown is indicated by a buzzer and a red light. Two retractable probes are used; one is applied to the winding and the other to the frame or iron core. An optional feature is a lockout that requires manual reset after rejection. A high-voltage dielectric leakage current tester is shown in Fig. 18-14. In addi-

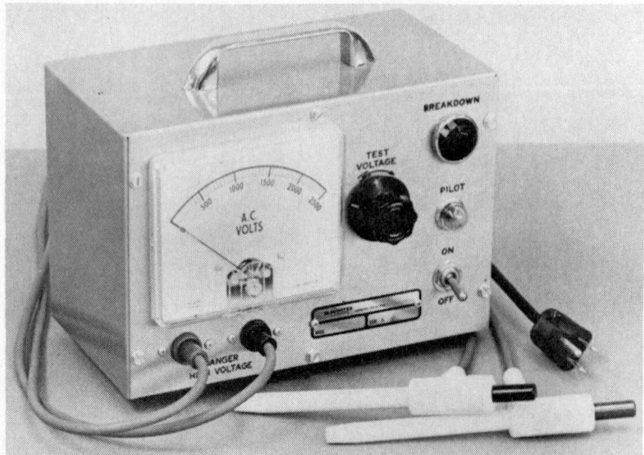

Fig. 18-13 A Model 106-2.5 dielectric tester with adjustable output voltage. Insulation breakdown is indicated by a buzzer and a red light. *(Slaughter Company.)*

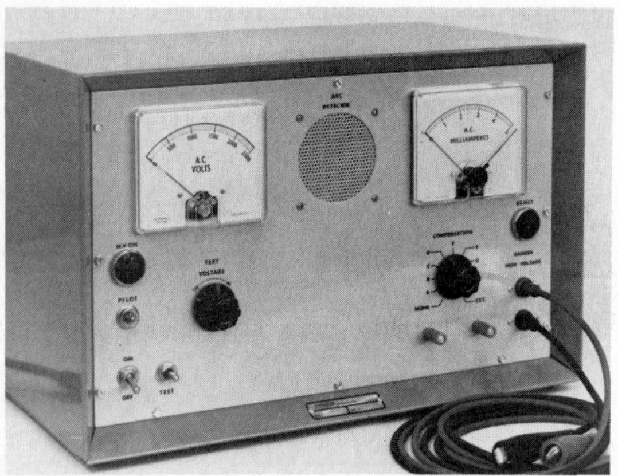

Fig. 18-14 A Model 103-2.5 high-voltage dielectric and leakage current tester. *(Slaughter Company.)*

tion to testing for dielectric withstand, this instrument measures the total leakage current of the part under test and by adjustment of a compensating circuit, measures resistive leakage current. Optional features include automatic control of the duration of time of application of the test voltage, dual-range voltmeter, and automatic reset of rejection signals.

SEGREGATION OF LOSSES

Segregation of losses means the breaking down of the total losses in a motor into their several components. A knowledge of the components of loss affords a means for checking the accuracy of input-output tests; also it is often of material aid to the design or application engineer.

18-23 Polyphase Induction Motors. Methods for segregating the losses of a polyphase induction motor are covered in detail by IEEE[7] and are generally so well known that they are not covered here. The components of loss are:

1. Friction and windage
2. Core loss, which is the loss in the magnetic-circuit iron at no load
3. Stator I^2R losses in the primary winding
4. Rotor I^2R losses in the secondary winding, which are equal to

   Slip $\times$ (stator input $-$ stator I^2R loss $-$ core loss)

5. Stray-load losses, which are those losses not otherwise accounted for

Total losses are taken as the difference between the input to the motor and its output; that part of the total loss not accounted for by 1, 2, 3, and 4 is considered to be stray-load loss. However, stray-load losses may be measured directly by a reverse-rotation test described in the IEEE test procedure.[7]

18-24 Single-Phase Induction Motors. Segregation of losses is more difficult in a single-phase than in a polyphase induction motor and is not covered in the IEEE test procedure. It was discussed by Veinott in *Electrical Engineering*[11] and in Chap. XIX of a previous edition,[12] as well as in a work by Suhr.[13] Space here permits explaining only how and not why.

A practical form for recording and analyzing the necessary tests is given in Fig. 18-15, which is applicable both to motors that do use a running capacitor and to those that do not. Use of this form is illustrated by analyses of tests taken on a two-value capacitor motor, both with and without the running capacitor in the circuit. The numbers in this figure that are underlined with a dotted line represent values taken directly from tests without computation, as follows:

E_m, I_m, W_m and *line 5* are taken from the locked reading on the main winding.

E (line 9) is the rated voltage.

I_0' is the current from the no-load saturation curve at E' volts.

Lines 17 and *19* are measured by bridge as indicated.

W_0' is the number of watts read from the no-load saturation curve at E' volts.

		WINDINGS ONLY			AUX. WDG.		CAPACITOR UNIT ONLY		
	MAIN		AUXILIARY		WITH CAP.			STG. C.	RNG. C.
1 Locked volts	E_m	110	E_a	110	E_{ac}	110	E_c	*120	**180
2 Locked amps., sec. shorted	I_m	16.6	I_a	8.3	I_{ac}	7.7	I_c	7.15	1.08
3 Locked watts, sec. shorted	W_m	1260	W_a	640	W_{ac}	710	W_c	147	10.1
4 Total res., $R = W/I^2$	R_m	4.57	R_a	9.28	R_{ac}	12.0	R_c	2.88	8.61
5 Pri. res. after locked rdg.		1.48		3.49	Z_{ac}	14.3	Z_c	16.8	166
6 ④ − ⑤		3.09		5.79	X_{ac}	7.75	X_c	16.55	166
7 Impedance, $Z = E/I$	Z_m	6.62	Z_a	13.26	E_a	102	Mfd	160.1	16.0
8 Sh. ckt. reactance $X = \sqrt{Z^2 - R^2}$	X_m	4.79	X_a	9.45	E_c	131	$E_c I_c$		

9 No-load volts	E	110
10 $\frac{1.68}{1.15 r_{1m}} \times \frac{1.36}{A}$ or $\frac{\quad}{1.15 r_{1m}} \times \frac{\quad}{I_1} \times \frac{\quad}{P.F.}$		2.3
11 Induced volts @ F.L. ⑨ − ⑩	E'	107.7
12 Mag. amps. @ E' volts	I'_o	2.73
13 $X_o = 2E'/I'_o - X$	X_o	74.2
14 $K_r = X_o - X/X_o$	K_r	0.935
15 r_2 cold (⑥/K_r) × (⑰/⑤)	r_2	3.26
16 $0.5 K_r r_2$		1.525
17 r_1 cold @ 25°C	r_1	1.463
18 r_o cold = ⑯ + ⑰	r_o	2.988
19 Pri. res. after rng. sat.		1.64
20 r_o hot = r_o × ⑲/⑰		3.34
21 No-load Watts @ E' volts	W'_o	56.0
22 I^2R @ E' volts = $I'^2_o\, r_o$ hot		24.9
23 Core loss + friction = ㉑ − ㉒		31.1
24 Friction [min. cu. loss = 1.9]		11.0
25 Core loss @ F.L.		20.1

Cap. element volts

*Cap. volts locked; ** Cap. volts @ F.L.

Wdg. ratio; $k = \sqrt{\frac{E'_a = 122.5}{E_m = 110} \times \frac{E_a = 145}{E'_m = 99}} = 1.274$

Actual k from design data =

	WATTS @ F.L.	AMPERES @ F.L. TOTAL	REAL	REACTIVE
Main ph.	W_{1m} 150	I_{1m} 2.31	$A = 1.362$	$B = -1.87$
Aux. ph.	W_{1a} 99	I_{1a} 1.09	$g = 0.90$	$h = 0.61$
Line	(246) 249	(2.55) 2.56	2.26	−1.26

26 I^2_{1m}	5.33	A_h	0.831
27 $(kI_{1a})^2$	1.93	B_g	−1.683
28 $I^2_{1m} + (kI_{1a})^2$	7.26	$A_h - B_g$	2.514
29 $2k(A_h - B_g)$	6.40		
30 ㉘ + ㉙	13.66		
31 ㉘ − ㉙	0.86		

$A = W_{1m}/E$

$B = \pm\sqrt{I^2_{1m} - A^2}$

	AS 1-Φ MOTOR		AS CAP. MOTOR	
32 r_2/X_o hot = $1.15 r_2/X_o$	r_2/X_o	0.0505	r_2/X_o	0.0505
33 Slip @ full load	s	0.0417	s	0.0333
34 Slip of backward field	$2-s$	1.958	$2-s$	1.967
35 sR_f		0.711		0.531
36 $(2-s)R_b$		1.753		1.753
37 Amperes @ F.L.	I_1	3.75		
38 Sec. $I^2R(f)$	I_1^2 × ㉟	10.0	㉚ × ㉟	7.25
39 Sec. $I^2R(b)$	I_1^2 × ㊱	24.6	㉛ × ㊱	1.51
40 Main wdg. cu. loss; r_{1m} hot = 1.68	$I_1^2 r_{1m}$	23.6	$I^2_{1m} r_{1m}$	9.0
41 Aux. wdg. cu. loss; r_{1a} hot = 3.97			$I^2_{1a} r_{1a}$	4.7
42 Cap. unit loss			$I^2_{1a} R_c$	10.2
43 Core loss + friction	㉓	31.1	㉓	31.1
44 Total losses @ F.L.		89.3		63.3
45 Output		186.4		186.4
46 Output + losses		275.7		249.7
47 Efficiency by losses		67.6		74.7

Note:

$sR_f = \frac{1.15 \times ⑯ = 1.753}{1 + \left(\frac{㉜}{㉝}\right)^2}$

1+1.465=2.465

1+2.30=3.30

$(2-s)R_b = \frac{1.15 \times ⑯ = 1.753}{1 + \left(\frac{㉜}{㉞}\right)^2}$

TYPE FT FRAME B145 CALC. REF.____ SERVICE____

HP 1/4 VOLTS 110 CYCLES 60 POLES 4 R.P.M.____ DATE 2-5-38 CAP. L.____

D-SPEC.____ S.O.____ SIG.____ MOT. L.____

Fig. 18-15 A form for analyzing the test results and segregating the losses of either a single-phase induction motor or a capacitor motor.

s, and I_l are from the load-performance test at the full-load point taken when the motor was operating without the running capacitor.

How to fill out the rest of the form is more or less self-explanatory. In line 10, if the motor is not a capacitor-run motor, the following equation should be used:

$$1.15 \times r_{1m} \times I_1 \times \text{power factor}$$

If it is a capacitor-run motor, for line 10 the following should be used:

$$1.15 r_{1m} \times A$$

The friction (line 24) is determined from the no-load saturation curve which is extended to zero volts, and the watts input at zero volts is read from the curve; from this value is subtracted the

$$\text{Minimum copper loss} = (\text{minimum current})^2 \times (\text{item } 20)$$

The remainder of the analysis of losses is given in lines 32 to 47, inclusive, under "As 1-ϕ motor." It is to be noted that the scratchwork calculations for lines 35 and 36 are carried out at the right under "Note."

The foregoing method of analysis is directly applicable to split-phase motors, reactor-start motors, capacitor-start motors, or repulsion-start motors.

18-25 Capacitor-Run Motors. The test analysis form of Fig. 18-15 is generally applicable to conventional capacitor-run motors, including the tapped-winding motors of Figs. 7-10 and 7-11 when they are connected for high-speed operation but not the motor of Fig. 7-9. In general, Fig. 18-15 applies where, on the running connection, the main winding is used only as a main winding, and not also as an autotransformer. In using the form, items 1 to 25, except item 10 as previously noted, are determined exactly as described in Art. 18-24, as if the motor had no running capacitor. The winding ratio a (called k in the figure) is determined from the winding-ratio test. W_{1m}, W_{1a}, I_{1m}, and I_{1a} are read from the load-performance test as a capacitor motor. Main-winding current I_{1m} is broken up into its real and reactive components A and B, by the formulas given. The auxiliary-winding current is similarly resolved into g and h. The measured value of watts, 246, is entered in parentheses to compare with 249, which is the sum of the two wattmeter radings. The measured line amperes, 2.55, is compared with the value 2.56, computed from

$$\text{Line amperes} = \sqrt{(A + g)^2 + (B + h)^2}$$

The close agreement between the two figures of watts, and also between the two figures of current, is a good check on the accuracy of the tests.

Items 26 to 31 are self-explanatory. The remainder of the analysis of losses is given in lines 32 to 47, under "As cap. motor." The scratch-work calculations for items 35 and 36 are shown in the blank space under "Note," together with the similar previous calculations.

The effect of the running capacitor is strikingly illustrated in the figure. Rotor and stator I^2R losses are greatly reduced, improving efficiency; this comparison was noted in Art. 6-3.

VIBRATION AND NOISE MEASUREMENTS

Vibration measurements may be made on a motor to determine how well the rotor is balanced. Noise measurements may be made for an objective comparison of the relative noisiness of different motors. Noise measurements are the more difficult of the two to make, and the results are even more difficult to interpret.

18-26 Vibration Measurement. To ensure that the vibrations measured are due to rotor unbalance and are not influenced by the mounting, the motor should be mounted on an elastic mounting so chosen that the up-and-down natural frequency of the motor on the base is not over one-quarter of the rotational speed of the motor. To accomplish this, it is necessary that the mounting be deflected downward because of the weight of the motor, by as much as or more than the amount given in the following table:

	Compression	
rpm	in	mm
900	1	25
1800	¼	6
3600	1/16	1.6

If a flexible pad is used, it should not be compressed more than half its thickness, lest the supports become too stiff.

A common instrument used is the Davy vibrometer, which measures motions down to 0.0001 in. A half key should be securely fastened in the keyway. With the motor running at normal speed, the amplitude of vibration (total excursion) should be measured on each bearing housing. Normal speed for induction motors is no-load speed; for dc motors it is the highest rated speed; for series and universal motors, it is the normal operating speed.

In general, for fractional horsepower motors, the total amplitude of vibration of either bearing housing, in any direction, when measured as described previously, should not exceed 0.001 in. For some applications, even less vibration is permissible.

18-27 Nature of Sound: Certain Terms Explained. Some terms commonly used in discussing noise problems are the following:

Noise. Any unwanted sound.

Sound (in air). A series of minute compressions and rarefactions of the air pressure, often called sound waves, traveling at a speed of approximately 1200 feet/s. When these sound waves strike a surface, they tend to make the latter vibrate. A given sound may consist of but a single pure tone, or it may contain many tones.

Sound pressure. The root-mean-square (rms) variation of the air pressure. Because the range of audible sound pressure is more than 1 million to 1 and because we are usually more interested in percent changes in sound pressure, the term sound-pressure level is used almost exclusively.

Sound-pressure level. Twenty times the logarithm to base 10 of the ratio of actual sound pressure to a standard reference level of 0.0002 microbar for airborne sound. Sound-pressure level is expressed in decibels, abbreviated dB. Each decibel increase in level means an increase in sound pressure of approximately 13 percent; each 6 dB means a doubling of the sound pressure, and each 20-dB increase represents a 10-fold increase in sound pressure. One decibel represents about as small a change in noise as the average human ear can notice.

A pure tone. The sound of a single frequency.

Frequency. The total number of cycles of variation of pressure per second, expressed in hertz. Pitch is often used synonymously with frequency. Most sounds contain many frequency components.

White noise. The sound containing all possible frequency components in the audible range, all of them being substantially equal in magnitude.

Sound level. An overall sound pressure where the different components of frequency are weighted differently. Sound-level meters generally have three weighting networks, called A, B, and C. The A network has the most weighting, the C the least; the weighting reduces the low-frequency components of sound.

Octave band. A band of frequencies with a total range of one octave; that is, the top of the band has a frequency twice that of the bottom. Similarly a one-third-octave band is a band of frequencies having a width of one-third of an octave; the highest frequency is 26 percent greater than the lowest.

Sound power. A term applied to the total power radiated acoustically by a noise source. It may be expressed in watts or microwatts. More commonly, it is expressed in decibels where the reference level is 1 pW (10^{-12} W). *Sound power level* is 10 times the logarithm to the base 10 of the ratio of the acoustic power to 1 pW. Each decibel increase represents an increase in power of 26 percent, 3 dB represents a doubling of the sound power, and each 10 dB represents a tenfold increase in sound power. Sound power level is commonly used when integral horsepower motors are involved, but less commonly when only fractional horsepower motors are under consideration.

18-28 Instrumentation and Noise Testing Procedures. Space does not permit covering this subject in any depth here; it is covered much more fully in an IEEE test procedure.[14]

(1) Sound-level meter. A sound-level meter consists of a microphone, an amplifier, and weighting networks A, B, and C. The A network weights the frequencies so that the meter reads a number more nearly correlated to what the human ear would sense at a 40-dB level than it would without weighting. The C network gives virtually no weighting at all. When a network is used, the fact should be so noted in reporting the reading.

A small, portable, sound-level meter is illustrated in Fig. 18-16. Regular laboratory instruments are usually somewhat larger.

(2) Bandpass analyzer. A bandpass analyzer is an instrument usually associated with the sound-level meter that passes only a selected band of frequencies. This band may have a width of one octave, one-half an octave, or one-third of an octave.

(3) Narrowband analyzer. A narrowband analyzer measures discrete components of sound and is often useful in tracking down the source of a troublesome noise by identifying the troublesome frequency component. There are two types in use: with one of them, the band width, expressed in hertz, is the same for all frequencies, and with the other, the band width is a constant percentage of the center frequency at all frequencies. The latter type is generally more convenient for testing rotating electrical machines.

(4) Test procedure. The motor to be tested is mounted on a resilient suspension, like the one described in Art. 18-26 and the whole put into a quiet environment, preferably quiet enough that the meter reads 10 dB or more higher with the motor running than with it idle.

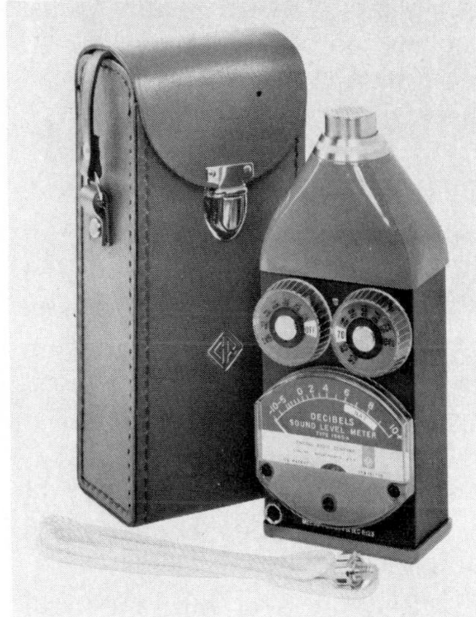

Fig. 18-16 A pocket-sized lightweight sound-level meter. *(General Radio Company.)*

For fractional horsepower motors, the microphone is positioned 12 in (0.3 m) from the major surfaces; for subfractional horsepower motors, 6 in (0.15 m) may be a more convenient distance. A single reading in one position may suffice, but it is better to take readings in a number of positions; four at shaft height around the motor and one directly above the motor have often been used. For most purposes, it is satisfactory to average the readings, taken in the different positions. Also, for most purposes, the A network is to be preferred. Readings can be taken with both A and C networks; if there is no difference between these readings, the noise is concentrated in frequencies of 1000 Hz or higher, but a larger difference would indicate a concentration of the noise in the lower frequencies. Readings should be taken with motor running and with motor not energized; the latter are ambient readings. A correction can be made for the effect of ambient, as follows:

Difference in readings	Correction, dB
Less than 3	Reading invalid
3	3
4	2
5	1.6
6	1.2
8	0.7
10	0.4
Over 10	0

"Difference" in this table is the difference between the motor-plus-ambient and ambient readings. The correction should be subtracted from the motor-plus-ambient reading to obtain the motor noise.

APPLICATION TESTS

The way to select the proper motor for any specific application is the principal subject of Chap. 19. After a preliminary choice has been made, it is a good idea to verify the selection by running an application test. When possible, it is desirable to run these tests on a new stiff machine, *and* on one that is well worn in.

18-29 Determination of Power Requirements. The test motor should be installed on the machine or appliance it is to drive and connected to a source of power through meters, connected as shown in Fig. 18-4. The load on the motor should be varied, subjecting the machine to all possible and conceivable operating conditions likely to be encountered in service; for each different load, the motor voltage should be adjusted to rated value, and readings taken of watts, amperes, and rpm. If a load-performance test on the test motor is available, the reading can be converted to horsepower or torque output. If no such load-performance test is available, one should be taken, or the output measured at the same inputs observed when testing the machine. It should be noted here that watts input is a much more reliable indicator of output than amperes. *In no case should one attempt to estimate the power output by comparing observed amperes under load with the amperes figure stamped on the nameplate.* (A careful study of Fig. 18-11, for example, will show that the output increases much more rapidly than the current input.)

If a Torque Table, such as the one described in Art. 18-14(4), is available, the torque required to drive the appliance can be measured directly without making electrical measurements on the motor itself.

18-30 Determination of Torque Requirements. Locked-rotor torque is almost never a problem with universal or with dc motors, and so this discussion is intended mainly for induction-type motors. With such motors, switching, pull-up, locked-rotor, and breakdown torques are often of importance. To find out what the machine requires, use a selected test motor and run tests to determine

1. Minimum voltage to start the load, in any position
2. Minimum voltage to bring the machine up above half speed

3. Minimum voltage at which the motor will transfer from the starting to the running connection

4. Minimum voltage at which the motor will carry the required load without an abrupt drop in speed

The driven machine, or appliance, should, of course, be adjusted for severe conditions, and the locked-rotor, switching, pull-up, and breakdown torques of the test motor must be known. Now, assuming that torque varies as the square of the voltage, the results of tests 1, 2, 3, and 4 give, respectively, the locked-rotor, pull-up, switching, and breakdown torques actually required by the appliance or device.

References*

1. *Master Test Code for Electrical Measurements in Power Circuits,* IEEE no. 120-1955.
2. *Standard Test Code for Resistance Measurements,* IEEE no. 118-1978.
3. *Recommended Practice for Testing Insulation Resistance of Rotating Machinery,* IEEE no. 43-1974.
4. *Recommended Practice for General Principles of Temperature Measurement as Applied to Electrical Equipment,* IEEE no. 119-1974.
5. Seely, R. E.: *A Circuit for Measuring the Resistance of Energized A-C Windings,* AIEE Paper no. 55-23, 1955 (available from IEEE).
6. *Test Procedure for Single-Phase Induction Motors,* IEEE no. 114-1982.
7. *Standard Test Procedure for Polyphase Induction Motors and Generators,* IEEE no. 112-1983.
8. *Trial-Use Test Procedure for AC 400 Hz Aircraft Induction Motors,* IEEE no. 137-1960.
9. *Standard Test Code for DC Machines,* IEEE no. 113-1973.
10. *Test Code for DC Aircraft Rotating Machines,* IEEE no. 132-1953.
11. Veinott, C. G.: "Segregation of Lossesx in Single-Phase Induction Motors," *AIEE Trans.,* 1935, p. 1302.
12. ————: *Fractional Horsepower Electric Motors,* 2d ed., 1948, Fractional- and Subfractional-Horsepower Electric Motors, 3d ed., 1970, McGraw-Hill Book Company, New York.
13. Suhr, F. W.: "Toward an Accurate Evaluation of Single-Phase Induction-Motor Constants," *Trans. AIEE,* pt. III, 1952, p. 221.
14. *Test procedure for Airborne Sound Measurements on Rotating Electrical Machinery,* IEEE no. 85-1973.

Bibliography

Veinott, C. G.: *Theory and Design of Small Induction Motors,* McGraw-Hill Book Company, New York, 1959. (Out of print, but available from University Microfilms, P. O. Box 1647, Ann Arbor, Mich. 48106.)

*The IEEE publications listed here are available from the Institute of Electrical and Electronics Engineers, 345 E. 47th St., New York, NY 10017.

CHAPTER

19

SELECTING THE RIGHT MOTOR FOR THE JOB

Selection of the best and most suitable motor for any particular job can involve a whole host of considerations, many of which conflict more or less with one another. Many questions need to be raised and answered; many of the answers are simple and obvious, but many are not. Often, a lot of the questions are best answered by the experience of others who have had a similar application problem, and a simple answer is to follow their lead. In this chapter, we shall review a number of the more important factors and considerations involved in the application of fractional-horsepower electric motors. The discussion will be mainly devoted to the more-or-less conventional types such as those discussed in the first 14 chapters; application considerations pertaining to the more special kinds of machines, such as those discussed in Chaps. 15 and 16, are discussed in their respective chapters.

Perhaps it should here be emphasized that the various factors discussed in the following pages have to be considered *in parallel* rather than *in series;* that is, it is not usually possible to go through each question and answer it once and for all and arrive at the end with a single pat answer to the selection problem. Rather, one has to make tentative or multiple-choice answers to each question as encountered, and then go back and review them all. In short, application considerations have to be taken into account collectively and simultaneously rather than serially or sequentially.

The type and rating of a motor for a particular application are dependent on the expected duty cycle or annual hours of operation, total power consumed, method of connecting the motor to the load, and type of load. In general, for a given motor type and frame size, the larger the rating, the higher the efficiency. However, the motor horsepower rating should be selected to match the load so that the motor operates near its full-load rating. Likewise, the higher the synchronous speed, the greater the efficiency, but not all loads are suitable for coupling to a high-speed motor.

Vacuum cleaners, kitchen blenders, portable hand-held tools, and washing machines are examples of applications that operate intermittently or relatively few hours per year. Even a large increase in efficiency will have little effect on the total annual power consumption.

Direct- or belt-drive heating and air-conditioning systems, centrifugal pumps, business computer equipment, and refrigeration compressors are examples of appliances that operate continuously. A substantial increase in motor efficiency can be effected by using a capacitor-run motor as opposed to a shaded-pole or split-phase type. An electric clock, which has an input of 2 W even if driven by a more efficient motor, has little effect on total power consumption.

CLASSIFICATION OF MOTORS ACCORDING TO APPLICATION

Motors may be said to fall into three broad classes, according to the application for which they were designed: *general-purpose*, *definite-purpose*, and *special-purpose*. Although the use of a special motor is sometimes justified, first consideration should be given to the use of a general-purpose motor, or a definite-purpose motor. By so doing, the user usually obtains a lower cost, greater ease in obtaining new motors, more sources of supply, and more general all-round satisfaction.

19-1 General-Purpose Motors. A general-purpose ac motor is one which is built in standard ratings, with standard operating characteristics, a standard mechanical construction, and offered for general use without restriction to any particular application. It is of open construction, is continuously rated, and has a service factor. Standard split-phase motors are sometimes referred to as general-purpose motors, since they have many of the characteristics of general-purpose motors, though they do not meet all National Electrical Manufacturers Association (NEMA) requirements for general-purpose motors. As a rule, first consideration should be given to the possibility of using a general-purpose or standard split-phase motor before other varieties are considered.

19-2 Definite-Purpose Motors. A definite-purpose motor is one designed in certain standard ratings, with standard operating characteristics or mechanical construction to make it particularly suitable for a given application. That is, it is usually designed to meet the special requirements of a particular industry or type of application. NEMA standards cover not only general-purpose motors, but also a considerable variety of definite-purpose motors for a number of industries.[1,*] Such motors are described in some detail later in this chapter; in each of these particular industries, it is often preferable and more satisfactory to use the definite-purpose than a general-purpose motor. If neither fills the need, it may be necessary to have a special-purpose motor designed and built.

19-3 Special-Purpose Motors. A special-purpose motor is one especially designed and built to suit a combination of requirements not filled by a general-purpose or by any definite-purpose motor. The requirements can involve almost anything special: torques, enclosures, bearings, severe environmental requirements, quietness, built-in brakes, clutches or other mechanical or electrical devices. All we can do here is to point out that such special-purpose motors do exist.

19-4 Standard Ratings and Performance

(1) Horsepower and speed ratings. Horsepower and speed ratings for ac fractional and subfractional horsepower motors are given in Tables 19-1 and 19-2, respectively. In general, these also apply to dc motors.

(2) Voltage ratings. Voltage ratings are as follows:

Motor	Voltage rating, V
Universal	115 and 230
AC, single-phase	
60 Hz	115 and 230
50 Hz	110 and 220
AC, polyphase	
60 Hz	115, 200, 230, 460, and 575
50 Hz	220 and 380
DC	115 and 230

*For numbered references, see References at end of this chapter.

TABLE 19-1 Performance Characteristics of 60-Hz General-Purpose Capacitor-Start and Standard Split-Phase Motors

				Capacitor-start				Split-phase			
		Full-load		Torque, oz-ft		Locked-rotor-amperes		Torque, oz-ft		Locked-rotor-amperes	
Horse-power	Poles	rpm	Torque oz-ft	Break-down	Locked-rotor	115 V	230 V	Break-down	Locked-rotor	115 V	230 V
1/20	2	3450	1.22	3.7	5.0	20	12	3.1	1.7	20	12
	4	1725	2.44	7.1	10.2	20	12	6.1	3.5	20	12
	6	1140	3.69	10.4	14.1	20	12	8.9	4.9	20	12
1/12	2	3450	2.03	6.0	8.0	20	12	5.2	2.7	20	12
	4	1725	4.06	11.5	16.7	20	12	10.0	5.7	20	12
	6	1140	6.15	16.5	22.7	20	12	14.5	8.1	20	12
1/8	2	3450	3.05	8.7	11.5	20	12	7.8	4.1	20	12
	4	1725	6.09	16.5	24.0	20	12	14.8	8.5	20	12
	6	1140	9.22	24.1	32.0	20	12	21.6	12.0	20	12
1/6	2	3450	4.06	11.5	15.0	20	12	10.6	5.5	20	12
	4	1725	8.12	21.5	33.0	20	12	19.8	11.0	20	12
	6	1140	12.29	31.5	43.0	20	12	29.0	15.7	20	12
1/4	2	3450	6.09	16.5	21.0	26	15	14.8	7.7	26	15
	4	1725	12.18	31.5	46.0	26	15	28.2	16.3	26	15
	6	1140	18.44	44.0	59.0	26	15	39.8	23.1	26	15
1/3	2	3450	8.12	21.5	26.0	31	18	19.8	10.2	31	18
	4	1725	16.24	40.5	57.0	31	18	37.5	21.4	31	18
	6	1140	24.58	58.0	73.0	31	18	53.3	30.0	31	18
1/2	2	3450	12.18	31.5	37.0	45	25	28.2	14.9	45	25
	4	1725	24.37	58.0	85.0	45	25	52.2	31.7	45	25
	6	1140	36.87	82.5	100.0	45	25	74.3	44.3	45	25
3/4	2	3450	18.28	44.0	50.0	61	35				
	4	1725	36.55	82.5	110.0	61	35				
1	2	3450	24.37	58.0	61.0	80	45				

Note: The breakdown torque of a general-purpose polyphase squirrel-cage fractional horsepower motor with rated voltage and frequency applied shall be not less than 140 percent of the breakdown torque of a single-phase general-purpose fractional horsepower motor of the same horsepower and speed rating.

Source: From *NEMA Motor and Generator Standards*, NEMA Publ. no. MG1-1978, National Electrical Manufacturers Association, Washington, 1978, and original sources.

(3) Frequencies. Frequencies for ac motors are 60 and 50 Hz, and for universal motors they are 60 Hz to direct current.

(4) Torques. Full-load, breakdown, and locked-rotor torques for capacitor-start and split-phase fractional horsepower motors are given in Table 19-1. Full-load and breakdown torques for subfractional horsepower motors are given in Table 19-2. The values given for break-

TABLE 19-2 Performance Characteristics of 60-Hz Subfractional Horsepower Motors

Millihorsepower	Poles	Full-load rpm	Full-load Torque, oz-in	Breakdown torque, oz-in
1	2	3000	0.34	0.55
	4	1550	0.65	1.10
	6	1050	0.96	1.65
1.5	2	3000	0.50	0.70
	4	1550	0.98	1.45
	6	1050	1.44	2.20
2	2	3000	0.67	1.10
	4	1550	1.30	2.20
	6	1050	1.92	3.30
3	2	3000	1.01	1.80
	4	1550	1.95	3.60
	6	1050	2.88	5.40
5	2	3000	1.68	2.70
	4	1550	3.25	5.40
	6	1050	4.80	8.10
7.5	2	3000	2.52	3.60
	4	1550	4.88	7.20
	6	1050	7.21	11.0
10	2	3000	3.36	5.50
	4	1550	6.51	11.0
	6	1050	9.61	17.0
15	2	3000	5.04	9.50
	4	1550	9.76	19.0
	6	1050	14.41	29.0
25	2	3000	8.41	15.0
	4	1550	16.27	30.0
	6	1050	24.02	46.0
35	2	3000	11.77	24.0
	4	1550	22.78	48.0
	6	1050	33.63	72.0

Source: Compiled from *NEMA Motors and Generator Standards*, NEMA Publ. no. MG1-1978, National Electrical Manufacturers Association, Washington, 1978, and original sources.

down and locked-rotor torques are minimum values, and commercial motors usually have somewhat higher values than shown.

(5) Locked-rotor currents. Maximum values of locked-rotor currents for 60-Hz capacitor-start and split-phase motors are given in Table 19-1.

(6) Mounting dimensions. NEMA has established mounting dimensions for fractional horsepower motor frames; these are shown in Fig. 19-1. Frame sizes into which the general-purpose motors are put are given in Table 19-3, which is not a NEMA table.

ELECTRICAL CONSIDERATIONS

19-5 Selection of the Type. A comparison of the principal characteristics of 13 major types of motors is given in Table 19-4. Each of the characteristics listed may be a factor in selecting the type of motor. The kind of *power supply* available may narrow down the number of possible types to be considered. Single-phase motors can be used on poly-

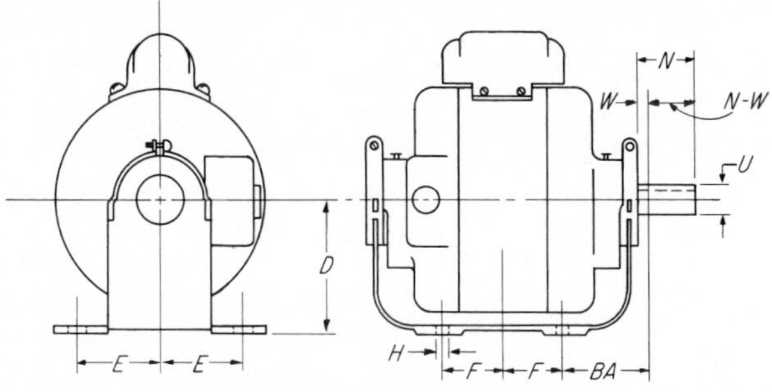

Frame number	Key Width	Key Thickness	Key Length	D	E*	F*	BA	H†	$N-W$‡	U
				Dimensions in inches						
42		3/64 flat		2 5/8	1 3/4	27/32	2 1/16	9/32 slot	1 1/8	3/8
48		3/64 flat		3	2 1/8	1 3/8	2 1/2	11/32 slot	1 1/2	1/2
48H		3/64 flat		3	2 1/8	2 3/8	2 1/2	11/32 slot	1 1/2	1/2
56	3/16	3/16	1 3/8	3 1/2	2 7/16	1 1/2	2 3/4	11/32 slot	1 7/8	5/8
56H	3/16	3/16	1 3/8	3 1/2	2 7/16	2 1/2	2 3/4	11/32 slot	1 7/8	5/8
				Dimensions in millimeters						
42		1.19 flat		66.7	24.4	21.4	52.4	7.1 slot	28.6	9.6
48		1.19 flat		76.2	54.0	34.9	63.5	8.7 slot	38.1	12.7
48H		1.19 flat		76.2	54.0	60.3	63.5	8.7 slot	38.1	12.7
56	4.8	4.8	34.9	88.9	61.9	38.1	69.9	8.7 slot	47.6	15.9
56H	4.8	4.8	34.9	88.9	61.9	63.5	69.9	8.7 slot	47.6	15.9

*Tolerance for the $2E$ and $2F$ dimensions is ±1/32 in (±0.79 mm).
†Tolerance is +3/64, −0 in (+1.19, −0 mm).
‡$N-W$ is the length of the shaft extension from the shoulder (if any) to the end of the shaft, i.e., the usable shaft extension.

Fig. 19-1 Mounting dimensions of NEMA frame sizes of fractional horsepower motors.

TABLE 19-3 Frame Sizes for 60-Hz General-Purpose Fractional Horsepower Motors

Horsepower	rpm	Frame
1/8	3450	48
	1725	48
1/6	3450	48
	1725	48
	1140	48
1/4	3450	48
	1725	48
	1140	56
1/3	3450	48
	1725	56
	1140	56
1/2	3450	48
	1725	56
	1140	56
3/4	3450	56
	1725	56
	1140	56
1	3450	56

phase circuits, but polyphase motors are not generally suitable for operation on single-phase circuits. (There are on the market a number of static devices intended to make polyphase motors operate on single-phase circuits. Although such a device may solve an isolated problem, a polyphase motor, operated from a single-phase supply through such a device, does so at some sacrifice in locked-rotor and breakdown torques, as well as at reduced efficiency and increased temperature rise.) *Horsepower* and *speed* ratings required are important because most types are not available over the whole range of either; however, by using a belted or a geared drive, almost any speed can be obtained from any type. Horsepower needs can be determined by an application test. (See Art 18-29.)

What *speed classification* is required: constant-speed, two-speed, multispeed, varying, adjustable-varying, or absolutely constant? Not to be forgotten, however, is that variability of speed can also be achieved by a variable-speed drive system, such as discussed in Chap. 13, for example, or by use of a mechanical variable-speed transmission such as the Reeves Moto drive, manufactured by the Reeves Division of the Reliance Electric Company.

Torques required by the driven device, both starting and running, play an important role in the selection of the type. How to determine torque requirements is discussed in Art. 18-30.

Reversibility may play a role in the selection of the type. For a discussion of *reversing motors* see Art 5-21.

Other factors influencing selection of the best type for a particular application are discussed in Table 19-5 under "General Remarks."

19-6 What Full-Load Speed? In usual fractional horsepower sizes, motors with a rated full-load speed of 1725 rpm are usually less costly and easier to obtain than those with a rated speed of 3450, 1140, or 850 rpm; this statement applies to dc and to 60-Hz motors, both single-phase and polyphase. Operating speeds of permanent-split capacitor and shaded-pole motors are somewhat lower, as listed in Table 19-5. Where low and very low speeds are needed, gearmotors are available. (Refer to Arts. 17-22 to 17-24.) Universal motors may be operated at any speed from 3500 up to 10,000 or 15,000 rpm.

19-7 Choosing the Horsepower Rating. First, the actual power requirements of the load under different conditions should be determined by means of an application test (see Art. 18-29). When taking this test, it is best to determine how much the load can be increased in service; for example, if a V belt is used, the load can be increased considerably by tightening the belt, which an electrician or service mechanic is likely to do; tightening the packing gland of a pump can increase the load considerably; on an air compressor, the load can sometimes be increased by tampering with the setting of the pressure-limiting switch. After the true horsepower requirements of the load have been determined, it is generally best to select a motor of the nearest standard horsepower rating. Although fractional horsepower motors are generally capable of carrying small overloads continuously, if long life is desired, it is best to select a motor large enough to handle the job under normal *and* adverse conditions.

Motors having a service factor stamped on the nameplate may be safely operated at service-factor load without injurious overheating provided rated voltage is applied at rated frequency. However, if a motor is operated continuously at service-factor load, the insulation life may be less than half of what it would have been had the motor been operated continuously at rated load.

Thermal protection offers additional assurance of trouble-free operation, but as thermal protectors are generally applied so as to permit continuous operation at service-factor load, they do not prevent reduced insulation life if the motor is continuously operated at over-

TABLE 19-4 Characteristics of Fractional and Subfractional Horsepower Motors

	Alterna					
	Single-phase motors					
	Split-phase types		Capacitor types			
	Standard	Special-service	Split-phase capacitor	Permanent-split	Capacitor-start	Two-value
Schematic diagram of connections: Arrangements shown are typical or representative; most of the types illustrated have numerous other arrangements which are also used.						
Characteristic speed-torque curves: Ordinates are speed. For all ac motors, 1 division = 20% of synchronous rpm; for universal motors, 1 division = 1000 rpm; for dc motors, 1 division = 20% of full-load rpm. Abscissas are torque. For all motors, 1 division = 100% of full-load torque.						
Rotor construction	Squirrel-cage	Squirrel-cage	Squirrel-cage	Squirrel-cage	Squirrel-cage	Squirrel-cag
Built-in automatic starting mechanism	Centrifugal switch	Centrifugal switch	Centrifugal switch	None required	Centrifugal switch	Centrifugal switch
Horsepower ratings commonly available	1/20–1/2	1/6–1/2	1/6–1/2	1/20–3/4	1/8–1	1/4–1
Usual rated full-load speeds (for 60-Hz ac motors; also for dc motors)	3450, 1725, 1140, 850	1725	1725	1550, 1050, 825	3450, 1725, 1140, 850	3450, 1725
Speed classification	Constant	Constant	Constant or adjustable varying	Constant or adjustable varying	Constant	Constant
Means used for speed control			Tapped winding	Tapped winding		
Comparative torques: Locked-rotor	Moderate	High	Moderate	Low	Very high	Very high
Comparative torques: Breakdown	Moderate	High	High	Moderate	High	High
Radio interference: Running	None	None	None	None	None	None
Radio interference: During Acceleration	One click	One click	One click	None	One click	None
Approximate comparative costs between types for same horsepower rating: Below 1/20						
Approximate comparative costs between types for same horsepower rating: 1/20–1/4 hp	100	85	130	75	140	160
Approximate comparative costs between types for same horsepower rating: 1/3–3/4 hp	115	100	145	100	150	170
General remarks: Standard motors are ordinarily designed to operate in ambient temperatures of 10–40°C (50–104°F). Variations in line voltage of ±10% or variations in frequency of ±5% are allowable. Split-phase and capacitor-start motors are available in two-speed pole-changing ratings, 1725/1140 and 1725/850 rpm. Permanent-split capacitor and shaded-pole motors are available in tapped-winding multispeed ratings. Fractional horsepower motors are built for across-the-line starting. The standard direction of rotation is counterclockwise, facing the end opposite the shaft extension.	For constant-speed operation, even under varying load conditions, in which moderate torques are desirable or mandatory, this type is often used in preference to the more costly capacitor-start motor. This type meets NEMA starting currents. Typical applications are as motors for blowers; centrifugal pumps, duplicating machines, and oil burners.	High locked-rotor currents (Design O) that have a tendency to cause lights to flicker limit the use of this type on lighting circuits to only those applications in which the motor starts very infrequently. Principal applications are as motors for washing machines, dryers, dishwasers, sump pumps, and tools for a home workshop.	This type is used in place of split-phase motors when higher operating efficiencies are desired. It has the starting characteristics of the split-phase motor and the running characteristics of the two-value capacitor motor. Common applications are as motors for belted blowers in warm-air furnaces, for attic ventilators, and for air-conditioning.	This motor is primarily used for shaft-mounted fans and blowers. It is essentially a constant-speed motor, but by means of tapped windings is often multispeed. It can also be made for intermittent duty for plug-reversal service. Typical applications are as motors for air-conditioners, ceiling fans, attic ventilators, and hospital beds.	This type is general-purpose motor suitable for most applications requiring constant speed under varying loads, high starting and running torques, and high overload capacity. It is also available as a two-speed pole-changing motor above ¼ hp. A few important applications are as motors for air compressors, gasoline pumps, jet pumps, table saws, and jointer-planers,	This motor has starting characteristics similar to a capacitor-start motor, with higher efficiency at operating loads. It is used in applications that run continuously, such as conveyor belts, elevators, chain drives, and large refrigeration equipment.

Alternating current				Direct current		
	1-, 2-, or 3-phase motors	Polyphase motor				
Shaded-pole	Synchronous (reluctance, permanent-magnet, or hysteresis)	Squirrel-cage induction	DC or ac (60 H or less), universal	Shunt or compound	Series	Permanent-magnet
	Stator winding may be split-phase, capacitor-type, or polyphase					
Squirrel-cage	Reluctance, permanant-magnet, or hysteresis	Squirrel-cage	Drum-wound; commutator	Drum-wound; commutator	Drum-wound; commutator	Drum-wound; commutator
None required	Depends on stator winding	None	None	None	None	None
1/2000–1/4	1/3000–1/2	1/6–1	1/150–1	1/20–3/4	1/125–1/30	1/125–1
3000, 1550	3600, 1800, 1200, 900	3450, 1725, 1140, 850	3000–11,000	3450, 1725, 1140, 850	900–2000	200–14,000
Constant or adjustable varying	Absolutely constant	Constant	Varying or adjustable varying	Constant or adjustable varying	Varying or adjustable varying	Constant or adjustable varying
Tapped winding			Resistor or tapped field	Armature resistance	Resistor	Armature resistance
Low Low	Low Moderate	Very high Very high	Very high	Very high	Very high	Very high
None None	None	None None	Continuous Continuous	Continuous Continuous	Continuous Continuous	Continuous Continuous
100 110	300 200	180 150	80 140	250 250	200	200 250 275
For ratings below 1/20 hp, this motor is the most widely used. For fan applications speed control can be effected by tapped windings. Applications are as motors for humidifiers, hair dryers, and damper controls.	Synchronous speed results from rotor construction; the rotor has magnetic poles that are fixed in place by notches or permanent magnets or by a hysteresis effect. Pull-in ability is affected by the inertia of the connected load. For very low speeds, synchronous inductors are used. Common applications are as motors for teleprinters, graphic instruments, clocks, and timing devices.	This motor is a companion motor to capacitor-start-type motors with comparable torques and is generally suited to the same applications if polyphase power is available. It is inherently plug-reversible and suitable for door openers, hoists, and elevators. High-frequency motors are used for high-speed applications such as motors for woodworking machinery, rayon spinning, and portable tools.	This motor's light weight for a given output, high speeds, and varying-speed and universal characteristics make this type very popular for hand tools of all kinds, vacuum cleaners, and kitchen food processors. Some speed control can be effected by a resistor or by use of a tapped field. It is used with reduction gearing for slower-speed applications.	This type is a constant-speed companion motor for the capacitor-start or split-phase motor for use where only dc power is available. Resistance in the armature circuit is used to obtain speed control. It is not usually designed for field control. Most are designed for and used on rectified ac power. Applications include motors for air compressors, pumps, and blowers.	This motor is principally used as the dc companion motor to the shaded-pole motor for fan applications. It is used in the smaller ratings in place of shunt motors to avoid using extremely small wire. Speed is varied by adding series resistance.	This motor has a high efficiency because permanent magnets, in lieu of field windings, supply the excitation. Its characteristics are similar to those of a shunt motor. It is very popular for battery-operated applications and for applications in which easy reversibility is needed.

TABLE 19-5 Performance Characteristics of 60-Hz Shaded-Pole and Permanent-Split Capacitor Motors for Fan and Pump Applications

				Torque, oz-ft			
		Full-load		Permanent-split capacitor		Shaded-pole	
Horse-power	Poles	rpm	Torque oz-ft	Break-down	Locked-rotor	Break-down	Locked-rotor
1/20	4	1550	2.71	4.13	2.44	3.5	1.6
	6	1050	4.0	6.09	3.92	5.2	2.8
	8	825	5.1	8.0	5.35	6.6	3.8
1/15	4	1550	3.62	5.23	3.22	4.5	2.1
	6	1050	5.34	7.72	5.23	6.8	3.7
	8	825	6.79	10.1	7.13	8.8	5.1
1/12	4	1550	4.52	6.39	3.97	5.7	2.7
	6	1050	6.67	9.42	6.47	8.6	4.6
	8	825	8.49	12.4	8.82	11.0	6.3
1/10	4	1550	5.42	8.0	4.72	6.9	3.1
	6	1050	8.01	11.8	7.69	10.4	5.4
	8	825	10.19	15.5	10.5	13.2	7.5
1/8	4	1550	6.78	10.4	5.83	8.8	3.8
	6	1050	10.01	15.3	9.41	13.0	6.8
	8	825	12.74	20.1	13.0	16.5	9.4
1/6	4	1550	9.04	12.7	7.68	11.3	5.2
	6	1050	13.34	18.8	12.28	17.3	8.9
	8	825	16.98	24.6	17.06	22.0	12.4
1/5	4	1550	10.85	16.0	9.11	13.8	6.1
	6	1050	16.01	23.6	14.41	20.8	10.7
	8	825	20.38	31.0	20.17	26.5	14.7
1/4	4	1550	13.56	21.0	11.25	17.3	7.5
	6	1050	20.02	31.5	17.8	26.0	13.2
	8	825	25.48	41.0	24.97	33.1	18.1
1/3	4	1550	18.08	31.5	14.82		
	6	1050	26.69	47.0	23.49		
	8	825	33.97	61.0	32.95		
1/2	4	1550	27.12	47.5	21.96		
	6	1050	40.03	70.8	32.02		
3/4	4	1550	40.68	63.5	32.54		

Source: Compiled from *NEMA Motor and Generator Standards,* NEMA Publ. no. MG1-1978, National Electrical Manufacturers Association, 1978, and original sources.

loads. Thermal protection should be used to protect against the unusual condition, not as a means for using underrated motors.

Sometimes, the horsepower rating is selcted on the basis of torque requirements, which will be discussed next.

19-8 Torque Requirements of the Application. The actual torque requirements of the load should be determined by an application test such as described in Art. 18-30 or by other appropriate means. Then, if the horsepower rating has been tentatively chosen, the torque characteristics referenced in Art. 19-4 should be compared with the needs of the application. Sometimes it is desirable to select a motor of higher horsepower rating to obtain more torque, rather than attempt to buy a specially designed motor to get the higher torque. Sometimes, also, the device itself can be altered so as to reduce slightly the torque needed by it.

19-9 Duty. In many applications, the motor is not required to run continuously and a short-time rated motor can be used. It may be rated for 5, 15, 30, or 60 min. These ratings mean that the motor may operate at full load for a period not to exceed the time rating; the ratings assume that the rest period between each such operation is long enough for the windings and other parts of the machine to cool down to within 5°C of the ambient temperature. If the motor is subjected to a complicated duty cycle, the best procedure is to make a heat run while the motor is driving its load through a typically heavy duty cycle, running the motor long enough to find the peak winding temperature.

MECHANICAL CONSIDERATIONS

Mechanical considerations involved in the selection and application of fractional horsepower motors are the following: enclosure and method of cooling, mounting arrangement, bearing construction, frame size and shaft extension, and reduction gearing.

19-10 Enclosure and Method of Cooling. Enclosures for fractional horsepower motors fall into two broad categories, open and totally enclosed, with several variations of each.

(1) Open machines. An open machine has ventilating openings through which external cooling air passes to take away the heat generated by the motor losses. An open machine may be drip-proof or splashproof. It may also be guarded, that is, have the ventilating openings so limited in size as to prevent accidental insertion of a finger into the motor where it might encounter live or rotating parts. Or, an open machine may be semiguarded, which means that part of the ventilating openings, usually those in the top half, are guarded.

(2) Totally enclosed machines. A totally enclosed machine is so enclosed as to prevent the free exchange of air between the inside and the outside of the frame, but it is not airtight. A totally enclosed motor may be nonventilated, fan-cooled (with a cooling fan external to the housing, covered by a fan cover or shroud), or explosionproof.

19-11 Mounting Arrangements. Rigid and resilient mountings are both standard and commonly available, but the latter is much to be preferred if quietness of operation is an important consideration (see Art. 17-14). Floor mounting is the standard arrangement for foot-mounted motors. Other arrangements of foot-mounted motors are wall mounting and ceiling mounting. Right-hand wall mounting means that, as the observer looks squarely at the front end shield, the mounting wall is at his right. Face-mounted motors use a machined rabbet fit on the back end shield which is bolted to the driven device; ordinarily it has no foot since it is hung by its back end shield, but sometimes a foot is used to mount the motor and a pump or other driven device is mounted on the back end shield. Examples of face-mounted motors are Figs. 19-4 and 19-6. A face-mounting end shield is also often used as the front end shield and a brake attached to it. Another form of mounting is by means of extended end-shield clamp bolts [see Art. 19-19(5)].

19-12 Bearing Constructions. Both sleeve and ball bearings are standard. Both are now commonly designed so as to permit operation of the motor with the shaft in any position, though sleeve bearings were formerly designed for operation in but one position of the shaft, which was usually horizontal. These are discussed in greater detail in Chap. 17. If there is to be an abnormal thrust or overhung load on the shaft, the problem should be taken up with the motor manufacturer.

19-13 Shaft Extension. Standard construction for extension diameters ½ in or less is to use a milled flat; for larger shafts, a keyway is standard. The number of other forms is almost limitless. Aircraft motors, for example, commonly use a spline on the shaft extension. Shaft-extension sizes are given in Fig. 19-1.

19-14 Mechanical Drives. A wide variety of mechanical drives is discussed in a single issue of the magazine *Machine Design.*[2] For the most part, the material concerns itself with mechanical drive arrangements for motors larger than fractional horsepower, but a lot of discussion is devoted to basic principles which could apply to smaller versions; many of the drives discussed are used in fractional horsepower sizes. The issue discusses chains of various types and sprockets; types of V belts and their characteristics and sheaves for fixed or adjustable speeds; flat, conventional, and grooved or poly-V belts; gears of all types including epicyclic and harmonic drives and gear-tooth forms, precision, quality, and rating; packaged adjustable-speed drives including gear drives, belt and chain drives, friction and traction drives, variable-stroke drives, and torque converters; speed reducers including base-mounted and shaft-mounted arrangements; clutches of many kinds including mechanical clutches of the positive-drive type as well as friction clutches, overrunning clutches, and centrifugally actuated clutches; electric clutches of many types including eddy-current couplings and magnetic-particle clutches; fluid couplings; brakes, both mechanical and electric; mechanical couplings of many types; universal joints and flexible shafts.

By use of the proper gearing, almost any output speed may be obtained, higher or lower than motor speed. The gearing may be designed integral with the motor or supplied as a separate device; both such arrangements are discussed further in Arts. 17-22 to 17-24.

DEFINITE-PURPOSE MOTORS

Definite-purpose motors, it was explained in Art. 19-2, are motors designed and built to suit the special requirements of a particular industry or type of application. We will now proceed to examine these motors, and the material that follows is mostly extracted or adapted from NEMA standards. For more complete and more-up-to-date information, the standards themselves should be consulted.[1]

19-15 Permanent-Split Industrial Instrument Capacitor Motors and Gearmotors. These are subfractional horsepower motors intended for application in such instruments as recorders and timing devices.

(1) Types. Synchronous capacitor, nonsynchronous capacitor with normal slip, and nonsynchronous capacitor with high slip.

(2) Voltage ratings. 115 V.

(3) Frequencies. 60 and 50 Hz.

(4) Speed and torque ratings. For 60-Hz motors, see Table 19-6. Comparable gear motors are built with gear ratios from 6:1 to 1800:1.

(5) Terminal markings. See Figs. 7-3, 7-4, and 16-3 (servomotor).

(6) General mechanical features. Totally enclosed, ball-bearings, face-

TABLE 19-6 Characteristics of Permanent-Split Capacitor Motors for Industrial Instruments

	Axial length of motor body, in		
	1.75	2.00	2.50
Synchronous capacitor motors:			
Full-load rpm	1800	1800	1800
Full-load and pull-in torque, oz-in	0.25	0.33	0.60
Locked-rotor torque, oz-in	0.25	0.35	0.60
Nonsynchronous capacitor motors: normal slip:			
Approx. full-load rpm	1550	1550	1550
Full-load torque, oz-in	1.0	1.4	2.4
Breakdown torque, oz-in	1.7	2.4	4.2
Locked-rotor torque, oz-in	1.0	1.4	2.4
Nonsynchronous capacitor motors: high slip			
Approx. full-load rpm	1200	1200	1200
Full-load torque, oz-in	0.75	1.10	1.85
Breakdown torque, oz-in	1.5	2.2	3.7
Locked-rotor torque, oz-in	1.5	2.2	3.7

type mounting using two no. 8-32 tapped holes 180° apart. The shaft extension has a 0.1875-in (4.76-mm) diameter with a milled flat.

19-16 Low-Inertia Industrial Instrument Servomotors and Gearmotors. These are subfractional horsepower motors intended for applications in such instruments as self-balancing recorders and remote positioning devices. These motors are discussed in some detail in Chap. 16.

19-17 Universal Motor Parts. These are discussed in Chap 11.

19-18 Motors for Hermetic Refrigeration Compressors. A hermetic motor consists of a stator and rotor without a shaft, end shields, or bearings for installation in refrigeration compressors of the hermetically sealed type.

(1) Types. Split-phase, capacitor-start, two-value capacitor, permanent-split capacitor, and split-phase capacitor motors.

(2) Output ratings. Unlike most motors, hermetic motors are rated in terms of breakdown torque instead of horsepower. They are made in two- and four-pole speeds. Breakdown torque ratings, together with their respective locked-rotor currents for 60-Hz motors, are given in Table 19-7.

TABLE 19-7 Breakdown Torques and Locked-Rotor Currents of Single-Phase Motors for Hermetic Refrigeration Compressors

Four-pole motors			Two-pole motors	
Breakdown torque, oz-ft	Locked-rotor amperes at 115 V		Breakdown torque, oz-ft	Locked-rotor amperes at 115 V
10.5	20		5.25	20
12.5	20		6.25	20
15	20		7.5	20
18	20		9.0	20
21.5	20		10.75	21
26	21.5		13.0	23
31	23		15.5	26
37	28	23*	18.5	29
44.5	34	23*	22.0	33
53.5	40		27.0	38
64.5	48	46*	32.0	43
77	57	46*	38.5	49
92.5	68	46*	46.0	56

*Motors having these locked-rotor currents usually have lower locked-rotor torques than those with the higher locked-rotor amperes.

Source: From *NEMA Motor and Generator Standards,* NEMA Publ. no. MG1-1978, National Electrical Manufacturers Association, Washington, 1978.

(3) Voltage ratings. 115, 200, and 230 V for 60 Hz and 110 and 220 V for 50 Hz.

(4) Frequencies. 60 and 50 Hz.

(5) Terminal markings. Terminal markings are as follows: the auxiliary winding is white, the main winding is white with a red tracer, and the common lead is white with a black tracer.

(6) Dimensional standards. Special dimensional standards have been developed and published by NEMA.[1]

19-19 Motors for Shaft-Mounted Fans and Blowers. These motors are either totally enclosed or open and are designed for propeller fans or centrifugal blowers mounted on the motor shaft, with or without air drawn over the motors. They are generally not suitable for belted loads because many motors of this type do not develop enough locked-rotor torque for this purpose.

(1) Types. Single-phase, split-phase, permanent-split capacitor, or shaded-pole motors with horsepower ratings of 1/20 hp and larger. Also available are polyphase squirrel-cage induction motors with horsepower ratings of ⅛ hp and larger. DC motors with frame diameters up to approximately 6 in are shunt-wound, and DC motors with frame diameters of 6 in and larger are compound-wound.

(2) Horsepower and speed ratings. Single-speed motors are available in all the horsepower and speed ratings given in Table 19-1 except that there are few two-pole (3450-rpm) motors available.

(3) Voltage ratings. AC single-phase motors, 115 and 230 V; polyphase motors, 115, 200, 230, 460, and 575 V; dc motors, 115 and 230 V.

(4) Two-speed and adjustable varying-speed motors. Refer to Chap. 7.

(5) General mechanical features. Totally enclosed or open. Horizontal motors have sleeve bearings with a provision in the front bearing for taking the axial thrust of the driven fan. Vertical motors have sleeve or ball bearings. The rear-end oiler on sleeve-bearing motors may extend the length of the motor. End-shield clamp bolts have a threaded extension of ⅜ in (9.5 mm) or more beyond the nut on the back end of the motor for mounting the motor or for attaching a fan guard to the motor. Permanent-split capacitor and shaded-pole motors may sometimes be provided wtih blade terminals for line connections.

19-20 Motors for Belted Fans and Blowers. These motors are intended for operating belt-driven fans or blowers that are commonly used in conjunction with heating and air-conditioning installations and whole-house attic fans.

(1) Types. Single- or two-speed capacitor-start, split-phase, or split-phase capacitor motors.

(2) Speed ratings. Single-speed motors are rated 1725 rpm, and two-speed motors are rated 1725/1140 rpm.

(3) Horsepower ratings for single-speed motors. Horsepower ratings for split-phase and split-phase capacitor motors are ⅙, ¼, ⅓, ½, and ¾ hp. Horsepower ratings for capacitor-start motors are ⅓, ½, ¾, and 1 hp.

(4) Horsepower ratings for two-speed motors. On the higher speed (1725 rpm) these motors have the same horsepower ratings as single-speed motors. The horsepower ratings on the lower speed (1140 rpm) are about one-third the ratings of the higher speed.

(5) Torque characteristics. Split-phase and split-phase capacitor motors have the same torque characteristics as standard split-phase motors in Table 19-1. Capacitor-start motors have the same torque characteristics as the general-purpose capacitor-start motors in Table 19-1. Two-speed motors start on the high-speed connection and have about one-half as much breakdown torque on the lower speed as on the higher speed.

(6) Locked-rotor current. See Table 19-1.

(7) General mechanical features. Open, drip-proof, sleeve bearings, resilient mounting, automatic-reset thermal overload protection. Split-phase capacitor motors operate as split-phase motors on the lower speed connection.

19-21 Motors for Air-Conditioning Condensers and Evaporator Fans.

(1) Types. Shaded-pole and permanent-split capacitor motors. Very few shaded-pole motors are used today.

(2) Horsepower and speed ratings. See Table 19-5.

(3) Breakdown torques. See Table 19-5.

(4) Variations from rated speed. Since these motor are high-slip motors, as can be seen from Table 19-5, they are subject to wider variations in operating speed than general-purpose fractional horsepower motors. This point is discussed in Art. 9-8.

(5) Terminal markings for multispeed motors. For shaded-pole types, see Fig. 9-13; for permanent-split capacitor motors, see Chap. 7

(6) General mechanical features. Open or totally enclosed, sleeve bearings. Vertical shaft-up motors have a plastic rain shield pressed on the shaft to retard water flow into the bearing cavity.

19-22 Motors for Cellar Drainers and Sump Pumps. A cellar-drainer motor furnishes power for operating a pump for draining cellars, pits, or sumps.

(1) Type. Split-phase.

(2) Horsepower and speed rating. The horsepower rating is ⅓ hp at 1725 rpm.

(3) Torque characteristics. Breakdown torque is 32 oz-ft and locked-rotor toque is 20 oz-ft.

(4) General mechanical features. Vertical operation, open construction with the top end shield totally enclosed or protected by a canopy, louvres, or the equivalent. The bottom end shield has a hub machined so that the motor can be mounted directly on a support pipe. The motor may or may not have an automatic-reset thermal protector. The frame number bears the suffix K. The dimensions of such a motor are given in Fig. 19-2.

19-23 Motors for Gasoline-Dispensing Pumps. A gasoline-dispensing pump motor is commonly used in automobile service stations. It is built to meet the rigid requirements of Class I Group D explosionproof construction as established by the Underwriters Laboratories and is so certified on the nameplate.

(1) Types. Capacitor-start, polyphase squirrel-cage, and dc com-

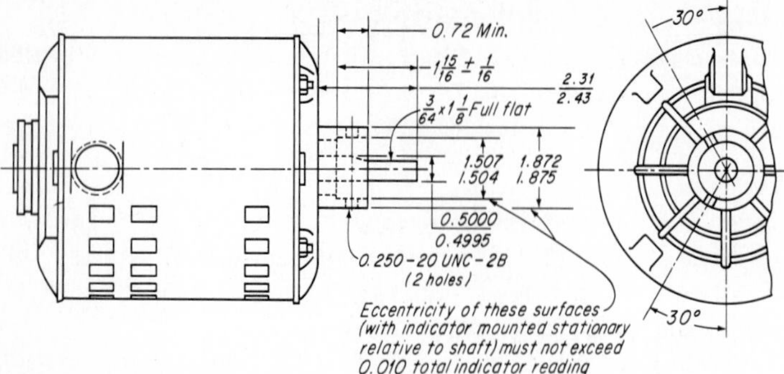

Fig. 19-2 Dimensions (in inches) for cellar-drainer and sump-pump motors. *(From NEMA Motor and Generator Standards, NEMA Publ. no. MG1-1978, National Electrical Manufacturers Association, Washington, 1978.)*

pound-wound motors.

(2) Horsepower and speed ratings. The horsepower rating is ½ hp. For a 60-Hz motor, the speed rating is 1725 rpm; for a 50-Hz motor, 1425 rpm; and for a dc motor, 1725 rpm.

(3) Torque characteristics. For 60-Hz single-phase motors, the breakdown torque is 46-oz-ft or more and the locked-rotor torque is 48.8 oz-ft or more.

(4) Locked-rotor current. See Table 19-1.

(5) General mechanical features. Totally enclosed, explosionproof Class 1 Group D, sleeve bearings, rigid base mounting. It has a built-in line switch with an external operating lever, a built-in voltage-selector switch on the same end as the line switch (see Fig. 19-3), and may have a built-in thermal protector. Line leads 36 in long are brought out through a swivel connector.

(6) Dimensions and frame size. Outline dimensions are given in Fig. 19-3. This frame size is designated as 61G.

19-24 Motors for Domestic Oil Burners. The motors described next are intended for operating mechanical-draft oil burners for domestic applications.

(1) Type. Split-phase.

(2) Horsepower and speed ratings. Horsepower ratings are 1/12, ⅛, and ⅙ hp at 1725 rpm and ⅛, 1/7, ⅙, and ¼ hp at 3450 rpm.

(3) Direction of rotation. Clockwise, facing the end opposite the shaft.

(4) Breakdown torque. The breakdown torque is slightly less than the values given in Table 19-1 for standard split-phase motors. Approximate values for 1725-rpm motors are 1/12 hp, 10 oz-ft; ⅛ hp, 14 oz-ft; and ⅙ hp, 18 oz-ft. For 3450-rpm motors the approximate values are ⅛ or 1/7

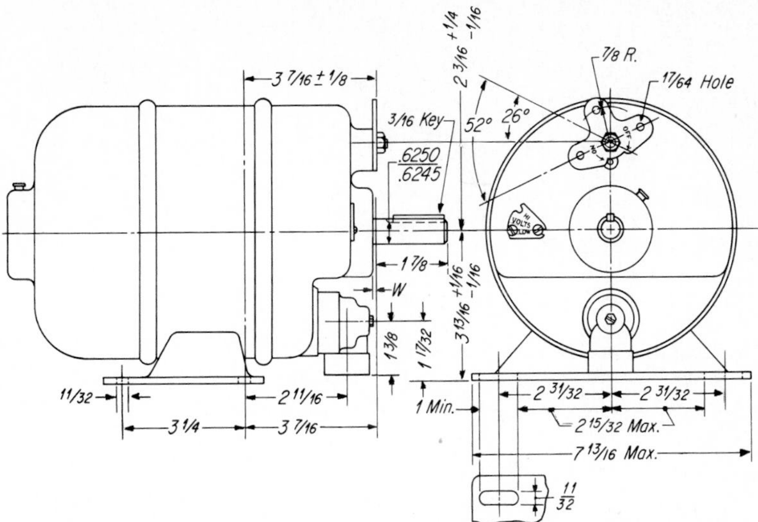

Fig. 19-3 Outline dimensions (in inches) for gasoline-dispensing pump motors. *(From NEMA Motor and Generator Standards, NEMA Publ. no. MG1-1978, National Electrical Manufacturers Association, Washington, 1978.)*

hp, 7.5 oz-ft; ⅙ hp, 10 oz-ft; and ¼ hp, 14 oz-ft.

(5) Locked-rotor current. Locked-rotor current is not more than 20 A for 115-V motors and not more than 12 A for 230-V motors up to ⅙ hp. Locked-rotor current for ¼-hp motors is not more than 26 A for 115-V motors and not more than 15 A for 230-V motors.

(6) General mechanical features. Totally enclosed or vent slots in the face of the flange end shield, manual-reset thermal overload protector. Two 20-in-long leads of flexible single-conductor wire exit the motor through a hole tapped for a ½-in conduit.

(7) Dimensions and frame size. Outline dimensions for oil-burner motors are given in Fig. 19-4. Note that these are designed for face mounting and there are two flange sizes, which are distinguished by the suffix letter following the frame number.

19-25 Motors for Home-Laundry Equipment.

For a discussion of these motors, see Arts. 4-27 and 4-29.

19-26 Motors for Jet Pumps.

A jet-pump motor is an open-type ball-bearing motor built for horizontal or vertical operation for direct connection to direct-driven centrifugal ejector pumps. A cutaway view of a jet-pump motor is illustrated in Fig. 19-5.

(1) Types. Split-phase, capacitor-start, polyphase squirrel-cage induction, and dc compound-wound motors.

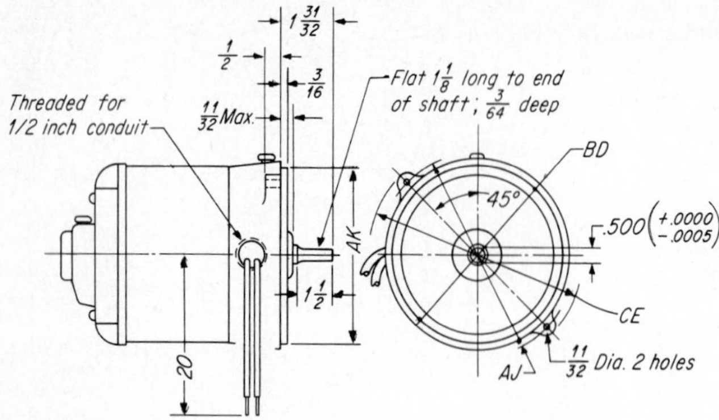

Frame suffix	AK	AJ	CE*	BD*
M	5½	6¾	7¾	6¼
N	6⅜	7¼	8¼	7

*Maximum dimensions.

Fig. 19-4 Outline dimensions (in inches) for motors for domestic oil burners. *(From NEMA Motor and Generator Standards, NEMA Publ. no. MG1-1978, National Electrical Manufacturers Association, Washington, 1978.)*

Fig. 19-5 Cutaway view of capacitor-start jet-pump motor. Note the large internal cooling fan to provide high service factors. Note also the internal capacitor. *(Emerson Motor Division.)*

(2) Horsepower and speed ratings. Horsepower ratings are 1, ¾, ½, ⅓, and ¼ hp and are all two-pole. The speed rating for 60-Hz and dc motors is 3450 rpm. Split-phase motors generally are not built for more than ⅓ hp.

(3) Direction of rotation. Clockwise, facing the end opposite the shaft extension.

(4) Breakdown torques. See Table 19-1.

(5) Locked-rotor current. See Table 19-1.

(6) Service factors. Jet-pump motors are customarily designed for higher service factors than the standard NEMA values given in Table 1-3. Common values for single-phase motors, for example, are 1.8 for ⅓-hp motors, 1.6 for ½-hp motors, 1.5 for ¾-hp motors, and 1.4 for 1-hp motors. For exact values in any particular case, refer to the motor nameplate or to the manufacturer.

(7) General mechanical features. Open construction, a ball bearing in the back end shield and a ball or sleeve bearing in the front end shield, suitable for horizontal or vertical operation, capable of taking axial end thrust. The back end shield is machined as in Fig. 19-6, and the front end shield is provided with a tapped hole or other means for holding a drip cover when used vertically. They have a standard shaft extension that is straight with a keyway and an alternate shaft extension that is threaded as in Fig. 19-6*b*. The terminals or leads are located in the front end shield or near it. The capacitor unit is mounted externally, as shown in Fig. 19-6, or internally, as shown in Fig. 19-5. These motors are provided with an automatic-reset thermal overload protector. When the shaft extension shown in Fig. 19-6*b* is provided, a screwdriver slot or other means for holding the shaft during assembly or removal of the impeller is provided in the end opposite the shaft extension.

(8) Mounting dimensions and frame size. Jet-pump motors are usually built in the same frame sizes as general-purpose motors, as given in Table 19-3. When the motor has a threaded shaft extension as in Fig. 19-6*b*, the frame number is followed by the suffix letter J; when the motor has a standard shaft extension as in Fig. 19-6*a*, the frame number is followed by the suffix letter C.

19-27 Motors for Carbonator Pumps. A carbonator-pump motor is an open-type sleeve bearing motor built for horizontal operation for direct connection to a carbonator pump. One such motor is illustrated in Fig. 19-7.

(1) Type. Split-phase.

(2) Horsepower and speed ratings. Horsepower ratings are ⅙, ¼, and ⅓ hp. The speed rating for a 60-Hz motor is 1725 rpm.

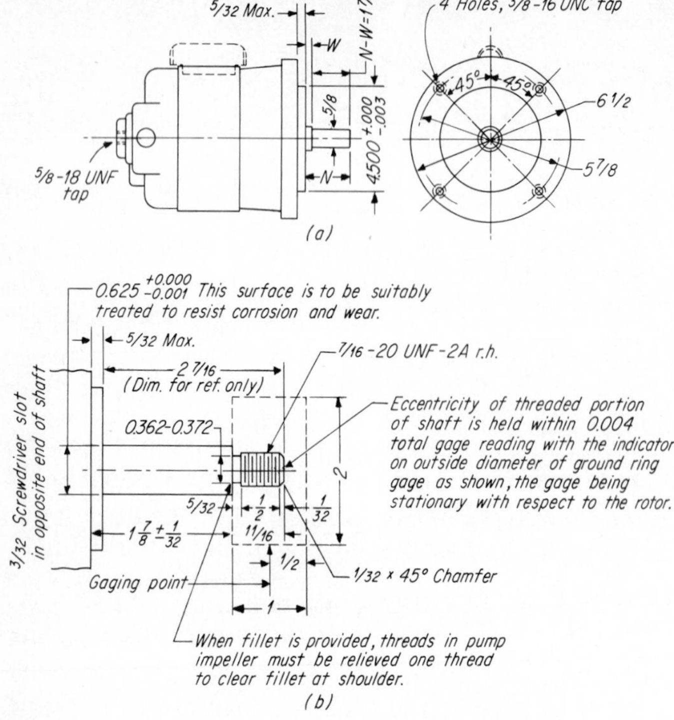

Fig. 19-6 Motor outline and shaft-exension dimensions (in inches) for face-mounted motors for jet pumps. (*a*) Motors with a straight shaft extension; (*b*) motors with a threaded shaft extension. *(From NEMA Motor and Generator Standards, NEMA Publ. no. MG1-1978, National Electrical Manufacturers Association, Washington, 1978.)*

Fig. 19-7 A carbonator pump motor with close-coupled carbonator pump attached to the specially machined hub. *(Emerson Motor Division.)*

(3) Direction of rotation. Counterclockwise, facing the end opposite the shaft extension.

(4) Breakdown torques. See standard split-phase motors in Table 19-1.

(5) Locked-rotor currents. These are not more than 50 A for 115-V motors and not more than 25 A for 230-V motors.

(6) General mechanical features. Open construction, sleeve bearings, resilient mounting, automatic-reset thermal overload protector. The extended hub on the shaft extension end shield is machined as in Fig. 19-8 to accept a pump clamped onto this hub. The shaft extension is shorter than the hub and has a cross-milled slot as in Fig. 19-8 to accept the pump shaft tang.

(7) Mounting dimensions and frame size. Carbonator-pump motors are normally built in the NEMA 48 frame, with dimensions as given in Fig. 19-8. When the motor has dimensions as given in Fig. 19-8, the frame number is followed by the suffix letter B.

19-28 Motors for Coolant Pumps. These are enclosed ball-bearing motors built for horizontal or vertical operation for direct connection to direct-driven centrifugal coolant pumps.

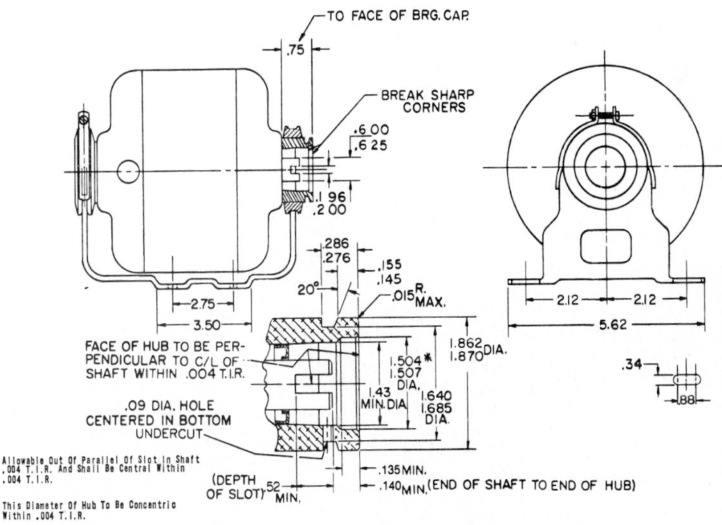

Fig. 19-8 Motor outline, shaft extension, and extended hub dimensions for carbonator pump motors. *From NEMA Motor and Generator Standards, NEMA Publ. no. MG1-1978, National Electrical Manufacturers Association, Washington, 1978.)*

(1) Types. Split-phase, capacitor-start, polyphase squirrel-cage induction, and dc compound-wound.

(2) Horsepower and speed ratings. These are the same as those shown for the two-pole and four-pole motors in Table 19-1.

(3) Direction of rotation. Clockwise, facing the end opposite the shaft extension.

(4) Breakdown torques. See Table 19-1.

(5) Locked-rotor current. See Table 19-1.

(6) General mechanical features. Generally the features are the same as for jet-pump motors, except that the motor is totally enclosed.

19-29 Submersible Motors for Deep-Well Pumps. A submersible motor for deep-well pumps is designed for operation totally submerged in water having a temperature not exceeding 25°C. Such fractional horsepower motors are built in a 3¾-in diameter frame, are two-pole, and have a splined shaft extension. One such motor is illustrated in Fig. 19-9. Motors of similar construction are also used in a variety of applications such as pumping oil, gasoline, chemicals, or solvents and operate submerged in the fluid.

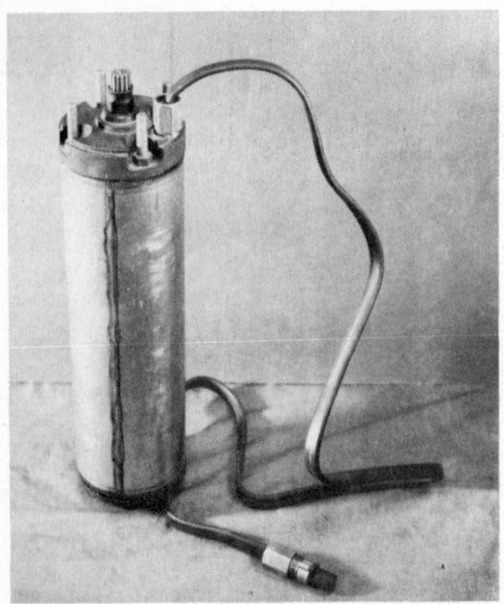

Fig. 19-9 A 4-in submersible motor. It is capable of withstanding thrust loads up to 300 lb and fluid pressures up to 1000 lb/in^2. It is built in ratings from ⅙ to 1½ hp in capacitor-start and polyphase types and up to ¾ hp in split-phase and permanent-split capacitor types. *(Franklin Electric Company.)*

19-30 DC Motors Intended for Use on Adjustable-Voltage Electronic Power Supplies. These motors are discussed in Art. 13-6, and more generally in Arts. 13-2 through 13-7.

MISCELLANEOUS

19-31 Service Conditions. Service conditions are classified by NEMA as *usual* or *unusual;* both of these two categories are subdivided into environmental conditions and operating conditions.

(1) Usual service conditions: environmental. These are exposure to an ambient temperature from 0 to 40°C, an altitude not exceeding 3300 feet (1000 m), and installation on a rigid mounting surface and in areas or supplementary enclosures which do not seriously interfere with the ventilation of the machine.

(2) Usual service conditions: operational. These service conditions include voltage variations that are not over plus or minus 10 percent for dc or induction motors and not over plus or minus 6 percent for universal motors (except fan motors), frequency variation that is not over plus or minus 5 percent, combined voltage and frequency variation of not over 10 percent, and belt, chain, or gear drives in accord with NEMA practices.

(3) Unusual service conditions: environmental. In general, these are all conditions that are not specified as usual. Some typical ones requiring special consideration and consultation with the motor manufacturer are exposure to chemical fumes, lint, steam, flammable or explosive gases, oil vapor, salt air, radiant heat, nuclear radiation, excessive dampness or dryness; exposure to dust, particularly if it is combustible, explosive, abrasive, or conducting; exposure to ambient temperatures below 0°C or above 40°C; exposure to abnormal shock or vibration, or mechanical loads involving abnormal thrust or overhung loads; and exposure to vermin infestation.

(4) Unusual service conditions: operational. These may be succinctly defined as operating conditions other than usual. Specific illustrative examples include operation at speeds above the highest rated, operation in poorly ventilated rooms or in pits or enclosures that restrict ventilation, operation where voltage and frequency are outside the usual range, operation where low noise levels are required, operation of a dc motor on rectified alternating current unless the motor is specifically designed for it, operation when subjected to torsional impact loads, continuous operation of a short-time-rated machine, and operation of a polyphase motor on an unbalanced-voltage supply.

19-32 Effects of Variation of Voltage and Frequency on the Performance of Induction Motors. Induction motors are normally designed to operate on a voltage range of plus or minus 10 percent of rated, a frequency variation of 5 percent, or a combined voltage and frequency variation of not over 10 percent. Performance characteristics are, however, affected somewhat by such variations.

Locked-rotor, pull-up, breakdown, and switching torques will be proportional to the square of the applied voltage when the frequency is held constant. At any speed, the torque developed by the motor will be proportional to the square of the applied voltage. The full-load slip will be nearly inversely proportional to the square of the applied voltage. A small variation in frequency, with voltage held constant, will make the torques, expressed in ounce-feet, nearly inversely proportional to the square of the frequency. Efficiency will usually be poorer at one end of the permissible voltage variation range, but not enough to cause concern. Full-load power factor usually improves at reduced voltage and becomes worse at higher voltages. If both voltage and frequency are varied together in the same direction over a small range, torques in ounce-feet remain constant. See Art. 19-33.

19-33 Operation of 60-Hz Motors on 50 Hz. In general, 60-Hz single-phase induction motors cannot be recommended for operation on 50-Hz circuits. One reason is that the starting switch would not operate at the correct speed; nor would a starting relay. Another reason is the capacitor, if one is used; it would need to have about 40 percent more capacitance. If, however, the proper starting device and capacitor for 50-Hz operation were obtained and installed, the effect on performance would be about the same as for a polyphase motor.

If a 60-Hz polyphase induction motor is to be operated on a 50-Hz circuit, the applied voltage should be reduced to five-sixths of its 60-Hz voltage rating: the torque loading on the motor should be kept the same, which means that the horsepower load is reduced by one-sixth. If these conditions are observed, the following results can be expected: locked-rotor, breakdown, and full-load torques remain unchanged; speed becomes five-sixths of the 60-Hz speed; locked-rotor amperes will be decreased by approximately 5 percent; service factor will be 1.0; temperature rise will not be excessive.

19-34 Relationships between Power, Torque, and Speed. The basic relationship for calculating torque from power and speed is

$$\text{Torque} = \frac{K \times \text{power}}{\text{rpm}} \tag{19-1}$$

For calculating power from torque and speed, the equation may simply be arranged in the form

$$\text{Power} = \frac{\text{torque} \times \text{rpm}}{K} \qquad (19\text{-}2)$$

The value of K in the preceding equations depends upon the units in which torque and power are expressed; proper values of K are given in Table 19-8.

19-35 Radio Interference and Suppression Devices. Commutator-type motors may give trouble as a result of radio interference, because the commutation, or repeated interruption, of the currents produces components in the radio-frequency ranges, which may interfere with reception. These interfering currents are conducted along the power wires from the motor, and the fields set up may affect the radio aerials in the vicinity. There may even be some radiation from the motor itself, though this is usually slight except within a few feet of the motor. The interference voltages exist across the power leads (symmetrical component) and from the power leads jointly to the motor frame (asymmetrical component).

In order to suppress the interference it is necessary to use either or both of the following methods:

1. Connect a capacitor from each motor lead to the frame. This tends to put a short circuit on the motor for the high frequencies. (On motors that are operated with their frame insulated from ground, the microfarads of capacitance used are limited to approximately 0.0001 mfd per

TABLE 19-8 Values of *K* for Power-Torque-Speed Calculations*

Units in which torque is expressed	Units in which power is expressed			
	Watts	Horsepower (746 W)	Chevauex-vapeur (735.5 W)	Millihorsepower (0.746 W)
Pound-feet	7.043	5,254	5,180	5.254
Pound-inches	84.52	63,050	62,160	63.05
Ounce-feet	112.69	84,070	82,890	84.07
Ounce-inches	1,352	1,008,800	994,600	1,008.8
Centimeter-grams	97,376	71,643,000	71,620,000	72,643
Centimeter-kilograms	97.376	72,643	71,620	72.643
Meter-kilograms	0.97376	726.43	716.2	0.72643
Newton-meters	9.549	7,124	7,024	7.124

*$\text{Torque} = \frac{K \times \text{power}}{\text{rpm}} \qquad \text{Power} = \frac{\text{torque} \times \text{rpm}}{K}$.

volt of motor rating; capacitance is limited here because of the danger of electric shock.)

2. Connect a choke coil in series with the power leads. This tends to open-circuit the motor for the high frequencies.

Capacitor methods generally used with quite satisfactory results are shown in Figs. 19-10 and 19-11. It is very important that the leads con-

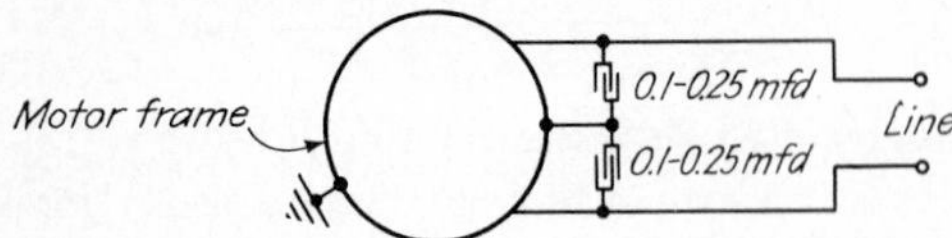

Fig. 19-10 Circuit for suppressing radio noise from a small motor with a solidly grounded frame.

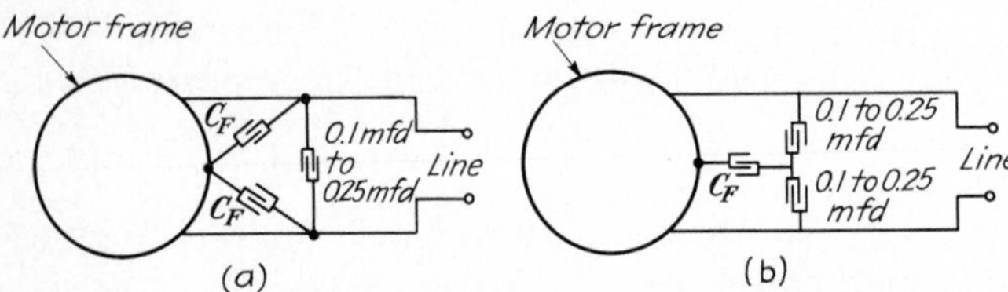

Fig. 19-11 Circuits for suppressing radio noise from a small motor with a frame insulated from ground. For 115-V motors, C_F = 0.10 mfd. For 230-V motors, C_F = 0.005 mfd. (*a*) "Pi" network; (*b*) T network.

necting the capacitors be as short as possible. In most cases, it is practicable to wire a capacitor to a power lead with 2 to 3 in of wire. Fuses are not commonly used in series with capacitors since they will reduce the effectiveness of a capacitor-type suppressor. One should use capacitors with sufficient insulation strength.

References

1. *NEMA Motors and Generator Standards,* NEMA Publ. no. MG1–1978, National Electrical Manufacturers Association, 2101 L Street N.W., Washington ,DC 20037, 1978.
2. "Mechanical Drives Reference Issue." *Mach. Des.,* Sept. 21, 1967.

Bibliography

Carville, T. E. M.: Selecting and Applying Fractional-Horsepower Motors, *Mach. Des.,* July 24, 1958, pp. 127–133.

"Electrical Motor Selector," *Mach. Des.,* July 24, 1958, pp. 141–150.

Electric Motors Reference Issue, Mach. Des., Dec. 16, 1965.

Electric Motors Reference Issue, Mach. Des., April 9, 1970.

Kordatzky, Robert W.: "Subfractional-Horsepower A-C Motors for Instrument Applications," *Mach. Des.,* Feb. 1, 1968, pp. 96–99.

Morgan, N. L.: "Which Fhp motor for the Job?" *Prod. Eng.*, Jan. 19, 1959, pp. 76–81.

Energy Management Guide for Selection and Use of Single-phase Motors, NEMA Standards Publ. no. MG11-1977, NEMA, 2101 L Street N. W., Washington, DC 20037, 1977.

National Electrical Manufacturers Association: *Energy Management Guide for Selection and Use of Polyphase Motors*, NEMA Standards Publ. no. MG10-1977, NEMA, 2101 L Street N.W., Washington, DC 20037, 1977.

National Electric Manufacturers Associatioh: *Guide for the Development of Metric Standards for Motors and Generators*, NEMA, 2101 L Street N. W., Washington, DC 20037, November 1980.

Matthews, Robert W.: "Selecting A-C Motors for Instrument Service," *Mach. Des.*, Aug. 20, 1959, pp. 166–171.

APPENDIX

TABLE A-1 Trouble Diagnosis Chart for Fractional Horsepower Motors

Problem	Probable causes	Test and remedy
A. Failure to start	1. Blowing of fuses	1. Check capacity of fuses; they should be rated at not less than 125 percent of full-load current. Disconnect motor from line, replace fuses, and reconnect motor to line. If fuses blow, investigate and remedy cause.
	2. Overload device open-circuited. (Applies to manual-reset devices)	2. Reset overload device. If it trips out repeatedly, investigate further to find out why.
	3. Improper current supply, incorrect voltage or frequency	3. Make certain that voltage, frequency, and number of phases of supply circuit agree with the motor nameplate stamping. Correction involves getting the proper motor. Single-phase motors can be operated on polyphase

TABLE A-1 Trouble Diagnosis Chart for Fractional Horsepower Motors (*Cont.*)

Problem	Probable causes	Test and remedy
A. Failure to start *(Continued)*		circuits, but not vice versa. AC motors—except universal motors—will not operate on direct current, and vice versa.
	4. No voltage or low voltage	4. Measure volts at motor terminals with line switch closed. It should be within 10 percent of rating on nameplate; if not, larger feeder circuits, or shorter connections to motor, may be required.
	5. Improper line connections	5. See that connections are exactly in accordance with the connection diagram or connection instructions furnished with the motor; if no wiring diagram is available, a suitable one may be found in this book. However, terminal markings of the motor at hand may be different from the markings on the wiring diagram given for a motor of the type concerned. If a wiring diagram that apparently fits is found, the first step should be a systematic check to see how the terminal markings differ, if at all, from those of the diagram selected.
	6. Mechanical failure in load	6. Check the driven appliance to see that it is not jammed and that it turns freely. Disconnect motor from the load, and determine whether it will start idle. Watch out for excessive belt tension.
	7. Motor bearings tight or seized	7. When bearings seize in a motor having high torques, such as a capacitor-start or repulsion-start motor, new bearings are usually required. In a motor having low torques, such as a shaded-pole motor, it is often sufficient to

TABLE A-1 Trouble Diagnosis Chart for Fractional Horsepower Motors (*Cont.*)

Problem	Probable causes	Test and remedy
A. Failure to start *(Continued)*		clean and relubricate the bearing. The bearing can be cleaned with a clean cloth soaked in gasoline or kerosene.
	8. Excessively worn or eccentric bearings	8. Motor shaft turns freely by hand when motor is not energized but locks when power is applied so that it cannot be turned by hand. Remedy is to install new bearings.
	9. Dirt or foreign matter in air gap	9. Remove the obstruction.
	10. Excessive load	10. If motor starts idle and has no apparent defect, it may not have enough locked-rotor torque for the application, or may otherwise be misapplied. Refer to Chap. 19 for guidance in selecting the right motor.
	11. Open circuit in starting phase, main phase, or both	11. This condition is often indicated by a humming sound when the line switch is closed. It may be due to a broken lead, improper or poorly soldered connection, defective starting switch (see Arts. 17-5 to 17-9), burned-out starting winding, defective capacitor, loose connection on thermal overload device.
	12. Short-circuited stator	12. This condition is likely to cause fuses to blow; rewinding is required. Similar condition can be caused by a grounded winding if motor is operated on a grounded circuit.
	13. Defective capacitor	13. If an electrolytic capacitor is used, disconnect it from the windings and test it as explained in Art. 8-4; if it is defective, it should be replaced (see Art. 8-3).
	14. Defective controller	14. If motor uses an external

TABLE A-1 Trouble Diagnosis Chart for Fractional Horsepower Motors (*Cont.*)

Problem	Probable causes	Test and remedy
A. Failure to start *(Continued)*		controller, the latter should be checked for defects.
	15. Faulty starting switch	15. See Arts. 17-5 to 17-9.
	Items applying especially to repulsion-type motors	
	16. Worn or sticking brushes	16. Brushes may be worn so that they do not touch commutator, or only very lightly, in which case new brushes are required. If the brushes are sticking in the holders, clean brushes and brush holders so that brushes move freely. Brush springs may be weak and require replacing.
	17. Incorrect setting of brushes	17. Check to see that brushes are set the proper distance off neutral.
	18. Defective short-circuiter, defective brush lifter	18. See whether mechanism works freely, whether commutator is clean opposite short-circuiter. Repair or replace the defective mechanism.
	19. Short-circuited armature	19. Remove brushes from commutator, impress full voltage on the stator winding, and turn armature by hand. If there are one or more points at which the rotor "hangs," the armature is short-circuited. By forcing the rotor to the position where it is most difficult to hold, the short circuit can be located, for the short-circuited coil will become hot. Sometimes the short circuit can be located and repaired; frequently, rewinding the armature is required.
	Items applying especially to universal motors	
	20. Worn or sticking brushes	20. See item 16 in this section.

TABLE A-1 Trouble Diagnosis Chart for Fractional Horsepower Motors (*Cont.*)

Problem	Probable causes	Test and remedy
A. Failure to start *(Continued)*	21. Defective speed governor	21. Inspect governor contacts, which may require cleaning. If contacts are badly burned, test capacitor for opens or shorts. Also inspect governor mechanism. Replace defective parts.
	22. Too much external resistance	22. See Art. 11-20.
B. Excessive bearing wear	1. Belt tension too great; misalignment of belt, coupling, or drive gears; unbalanced coupling; eccentric or too closely meshed gears; excessively heavy fan, flywheel or other load hung on motor shaft	1. Correct the faulty mechanical condition. Belts, whether flat or V-shaped, should have only enough tension to prevent slipping.
	2. Dirty bearings, particularly if dirt is abrasive	2. If condition is bad, provide means for shielding motor from the dirt or consult motor manufacturer and obtain a specially designed motor for the application.
	3. Insufficient or inadequate lubrication	3. Sleeve bearings should be lubricated with a good grade of light machine oil about once every year for normal applications and more frequently if the motor runs continuously (see Art. 17-2). For care of ball bearings, refer to Art. 17-4. See also Sec. A, item 7.
	4. Excessive thrust load	4. Reduce the thrust load or obtain a motor designed to handle the required thrust.
	5. Bent shaft (sleeve-bearing motors)	5. Remove rotor and check with a dial indicator; straighten as required.
C. Motor runs hot. *Do not judge motor temperature by hand.* Measure the temperature by thermometer,	1. Improper line connections	1. See Sec. A, item 5.
	2. Excessive load	2. Inspect the driven device and the belt, coupling, or other mechanical connection between the motor and its load. Run an application test, as explained in Art. 18-29,

TABLE A-1 Trouble Diagnosis Chart for Fractional Horsepower Motors (*Cont.*)

Problem	Probable causes	Test and remedy
thermocouple, or rise of resistance [see Arts. 18-10 and 18-16(9)]		and determine whether a motor of the right horsepower rating is used (see Art. 19-7); if not, it may be necessary to obtain a motor of larger rating. Continuous overloading shortens the life of a motor.
	3. Improper circuit voltage	3. Measure voltage at motor terminals with motor driving its load. If voltage differs more than 10 percent from nameplate voltage, take full-load saturation test as described in Art. 18-16(8). If watts input at operating voltage is materially greater than at rated voltage, this is the probable cause of the difficulty, but if there is little difference in inputs at these two voltages, a further search for the trouble should be made. See also Art. 19-32 for further discussion of this subject.
	4. Wrong frequency	4. Compare circuit frequency with nameplate frequency. See Art. 19-32 for a discussion of this subject.
	5. Too frequent starting	5. This condition may occur if motor is automatically controlled. Adjust the control to lengthen the cycle, if possible; reduce the WR^2 of the connected load; or obtain manufacturer's recommendation for a new motor.
	6. Failure of ventilation	6. Clean motor, especially the ventilating passages. If motor has air-flow baffles to direct the air for effective cooling, make certain that these are correctly placed. Allow nothing to prevent free flow of necessary cooling air to and from motor.

TABLE A-1 Trouble Diagnosis Chart for Fractional Horsepower Motors (*Cont.*)

Problem	Probable causes	Test and remedy
C. Motor runs hot *(Continued)*	7. Ambient temperature too high	7. An ambient temperature above 40°C (104°F) is too high for standard motors. If motor is operated within an enclosure or restricted space, the temperature of this space may often rise several degrees above the room temperature. It is the temperature of the *immediate surroundings* of the motor that governs, not the room temperature. Sometimes this restricted space can be opened up or otherwise ventilated. If not, it may be necessary to obtain a special motor.
	8. Bearing trouble	8. See Sec A, items 7 and 8, and Sec. B, items 1 to 5.
	9. Short-circuited coils	9. This condition may be indicated by abnormal magnetic noise or excessive no-load watts. Rewind motor.
	10. Grounded winding or grounded switch	10. If the ground cannot be located and repaired, rewinding is required.
	11. Starting switch fails to open	11. This condition applies to split-phase and capacitor-start motors. Repair or replace faulty switch member (see Art. 17-8). This condition is apt to burn out the windings, particularly the starting winding.
	12. Poor soldering to commutator necks	12. This condition applies to dc and universal motors. Resolder connections and check armature (see Arts. 11-16 to 11-18).
	13. Short-circuiter fails to function correctly	13. This condition applies to repulsion-start motors. Clean and repair or replace the mechanisms (see Sec. *G*).
	14. Rotor rubbing stator	14. Best method of diagnosis is to dismantle motor and inspect stator bore and rotor surface for visual evidences of rubbing. If in doubt, apply

TABLE A-1 Trouble Diagnosis Chart for Fractional Horsepower Motors (*Cont.*)

Problem	Probable causes	Test and remedy
C. Motor runs hot *(Continued)*		Prussian blue to these surfaces, reassemble, run motor, dismantle, and reexamine. Remedy is to install new bearings (see Sec. A, item 8).
D. Motor burns out	1. Frozen bearings, misapplication or any of the causes listed in Sec. *C*	1. Any of the bad conditions listed in Sec. *C*, if prolonged, may lead to motor burnout. After the motor has actually burned out, the real source of the trouble is often very difficult to trace. Recondition the motor, run an application test, as described in Arts. 18-29 and 18-30, and investigate possibilities discussed in Sec. *C*.
E. Motor is noisy	1. Unbalanced rotor	1. An unbalanced rotor will set up a vibration that can be easily felt; the cure is to rebalance the rotor, dynamically, if equipment is available. Also, the shaft may be sprung slightly and may need straightening.
	2. Worn bearings	2. In single-phase induction motors, worn or dry bearings give rise to a characteristic noise; the noise is modulated at slip frequency and, at no load, is not unlike the sound of a purring cat. Sometimes oil will quiet the motor; at other times the bearings may have to be replaced (see Art. 17-2). If trouble is unduly frequent, see Sec. *B*.
	3. Switch rattles or rubs	3. Check switch for correct location and operation. Replace if necessary (see Art. 17-8).
	4. Excessive end play	4. Dismantle motor and add thrust washers to take up end play (Art. 17-2); add these to the side opposite the switch so as not to disturb the location of the latter with respect to

TABLE A-1 Trouble Diagnosis Chart for Fractional Horsepower Motors (*Cont.*)

Problem	Probable causes	Test and remedy
E. Motor is noisy (*Continued*)		the stator. Some motors are provided with end-play take-up plug; take up end play on these motors by tapping in the end-play take-up device, being careful not to increase the input to the motor by more than 10 W.
	5. Motor not properly aligned with driven machine	5. Correct mechanical condition.
	6. Motor not fastened firmly to mounting base	6. Correct mechanical condition.
	7. Loose accessories on motor	7. Such parts as oil-well covers, capacitor box or cover, oil pipe, conduit boxes or covers, and guards should be carefully checked and tightened so that they cannot rattle.
	8. Air gap not uniform	8. Bent shaft may cause this condition (see Sec. E, item 1). For extreme quietness, it may be necessary, after the shaft is straightened, to grind the rotor true. Only the barest minimum of material should be removed, as this increases the power and current consumption of the motor.
	9. Dirt in air gap	9. Noise is irregular, intermittent, and scratchy; dismantle and clean motor.
	10. Amplified motor noises	10. If this condition is suspected, uncouple motor from load and allow it to run idle; if noise persists, loosen the mounting bolts and lift the motor—while it is still running—off the mounting bolts. If motor is quiet, the mounting was acting as an amplifier. Use of resilient mounting will usually eliminate noise from this source (Art. 17-14). In some cases, it may be necessary to provide torsional flexibility in the coupling that connects the motor to the load (see Art. 17-15).

TABLE A-1 Trouble Diagnosis Chart for Fractional Horsepower Motors (*Cont.*)

Problem	Probable causes	Test and remedy
E. Motor is noisy *(Continued)*	11. Burrs on shaft shoulders, nicks on journals	11. Determine by inspection. Correct mechanical condition.
	12. Rough commutator	12. True up. See Art. 11-7.
F. Motor noisy when stopping	1. Regeneration (cap-start motors)	1. Motor becomes noisy while coasting to a stop. See Art. 5-27.
	Repulsion-start motors	
G. Motor runs but governor does not operate in a few seconds	1. Dirty commutator	1. Clean with fine sandpaper, never with emery.
	2. Governor mechanism sticking	2. Work governor mechanism by hand. Clean and install new parts or new governor as required.
	3. Worn or sticking brushes	3. See Sec. A, item 16.
	4. Frequency of supply circuit low	4. Check frequency of supply circuit. Get motor of proper frequency rating.
	5. Low line voltage	5. See Sec. A, item 4.
	6. Wrong connections	6. See Sec. A, item 5.
	7. Incorrect brush setting	7. See Sec. *A*, item 17.
	8. Excessive load	8. See Sec. A, item 10.
	9. Short-circuited stator	9. See Sec. A, item 12.
	10. Incorrect spring tension	10. Springs may lose temper or get out of adjustment; governor may not be of the proper speed rating.
	Commutator-type motors	
H. Excessive brush wear	1. Dirty or oily commutator	1. Clean with fine sandpaper, never with emery.
	2. Any of the troubles in Sec. *G*	3. See Sec. *G*.
	3. High mica, rough commutator	3. True up. See Art. 11-17.
	4. Excessive spring tension	4. Adjust to correct tension.
	5. Loose-fitting brushes	5. Get brushes or brush-holder boxes of the correct size.
I. Radio interference	1. Any of the causes in Sec. *H*	1. Correct mechanical condition. If necessary, add suppression filter, as explained in Art. 19-35.

TABLE A-2 Conversion Table for Units of Power[*]

Multiply	By	To obtain
Cheval-vapeur†	985.9	Millihorsepower
	735.5	Watts
	0.9859	Horsepower
	542.5	Foot-pounds per second
Foot-pounds per second	1.8174	Millihorsepower
	1.3558	Watts
	0.0018174	Horsepower
	0.0018436	Cheval-vapeur
Horsepower	1000	Millihorsepower
	746	Watts
	1.0143	Cheval-vapeur†
	550.3	Foot-pounds per second
Millihorsepower	0.746	Watts
	0.001	Horsepower
	0.0010143	Cheval-vapeur†
	0.5503	Foot-pounds per second
Watts	1.3405	Millihorsepower
	0.0013405	Horsepower
	0.0013598	Cheval-vapeur†
	0.7376	Foot-pounds per second

[*]Conversions may be made in inverse order by dividing by the conversion factor instead of multiplying by it. For example: divide watts by 746 to obtain horsepower.

†Metric horsepower; abbreviated cv.

TABLE A-3 Conversion Table for Units of Torque, Energy, or Work°

Multiply	By	To obtain
Kilogram-centimeters	1.157	Ounce-feet
	0.868	Pound-inches
	0.09807	Newton-meters
Newton-meters	11.801	Ounce-feet
	0.7376	Pound-feet
	10.197	Kilogram-centimeters
Ounce-feet	0.75	Pound-inches
	12	Ounce-inches
	864.1	Gram-centimeters
	0.8641	Kilogram-centimeters
	0.08474	Newton-meters
Ounce-inches	72.01	Gram-centimeters
	0.07201	Kilogram-centimeters
	0.007061	Newton-meters
Pound-feet	12	Pound-inches
	16	Ounce-feet
	192	Ounce-inches
	13,825	Gram-centimeters
	13.825	Kilogrma-centimeters
	1.3558	Newton-meters

°Units of energy and work have the same dimensions as units of torque, and hence the conversion factors in the table apply equally well to units of work or energy. Also, conversions can be made in inverse order by dividing by the factor. For example: divide newton-meters by 1.3558 to obtain pound-feet.

TABLE A-4 Table for Converting Degrees Fahrenheit to Degrees Centigrade

	0	1	2	3	4	5	6	7	8	9
−100	−73.3	−73.9	−74.4	−75	−75.6	−76.1	−76.7	−77.2	−77.8	−78.3
−90	−67.8	−68.3	−68.9	−69.4	−70	−70.6	−71.1	−71.7	−72.2	−72.8
−80	−62.2	−62.8	−63.3	−63.9	−64.4	−65	−65.6	−66.1	−66.7	−67.2
−70	−56.7	−57.2	−57.8	−58.3	−58.9	−59.4	−60	−60.6	−61.1	−61.7
−60	−51.1	−51.7	−52.2	−52.8	−53.3	−53.9	−54.4	−55	−55.6	−56.1
−50	−45.6	−46.1	−46.7	−47.2	−47.8	−48.3	−48.9	−49.4	−50	−50.6
−40	−40	−40.6	−41.1	−41.7	−42.2	−42.8	−43.3	−43.9	−44.4	−45
−30	−34.4	−35	−35.6	−36.1	−36.7	−37.2	−37.8	−38.3	−38.9	−39.4
−20	−28.9	−29.4	−30	−30.6	−31.1	−31.7	−32.2	−32.8	−33.3	−33.9
−10	−23.3	−23.9	−24.4	−25	−25.6	−26.1	−26.7	−27.2	−27.8	−28.3
−0	−17.8	−18.3	−18.9	−19.4	−20	−20.6	−21.1	−21.7	−22.2	−22.8
0	−17.8	−17.2	−16.7	−16.1	−15.6	−15	−14.4	−13.9	−13.3	−12.8
10	−12.2	−11.7	−11.1	−10.6	−10	−9.4	−8.9	8.3	−7.8	−7.2
20	−6.7	6.1	−5.6	−5	−4.4	−3.9	−3.3	−2.8	−2.2	−1.7
30	−1.1	−0.6	0	0.6	1.1	1.7	2.2	2.8	3.3	3.9
40	4.4	5	5.6	6.1	6.7	7.2	7.8	8.3	8.9	9.4
50	10	10.6	11.1	11.7	12.2	12.8	13.3	13.9	14.4	15
60	15.6	16.1	16.7	17.2	17.8	18.3	18.9	19.4	20	20.6
70	21.1	21.7	22.2	22.8	23.3	23.9	24.4	25	25.6	26.1
80	26.7	27.2	27.8	28.3	28.9	29.4	30	30.6	31.1	31.7
90	32.2	32.8	33.3	33.9	34.4	35	35.6	36.1	36.7	37.2
100	37.8	38.3	38.9	39.4	40	40.6	41.1	41.7	42.2	42.8
110	43.3	43.9	44.4	45	45.6	46.1	46.7	47.2	47.8	48.3
120	48.9	49.4	50	50.6	51.1	51.7	52.2	52.8	53.3	53.9
130	54.4	55	55.6	56.1	56.7	57.2	57.8	58.3	58.9	59.4
140	60	60.6	61.1	61.7	62.2	62.8	63.3	63.9	64.4	65
150	65.6	66.1	66.7	67.2	67.8	68.3	68.9	69.4	70	70.6
160	71.1	71.7	72.2	72.8	73.3	73.9	74.4	75	75.6	76.1
170	76.7	77.2	77.8	78.3	78.9	79.4	80	80.6	81.1	81.7
180	82.2	82.8	83.3	83.9	84.4	85	85.6	86.1	86.7	87.2
190	87.8	88.3	88.9	89.4	90	90.6	91.1	91.7	92.2	92.8
200	93.3	93.9	94.4	95	95.6	96.1	96.7	97.2	97.8	98.3
210	98.9	99.4	100	100.6	101.1	101.7	102.2	102.8	103.3	103.9
220	104.4	105	105.6	106.1	106.7	107.2	107.8	108.3	108.9	109.4
230	110	110.6	111.1	111.7	112.2	112.8	113.3	113.9	114.4	115
240	115.6	116.1	116.7	117.2	117.8	118.3	118.9	119.4	120	120.6
250	121.1	121.7	122.2	122.8	123.3	123.9	124.4	125	125.6	126.1
260	126.7	127.2	127.8	128.3	128.9	129.4	130	130.6	131.1	131.7
270	132.2	132.8	133.3	133.9	134.4	135	135.6	136.1	136.7	137.2
280	137.8	138.3	138.9	139.4	140	140.6	141.1	141.7	142.2	142.8
290	143.3	143.9	144.4	145	145.6	146.1	146.7	147.2	147.8	148.3
300	148.9	149.4	150	150.6	151.1	151.7	152.2	152.8	153.3	153.9
310	154.4	155	155.6	156.1	156.7	157.2	157.8	158.3	158.9	159.4
320	160	160.6	161.1	161.7	162.2	162.8	163.3	163.9	164.4	165
330	165.6	166.1	166.7	167.2	167.8	168.3	168.9	169.4	170	170.6

TABLE A-4 Table for Converting Degrees Fahrenheit to Degrees Centigrade (*Cont.*)

	0	1	2	3	4	5	6	7	8	9
340	171.1	171.7	172.2	172.8	173.3	173.9	174.4	175	175.6	176.1
350	176.7	177.2	177.8	178.3	178.9	179.4	180	180.6	181.1	181.7
360	182.2	182.8	183.3	183.9	184.4	185	185.6	186.1	186.7	187.2
370	187.8	188.3	188.9	189.4	190	190.6	191.1	191.7	192.2	192.8
380	193.3	193.9	194.4	195	195.6	196.1	196.7	197.2	197.8	198.3
390	198.9	199.4	200	200.6	201.1	201.7	202.2	202.8	203.3	203.9
400	204.4	205	205.6	206.1	206.7	207.2	207.8	208.3	208.9	209.4

TABLE A-5 Wire Table

Gauge no.	Diameter bare (nominal), in	Area		Diameter over insulation, in*		Heavy, lb/1000 ft	Resistance, Ω/1000 ft at 25°C
		Circular mils	mm²	Single	Heavy		
14	0.0641	4109	2.08	0.0666	0.0682	15.9	2.575
15	0.0571	3260	1.652	0.0594	0.0609	12.6	3.247
16	0.0508	2581	1.308	0.0531	0.0545	10.05	4.094
17	0.0453	2052	1.040	0.0475	0.0488	6.34	5.163
18	0.0403	1624	0.823	0.0424	0.0437	5.02	6.51
19	0.0359	1289	0.653	0.0379	0.0391	4.00	8.21
20	0.0320	1024	0.519	0.0339	0.0351	3.17	10.35
21	0.0285	812	0.411	0.0303	0.0314	2.51	13.05
22	0.0253	640	0.324	0.0270	0.0281	1.99	16.46
23	0.0226	511	0.259	0.0243	0.0253	1.58	20.76
24	0.0201	404.0	0.2047	0.0217	0.0227	1.26	26.17
25	0.0179	320.4	0.1621	0.0194	0.0203	0.998	33.00
26	0.0159	252.8	0.1282	0.0173	0.0182	0.793	41.62
27	0.0142	201.6	0.1024	0.0156	0.0164	0.630	52.48
28	0.0126	158.8	0.0806	0.0140	0.0147	0.501	66.17
29	0.0113	127.7	0.0649	0.0126	0.0133	0.396	83.4
30	0.0100	100.0	0.0507	0.0112	0.0119	0.316	105.2
31	0.0089	79.2	0.0400	0.0100	0.0108	0.251	132.7
32	0.0080	64.2	0.0324	0.0091	0.0098	0.198	167.3
33	0.0071	50.4	0.0253	0.0081	0.0088	0.158	211.0
34	0.0063	39.7	0.0203	0.0072	0.0078	0.126	266.0
35	0.0056	31.4	0.0157	0.0064	0.0070	0.0966	335.5
36	0.0050	25.0	0.0127	0.0058	0.0063	0.0791	423.0
37	0.0045	20.3	0.0101	0.0052	0.0057	0.0628	533.4
38	0.0040	16.0	0.0081	0.0047	0.0051	0.0498	672.6

*Insulations include formvar, Nyform, Nyleze, polyester, Polythermaleze, and ML.

Source: NEMA Motor and Generator Standards, NEMA Publ. no. MW 1000-1981, National Electrical Manufacturers Association, Washington, 1981 and data from Essex, Fort Wayne, IN 46804.

TABLE A-6 Locked-Rotor Amperes from Code Letter Markings

Code	1/20	1/12	1/8	1/6	1/4	1/3	1/2	3/4	1
	Horsepower rating hp								
	115-V Single-phase motor*								
A	0.00	0.00	0.00	0.00	0.00	0.00	0.00	0.00	0.00
B	1.37	2.28	3.42	4.57	6.85	9.13	13.70	20.54	27.39
C	1.54	2.57	3.86	5.14	7.72	10.29	15.43	23.15	30.87
D	1.74	2.90	4.35	5.80	8.70	11.59	17.39	26.09	34.78
E	1.96	3.26	4.89	6.52	9.78	13.04	19.57	29.35	39.13
F	2.17	3.62	5.43	7.25	10.87	14.49	21.74	32.61	43.48
G	2.43	4.06	6.09	8.12	12.17	16.23	24.35	36.52	48.70
H	2.74	4.57	6.85	9.13	13.70	18.26	27.39	41.09	54.78
J	3.09	5.14	7.72	10.29	15.43	20.58	30.87	46.30	61.74
K	3.48	5.80	8.70	11.59	17.39	23.19	34.78	52.17	69.57
L	3.91	6.52	9.78	13.04	19.57	26.09	39.13	58.70	78.26
M	4.35	7.25	10.87	14.49	21.74	28.99	43.48	65.22	86.96
N	4.87	8.12	12.17	16.23	24.35	32.46	48.70	73.04	97.39
P	5.43	9.06	13.59	18.12	27.17	36.23	54.35	81.52	108.70
R	6.09	10.14	15.22	20.29	30.43	40.58	60.87	91.30	121.74
S	6.96	11.59	17.39	23.19	34.78	46.38	69.57	104.35	139.15
T	7.83	13.04	19.57	26.09	39.13	52.17	78.26	117.39	156.52
U	8.70	14.49	21.74	28.99	43.48	57.97	86.96	130.43	173.91
V	9.74	16.23	24.35	32.46	48.70	64.93	97.39	146.09	194.78
	230-V three-phase motors†								
A	0.00	0.00	0.00	0.00	0.00	0.00	0.00	0.00	0.00
B	0.40	0.66	0.99	1.32	1.98	2.64	3.95	5.93	7.91
C	0.45	0.74	1.11	1.49	2.23	2.97	4.46	6.68	8.91
D	0.50	0.84	1.26	1.67	2.51	3.35	5.02	7.53	10.04
E	0.56	0.94	1.41	1.88	2.82	3.77	5.65	8.47	11.30
F	0.63	1.05	1.57	2.09	3.14	4.18	6.28	9.41	12.55
G	0.70	1.17	1.76	2.34	3.51	4.69	7.03	10.54	14.06
H	0.79	1.32	1.98	2.64	3.95	5.27	7.91	11.86	15.81
J	0.89	1.49	2.23	2.97	4.46	5.94	8.91	13.37	17.82
K	1.00	1.67	2.51	3.35	5.02	6.69	10.04	15.06	20.08
L	1.13	1.88	2.82	3.77	5.65	7.53	11.30	16.94	22.59
M	1.26	2.09	3.14	4.18	6.28	8.37	12.55	18.83	25.10
N	1.41	2.34	3.51	4.69	7.03	9.37	14.06	21.09	28.11
P	1.57	2.61	3.92	5.23	7.84	10.46	15.69	23.53	31.38
R	1.76	2.93	4.39	5.86	8.79	11.71	17.57	26.36	35.14
S	2.01	3.35	5.02	6.69	10.04	13.39	20.08	30.12	40.16
T	2.26	3.77	5.65	7.53	11.30	15.06	22.59	33.89	45.18
U	2.51	4.18	6.28	8.37	12.55	16.73	25.10	37.65	50.20
V	2.81	4.69	7.03	9.37	14.06	18.74	28.11	42.17	56.23

Note: Currents included in the range for any letter include the lower figure up to, but not including, the higher.

*For 230-V motors, divide table values by 2.

†For 115-V motors, multiply table values by 2.0; for 200-V motors, multiply table values by 1.15; for 460-V motors, divide table values by 2.0

TABLE A-7 Fractional Inches to Millimeters and Decimal Inches

Inches	Millimeters	Inches	Inches	Millimeters	Inches
1/64	0.3969	0.015625	33/64	13.0969	0.515625
1/32	0.7938	0.031250	17/32	13.4937	0.531250
3/64	1.1906	0.046875	35/64	13.8906	0.546875
1/16	1.5875	0.062500	9/16	14.2875	0.562500
5/64	1.9844	0.078125	37/64	14.6844	0.578125
3/32	2.3812	0.093750	19/32	15.0812	0.593750
7/64	2.7781	0.109375	39/64	15.4781	0.609375
1/8	3.1750	0.125000	5/8	15.8750	0.625000
9/64	3.5719	0.140625	41/64	16.2719	0.640625
5/32	3.9688	0.156250	21/32	16.6687	0.656250
11/64	4.3656	0.171875	43/64	17.0656	0.671875
3/16	4.7625	0.187500	11/16	17.4625	0.687500
13/64	5.1594	0.203125	45/64	17.8594	0.703125
7/32	5.5563	0.218750	23/32	18.2563	0.718750
15/64	5.9531	0.234375	47/64	18.6531	0.734375
1/4	6.3500	0.250000	3/4	19.0500	0.750000
17/64	6.7469	0.265625	49/64	19.4469	0.765625
9/32	7.1438	0.281250	25/32	19.8438	0.781250
19/64	7.5406	0.296875	51/64	20.2406	0.796875
5/16	7.9375	0.312500	13/16	20.6375	0.812500
21/64	8.3344	0.328125	53/64	21.0344	0.828125
11/32	8.7312	0.343750	27/32	21.4312	0.843750
23/64	9.1281	0.359375	55/64	21.8281	0.859375
3/8	9.5250	0.375000	7/8	22.2250	0.875000
25/64	9.9219	0.390625	57/64	22.6219	0.890625
13/32	10.3187	0.406250	29/32	23.0187	0.906250
27/64	10.7156	0.421875	59/64	23.4156	0.921875
7/16	11.1125	0.437500	15/16	23.8125	0.937500
29/64	11.5094	0.453125	61/64	24.2094	0.953125
15/32	11.9063	0.468750	31/32	24.6062	0.968750
31/64	12.3031	0.484375	63/64	25.0031	0.984375
1/2	12.7000	0.500000	1	25.4000	1.00000

TABLE A-8 Whole Inches to Millimeters°

	Millimeters									
Inches	0	1	2	3	4	5	6	7	8	9
0	000.0	25.4	50.8	76.2	101.6	127.0	152.4	177.8	203.2	228.6
1	254.0	279.4	304.8	330.2	355.6	381.0	406.4	431.8	457.2	482.6
2	508.0	533.4	558.8	584.2	609.6	635.0	660.4	685.8	711.2	736.6
3	762.0	787.4	812.8	838.2	863.6	889.0	914.4	939.8	965.2	990.6
4	1016.0	1041.4	1066.8	1092.2	1117.6	1143.0	1168.4	1193.8	1219.2	1244.6
5	1270.0	1295.4	1320.8	1346.2	1371.6	1397.0	1422.4	1447.8	1473.2	1498.6
6	1524.0	1549.4	1574.8	1600.2	1625.6	1651.0	1676.4	1701.8	1727.2	1752.6
7	1778.0	1803.4	1828.2	1854.2	1879.6	1905.0	1930.4	1955.8	1981.2	2006.6
8	2032.0	2057.4	2082.8	2108.2	2133.6	2159.0	2184.4	2209.8	2235.2	2260.6
9	2286.0	2311.4	2336.8	2362.2	2387.6	2413.0	2438.4	2463.8	2489.2	2514.6
10	2540.0	2565.4	2590.8	2616.2	2641.6	2667.0	2692.4	2717.8	2743.2	2768.6
11	2794.0	2819.4	2844.8	2870.2	2895.6	2921.0	2946.4	2971.8	2997.2	3022.6
12	3048.0	3073.4	3098.8	3124.2	3149.6	3175.0	3200.4	3225.8	3251.2	3276.6
13	3302.0	3327.4	3352.8	3378.2	3403.6	3429.0	3454.4	3479.8	3505.2	3530.6
14	3556.0	3581.4	3606.8	3632.2	3657.6	3683.0	3708.4	3733.8	3759.2	3784.6
15	3810.0	3835.4	3860.8	3886.2	3911.6	3937.0	3962.4	3987.8	4013.2	3038.6
16	4064.0	4089.4	4114.8	4140.2	4165.6	4191.0	4216.4	4241.8	4267.2	4292.6
17	4318.0	4343.4	4368.8	4394.2	4419.6	4445.0	4470.4	4495.8	4521.2	4546.6
18	4572.0	4597.4	4622.8	4648.2	4673.6	4699.0	4724.4	4749.8	4775.2	4800.6
19	4826.0	4851.4	4876.8	4902.2	4927.6	4953.0	4978.4	5003.8	5029.2	5054.6

°For example, 124 in = 3149.6 mm.

TABLE A-9 Millimeters to Inches°

	Inches									
Millimeters	0	1	2	3	4	5	6	7	8	9
0	0.000	0.039	0.079	0.118	0.157	0.197	0.236	0.276	0.315	0.354
1	0.394	0.433	0.472	0.512	0.551	0.591	0.630	0.669	0.709	0.748
2	0.787	0.827	0.886	0.906	0.945	0.984	1.024	1.063	1.102	1.142
3	1.181	1.220	1.260	1.299	1.339	1.378	1.417	1.457	1.496	1.535
4	1.575	1.614	1.654	1.693	1.732	1.772	1.811	1.850	1.890	1.929
5	1.969	2.008	2.047	2.087	2.126	2.165	2.205	2.244	2.283	2.323
6	2.362	2.402	2.441	2.480	2.520	2.559	2.598	2.638	2.677	2.717
7	2.756	2.795	2.835	2.874	2.913	2.953	2.992	3.031	3.071	3.110
8	3.150	3.189	3.228	3.268	3.307	3.346	3.386	3.425	3.465	3.504
9	3.543	3.583	3.622	3.661	3.701	3.740	3.780	3.819	3.858	3.898
10	3.937	3.976	4.016	4.055	4.094	4.134	4.173	4.213	4.252	4.291
11	4.331	4.370	4.409	4.449	4.488	4.528	4.567	4.606	4.646	4.685
12	4.724	4.764	4.803	4.843	4.882	4.921	4.961	5.000	5.039	5.079
13	5.118	5.157	5.197	5.236	5.276	5.315	5.354	5.394	5.433	5.472
14	5.512	5.551	5.591	5.630	5.669	5.709	5.748	5.787	5.827	5.866
15	5.906	5.945	5.984	6.024	6.063	6.102	6.142	6.181	6.220	6.260
16	6.299	6.339	6.378	6.417	6.457	6.496	6.535	6.575	6.614	6.654
17	6.693	6.732	6.772	6.811	6.850	6.890	6.929	6.969	7.008	7.047
18	7.087	7.126	7.165	7.205	7.244	7.283	7.323	7.362	7.402	7.441
19	7.480	7.520	7.559	7.598	7.638	7.677	7.717	7.756	7.795	7.835
20	7.874	7.913	7.953	7.992	8.031	8.071	8.110	8.150	8.189	8.228
21	8.268	8.307	8.346	8.386	8.425	8.465	8.504	8.543	8.583	8.622
22	8.661	8.701	8.740	8.780	8.819	8.858	8.898	8.937	9.876	9.016
23	9.055	9.094	9.134	9.173	9.213	9.252	9.291	9.331	9.370	9.409
24	9.449	9.488	9.528	9.567	9.606	9.646	9.685	9.724	9.764	9.803
25	9.843	9.882	9.921	9.961	10.000	10.039	10.079	10.118	10.157	10.197
26	10.236	10.276	10.315	10.354	10.394	10.433	10.472	10.512	10.551	10.591
27	10.630	10.669	10.709	10.784	10.787	10.827	10.866	10.906	10.945	10.984
28	11.024	11.063	11.102	11.142	11.181	11.220	11.260	11.299	11.339	11.378
29	11.417	11.457	11.496	11.535	11.575	11.614	11.654	11.693	11.732	11.772
30	11.811	11.850	11.890	11.929	11.969	12.008	12.047	12.087	12.126	12.165
31	12.205	12.244	12.283	12.323	12.362	12.402	12.441	12.480	12.520	12.559
32	12.598	12.638	12.677	12.717	12.756	12.795	12.835	12.874	12.913	12.953
33	12.992	13.031	13.071	13.110	13.150	13.189	13.228	13.268	13.307	13.346
34	13.386	13.425	13.465	13.504	13.543	13.583	13.622	13.661	13.701	13.740
35	13.781	13.819	13.858	13.898	13.937	13.976	14.016	14.055	14.094	14.134
36	14.173	14.213	14.252	14.291	14.331	14.370	14.409	14.449	14.488	14.528
37	14.567	14.606	14.646	14.685	14.724	14.764	14.803	14.843	14.882	14.921
38	14.961	15.000	15.039	15.079	15.118	15.157	15.197	15.236	15.276	15.315
39	15.354	15.394	15.433	15.472	15.512	15.551	15.591	15.630	15.669	15.709
40	15.748	15.787	15.827	15.866	15.906	15.945	15.984	16.024	16.063	16.102
41	16.142	16.181	16.220	16.260	16.299	16.339	16.378	16.417	16.457	16.496

TABLE A-9 Millimeters to Inches° (*Cont.*)

Millimeters	Inches 0	1	2	3	4	5	6	7	8	9
42	16.535	16.575	16.614	16.654	16.693	16.732	16.772	16.811	16.850	16.890
43	16.929	16.969	17.008	17.047	17.087	17.126	17.165	17.205	17.244	17.283
44	17.323	17.362	17.402	17.441	17.480	17.520	17.559	17.598	17.638	17.677
45	17.717	17.756	17.795	17.835	17.874	17.913	17.953	17.992	18.031	18.071
46	18.110	18.150	18.189	18.228	18.268	18.307	18.346	18.386	18.425	18.465
47	18.504	18.543	18.583	18.622	18.661	18.701	18.740	18.780	18.819	18.858
48	18.898	18.937	18.976	19.016	19.055	19.094	19.134	19.173	19.213	19.252
49	19.291	19.331	19.370	19.409	19.449	19.488	19.528	19.567	19.606	19.646
50	19.685	19.724	19.764	19.803	19.843	19.882	19.921	19.961	20.000	20.039

°For example, 237 mm = 9.331 in.

GLOSSARY

The definitions given are based on those of three organizations: (1) the Institute of Electrical and Electronics Engineers (IEEE), (2) the National Electrical Manufacturers Association (NEMA), and (3) the International Electrotechnical Commission (IEC). Numerals following the definitions indicate that the definition was abstracted, with or without editing, from the organization referenced; hence, the definition given may not have the exact wording of the definition in the reference. For terms not given, consult the Index, which lists many terms that are defined or explained in the text of the book but are not given in the Glossary.

Adjustable-speed motor A motor whose speed can be varied gradually over a considerable range but when once adjusted remains practically unaffected by the load, for example, a dc shunt-wound motor with field resistance control designed for a considerable range of speed adjustment. (1)

Adjustable varying-speed motor A motor whose speed can be adjusted gradually but when once adjusted for a given load will vary in considearble degree with change in load, for example, a dc compound-wound motor adjusted by field control or a wound-rotor induction motor with rheostatic speed control. (1)

Ambient temperature The temperature of the surrounding cooling medium, such as gas or liquid, which comes into contact with heated parts of the apparatus. (2) *Note:* Ambient temperature is commonly known as room temperature in connection with air-cooled apparatus not provided with artificial ventilation.

Apparent efficiency The ratio of the power output in watts to the total input in volt-amperes. It is also the product of the efficiency multiplied by the power factor. Used only in connection with ac machines.

Armature The member of an electric machine in which an alternating voltage is generated by virtue of relative motion with respect to a magnetic flux field. In dc, universal, ac-series, and repulsion-type machines, the term is commonly applied to the entire rotor. (1)

Asynchronous machine An ac machine in which the rotor does not turn at synchronous speed. (1)

Back The end of a motor or generator that carries the only or largest coupling or driving pulley. (1)

Base speed The lowest rated speed of an adjustable-speed motor obtained at rated load and rated voltage at the temperature rise specified in the rating. (2)

Bracket *See* End shield.

Breakdown torque The maximum shaft output torque that an induction motor (or a synchronous motor operating as an induction motor) develops when the primary winding is connected for running operation, at normal operating temperature, with rated voltage applied at rated frequency. (1) *Note:* A motor with a continually increasing torque as the speed decreases to standstill is not considered to have a breakdown torque. (See Fig. 11-1.)

Brush A conducting part, generally fixed, which provides electric connection through sliding contact with a part moving relative to it. (3)

Brush holder A structure which supports a brush and which enables it to be maintained in contact under pressure with the sliding surface. (3)

Capacitance (Capacity) That property of a system of conductors and dielectrics which permits the storage of electricity when potential differences exist between the conductors. Its value is expressed as the ratio of a quantity of electricity to a potential difference. A capacitance value is always positive. (1) *Note:* The introduction of capacitance into an ac circuit causes the current to lead the voltage in time; it is employed in motor circuits to accomplish a phase shift of the auxiliary winding relative to the main winding to increase the starting torque while decreasing the locked-rotor current. (See Art. 5-3.)

Capacitor A condenser consisting of two conducting bodies separated by insulation. If a positive charge is placed on one plate of a capacitor, an equal negative charge is induced on the other. The medium between the plates is called a *dielectric.* (1) *Note:* The primary purpose of a capacitor is to introduce capacitance into a circuit. Capacitors are usually classified according to their dielectrics, as air capacitors, paper capacitors, etc.

Capacitor motor A single-phase motor with a main winding arranged for direct connection to a source of power and an auxiliary winding connected in series with a capacitor. The capacitor may be directly in the auxiliary circuit or connected into it through a transformer. (1)

Capacitor-start motor A capacitor motor in which the auxiliary winding is energized only during the starting operation. (1) *Note:* The auxiliary-winding circuit is open-circuited during running operation.

Code letter A letter which appears on the nameplate of ac motors to show their locked-rotor kilovolt-amperes per horsepower. (2) (See Art. 1-15 and Table 1-2.)

Collector ring A conducting ring against which brushes bear, which is used to enable current to flow from one part of a circuit to another by sliding contact. (3) Also called *slipring.*

Commutator An assembly of conducting members insulated from one another, against which brushes bear, that is used to enable current to flow from one part of a circuit to another by sliding contact. (3)

Compensated series motor A series motor with a compensating field winding. (2) *Note:* The compensating field winding and the series field winding may be combined into one field winding.

Compound-wound motor A dc motor which has two separate field windings: one, usually the predominating field, connected in parallel with the armature circuit and the other connected in series with the armature circuit. (1) [See Art. 12-1(3).]

Constant-speed motor A motor in which the speed is constant or substantially constant over the normal range of loads, such as a synchronous motor, an induction motor with a small slip, or a dc shunt motor with constant excitation. (1)

Definite-purpose motor Any motor designed in standard ratings, with standard operating characteristics or mechanical construction, for use under service conditions other than usual or for use on a particular type of application. (2)

Design "N" motor A fractional horsepower single-phase motor designed for full-voltage starting with locked-rotor current not exceeding specified values, which are lower than for design "O" motors. (For values, see Table 19-1.) (1)

Design "O" motor A fractional horsepower single-phase motor designed for full-voltage starting with locked-rotor current not exceeding specified values, which are higher than for design "N" motors. (1)

Dripproof machine An open machine in which the ventilating openings are so constructed that successful operation is not interfered with when drops of liquid or solid particles strike or enter the enclosure at any angle from 0 to 15° downward from the vertical. (2)

Duty A statement of loads, including no-load and rest and deenergized periods, to which the machine is subjected, including their duration and sequence in time. (3) *Note:* Heavy-duty is often popularly used to denote an application requiring high locked-rotor torque and having high intermittent overloads.

Efficiency The ratio of useful power output to the power input expressed in the same units.

Either-rotation motor *See* Reversible motor.

Electric Containing, producing, arising from, actuated by, or carrying electricity and capable of so doing. Examples are an electric eel, energy, a motor, a vehicle, and a wave. (1)

Electrical Related to, pertaining to, or associated with electricity, but not having its properties or characteristics. Examples are an electrical engineer, a handbook, an insulator, a rating, and a school unit. (1)

End shield A shield secured to the frame and adapted to protect the windings and to support the bearing, but including no part thereof. Also called *end bell* or *bracket.*

Explosionproof machine A totally enclosed machine whose enclosure is designed and constructed to withstand an explosion of a specified gas or vapor which may occur within it and to prevent the ignition of the specified gas or vapor surrounding the machine by sparks, flashes, or explosions of the specified gas or vapor which may occur within the machine casing. (2) *Note:* See National Electrical Code, Article 500, "Hazardous (Classified) Locations."

Fractional horsepower motor A motor built in a frame smaller than that having an open continuous rating of 1 hp at 1700 to 1800 rpm. [See Art. 1-1(1).] In general, this is in NEMA frame size 56 or smaller.

Front The end of a normal motor or generator that is opposite the largest coupling or drive pulley. (1)

Full-load torque The torque necessary to produce a motor's rated horsepower at full-load speed. In ounces at a 1-ft radius, it is equal to the horsepower times 84,070 divided by the full-load speed. (2)

General-purpose ac motor An ac motor built in standard ratings, with standard performance and mechanical construction, offered for general use. (See Art. 19-1.)

Guarded machine An open machine in which all openings giving direct access to live metal or rotating parts (except smooth rotating surfaces) are limited in size by the structural parts or by screens, baffles, grilles, expanded metal, or other means to prevent accidental contact with hazardous parts. Openings giving direct access to such live or rotating parts should not permit the passage of a cylindrical rod of ¾-in diameter.

High-potential test A test which consists of the application of high voltage between windings and metal parts for a specified time for the purpose of determining the adequacy against breakdown of the insulating materials and spacings under normal conditions. (See Arts. 3-15, 3-16, and 18-22.) (1) *Note:* The test is used as a proof test of new apparatus, a maintenance test on older equipment, or one method of evaluating developmental insulation systems.

Hysteresis motor A synchronous motor without salient poles and without dc excitation that starts by virtue of the hysteresis losses induced in its hardened-steel secondary member by the revolving field of the primary and operates normally at synchronous speed as a result of retentivity of the secondary core. (1)

Induction generator An induction machine which, when driven above synchronous speed by an external source of mechanical power, is used to convert mechanical power to electric power. (1)

Induction motor An ac motor in which a primary winding on one member (usually the stator) is connected to the power source, and a polyphase secondary winding or a squirrel-cage secondary winding on the other member (usually the rotor) carries induced current. (1)

Inductor machine A synchronous machine in which one member, usually stationary, carries main and exciting windings effectively disposed in relation to each other and in which the other member, usually rotating, is without windings but carries a number of regular projections. (Permanent magnets may be used instead of the exciting windings.) (1) (See Arts. 10-18 to 10-21.)

Insulation Material or a combination of suitable nonconducting materials that provides electric isolation of two parts at different voltages. (1)

Insulation class A letter or number that designates the temperature rating of an insulation material or system with respect to thermal endurance. Four classes of insulation systems are used in motors: Classes A (105°C), B (130°C), F (155°C), and H (180°C).

Insulation system An assembly of insulating materials in association with the conductors and the supporting structural parts of a motor.

Integral horsepower motor A motor built in a frame having an open continuous rating of 1 hp at 1700 to 1800 rpm, or built in a larger frame. In general this is in NEMA frame size 140 or larger.

Journal That part of a shaft which is intended to rotate inside a bearing.

Leakage current The current that flows through or across the surface of insulation from motor windings to ground or frame and which consists of a resistive component and a capacitive component.

Locked-rotor current The steady-state current taken from the line with the rotor locked and with rated voltage and frequency applied to the motor. (1)

Locked-rotor torque The maximum torque that a motor will provide with locked rotor at any angular position of the rotor at a winding temperature of 25°C plus or minus 5°C with rated voltage applied at rated frequency. (1)

Machine with encapsulated windings An ac squirrel-cage machine having random windings filled with an insulating resin which also forms a protective coating. This type of machine is intended for exposure to more severe environmental conditions than usual varnish treatments can withstand. Other parts of the machine may require protection against such environmental conditions. (2)

Multispeed motor A motor which can be operated at any one of two or more definite speeds, each being practically independent of the load, for example, a dc motor with two armature windings or an induction motor with windings capable of various pole groupings. In the case of multispeed permanent-split capacitor and shaded-pole motors, the speeds are dependent on the load. (2)

Open machine A machine having ventilating openings which permit passage of external cooling air over and around the windings of the machine. (2)

Permanent-split capacitor motor A capacitor motor with the same value of effective capacitance for both starting and running conditions. (1)

Pull-in torque The maximum constant torque under which a synchronous motor will pull its connected inertia load into synchronism at rated voltage and frequency when its normal field excitation is applied. (2) *Note:* The speed to which a motor will bring its load depends on the power required to drive it, and whether the motor can pull the load into step from this speed depends on the inertia of the revolving parts, so that the pull-in torque cannot be determined without having the wk^2 as well as the torque of the load.

Pull-out torque The maximum sustained torque which a synchronous motor will develop at synchronous speed with rated voltage applied at rated frequency and with normal excitation. (2)

Pull-up torque The minimum torque developed by an ac motor during the period of acceleration from rest to the speed at which breakdown torque occurs with rated voltage applied at rated frequency. (1)

Radial magnetic pull The magnetic force on the rotor resulting from the radial (air-gap) displacement from magnetic center.

Rating The whole of the numerical values of the electrical and mechanical quantities with their duration and sequences that are assigned to the machine by the manufacturer and stated on the rating plate with the machine complying with the specified conditions.

Reactor A device whose primary purpose is to introduce reactance into an ac circuit.

Reactor-start motor A form of split-phase motor that uses an external reactor in series with the main winding for starting.

Regulation The amount of change in voltage or speed resulting from a load change. (1) *Note:* Usually the change is taken from no load to rated load, and the result expressed in percentage of the rated-load value.

Reluctance motor A synchronous motor similar in construction to an induction motor, in which the member carrying the secondary circuit has salient poles, without permanent magnets or dc excitation. It starts as an induction motor and is normally provided with a squirrel-cage winding but operates normally at synchronous speed. (1) (See Art. 10-3.)

Repulsion-induction motor A motor with repulsion-motor windings and short-circuited brushes, without an additional device for short-circuiting the commutator segments, and with a squirrel-cage winding in the rotor in addition to the repulsion-motor winding. (1)

Repulsion motor A single-phase motor that has a stator winding arranged for connection to a source of power and a rotor winding connected to a commutator. Brushes on the commutator are short-circuited and are so placed that the magnetic axis of the rotor windings is inclined to the magnetic axis of the stator winding. This type of motor has a varying-speed characteristic. (1)

Repulsion-start induction motor A single-phase motor with repulsion-motor windings and brushes that has a commutator-short-circuiting device that operates at a predetermined speed of rotation to convert the motor into the equivalent of a squirrel-cage motor for running operation. For starting operation, this motor performs as a repulsion motor. (1)

Resistance-start split-phase motor A split-phase motor using an external resistor in the starting-winding circuit.

Reversible motor A motor which is capable of being started from rest and operated in either direction of rotation.

Reversing motor A motor capable of being reversed, even while running at normal speed, by changing electric connections.

Rotor The rotating part of a machine. The shaft may or may not be included.

Semiguarded machine An open machine in which part of the ventilating openings in the machine, usually in the top half, are guarded, as in the case of a "guarded machine," but the others are left open. (2)

Series-wound motor A commutator machine in which the field circuit and armature circuit are connected in series. (1)

Service factor A multiplier which, when applied to rated power, indicates a permissible power loading which may be carried under the conditions specified for the service factor. (2) (See Table 1-3.)

Servomoter An ac induction motor with two primary windings mutually displaced in magnetic position from one another by 90 electrical degrees and a low-inertia high-resistance rotor that gives a speed-torque curve substantially linear in shape. It is designed for operation with a constant voltage applied to the fixed phase, and an adjustable voltage is impressed on the other phase, called the *control phase.* (See Art. 16-1.)

Shaded-pole motor A single-phase motor with a main winding and one or more short-circuited windings (or shading coils) disposed about the air gap. The effect of the winding combination is to produce a rotating magnetic field, which in turn induces the desired motor action. (1) (See Art. 9-1.)

Shunt-wound motor A dc motor in which the field circuit and armature circuit are connected in parallel. (1) [See Art. 12-1(1).]

Slipring *See* Collector ring.

Special-purpose motor A motor with special operating characteristics, or special mechanical construction, or both, designed for a particular application and not falling within the definition of a general-purpose or definite-purpose motor. (2)

Splashproof machine An open machine in which the ventilating openings are so constructed that successful operation is not interfered with when drops of liquid or solid particles strike or enter the enclosure at any angle not greater than 100° from the vertical. (2)

Split-phase motor A single-phase induction motor having a main winding and an auxiliary winding, which is designed to operate with no external impedance in either winding. The auxiliary winding is energized only during the starting operation and is open-circuited during the running operation. (1) (See Art. 4-1.)

Split-phase capacitor motor A single-phase induction motor which starts as a split-phase motor and switches a capacitor into the auxiliary circuit during running to operate as a permanent-split capacitor motor. [See Art. 5-1(4).]

Starting torque *See* Locked-rotor torque.

Stator The portion that includes and supports the stationary active parts. The stator includes the stationary portions of the magnetic circuit and the associated winding and leads. It may, depending on the design, include a frame or shell, winding supports, ventilation circuits, coolers, and temperature detectors. A base, if provided, is not ordinarily considered to be part of the stator. (1)

Stepper motor An electromagnetic device designed to convert a series of input power pulses into discrete angular movements, one for each power pulse. (See Art. 15-1.)

Subfractional horsepower motor Any motor rated at less than 1/20 hp. It is usually rated in millihorsepower.

Switching torque The minimum external torque developed by a motor as it accelerates through switch-operating speed. Applies only to motors with automatic connection-change means for starting. (See Fig. 18-9.)

Synchronous induction motor A cylindrical-rotor synchronous motor having a secondary coil winding similar to that of a wound-rotor induction motor. (1) *Note:* This winding is used for both starting and excitation purposes.

Synchronous machine A machine in which the average speed of normal operation is exactly proportional to the frequency of the system to which it is connected. (1)

Synchronous motor A synchronous machine that transforms electric power into mechanical power. Unless otherwise stated, it is generally understood to have field poles excited by direct current. (1)

Thermal protector A protective device for assembly as an integral part of the machine, which, when properly applied, protects the machine against dangerous overheating due to overload and, in a motor, failure to start. (2)

Thermally protected The words on the nameplate of a motor indicating that the motor is provided with a thermal protector. (2)

Torque A force which produces or tends to produce rotation. Common units are pound-feet, ounce-feet, ounce-inches, kilogram-centimeters, and newton-meters. A force of 1 ounce at a radius of 1 foot produces a torque of 1 ounce-foot.

Torque motor A motor designed primarily to exert torque through a limited travel or in a stalled position. (1) *Note:* Such a motor may be capable of being stalled continuously or for only a limited time.

Totally enclosed fan-cooled machine A totally enclosed machine equipped for exterior cooling by means of a fan or fans integral with the machine but external to the enclosing parts. (2)

Totally enclosed machine A machine so enclosed as to prevent the free exchange of air between the inside and the outside of the case but not sufficiently enclosed to be termed airtight. (2)

Totally enclosed nonventilated machine A totally enclosed machine which is not equipped for cooling by means external to the enclosing parts. (2)

Two-value capacitor motor A capacitor motor using different values of effective capacitance for the starting and running conditions. (1)

Universal motor A series-wound or compensated series-wound motor designed to operate at approximately the same speed and output on either direct current or single-phase alternating current with a specified frequency range and at the same root-mean-square voltage. (1)

Varying-speed motor A motor whose speed varies appreciably with the load, ordinarily decreasing when the load increases, for example, a series or repulsion motor.

Waterproof machine A totally enclosed machine constructed so that it excludes water applied in the form of a stream from a hose, except that leakage may occur around the shaft provided it is prevented from entering the oil reservoir. Automatic draining of the machine is provided. The means for automatic draining may be a check valve or a tapped hole at the lowest part of the frame which will serve for application of a drain pipe. (2) *Note:* A common form of the test for a waterproof machine is to play on the machine a stream of water from a hose with a 1-in nozzle delivering at least 65 gal/min from a distance of about 10 feet from any direction for a period of not less than 5 min.

Wound-rotor induction motor An induction motor in which a primary winding on one member (usually the stator) is connected to an ac power source and a secondary polyphase coil winding on the other member (usually the rotor) carries alternating current produced by electromagnetic induction. The terminations of the rotor winding are usually connected to collector rings. The brush terminals may be either short-circuited or closed through suitable adjustable circuits. (1)

INDEX

ABOUT THE AUTHORS

Cyril G. Veinott has an insider's knowledge of motors and motor construction. He was actively involved for over 40 years in the design of electric machinery and the development of design methods, first with the Westinghouse Electric Company and then with the Reliance Electric Company. He is a pioneer in the development of inherent thermal overheating protection, the design of ac machinery for military aircraft, and the application of digital computers to motor design. He has written scores of articles and technical papers and is the author of three books and parts of others. He is the recipient of many awards for his contributions to the motor industry including the Westinghouse Silver W, the IEEE Tesla as well as the Centennial Awards, medals from Czechoslovakia and Yugoslavia, and the Small Motor Manufacturers Association Hall of Fame Award. He is now a consultant, lecturer, and teacher.

Joseph E. Martin has been with the Emerson Motor Division for nearly 40 years, where he has been engaged in the design of small electric motors, in the application of motor designs to domestic and industrial appliances, and in the development of testing procedures. He also teaches in-house classes in motor design to engineers. He serves on standards writing committees of the Underwriters Laboratories, the Canadian Standards Association, the American Gas Association, and the International Electrotechnical Committee and represents Emerson Electric at meetings of the Motor and Generator Section of the National Electrical Manufacturers Association.